BIO
MATTER
TECHNO
SYNTHETICS
DESIGN FUTURES FOR THE MORE THAN HUMAN

FRANCA TRUBIANO
AMBER FARROW
MARÍA JOSÉ FUENTES
SUSAN KOLBER
MARTA LLOR

BIO MATTER TECHNO SYNTHETICS

DESIGN FUTURES FOR THE MORE THAN HUMAN

BIO

MATTER

TECHNO

SYNTHETICS

ACKNOWLEDGMENTS

It has been a work of joy to create this edited book alongside those who have given up their energy and imagination. B/M/T/S is the second publication created by women graduates of the Weitzman School of Design at the University of Pennsylvania. Like the earlier *Women [Re]Build: Stories, Polemics, Futures* (ORO - ar+d, 2019), ideas for this publication were incubated amid academic initiatives organized by the student group Penn Women in Design/Architecture (PWID/A). The brainchild of four of the group's leaders, Amber Farrow, María José Fuentes, Susan Kolber, and Marta Llor, B/M/T/S celebrates an empowered vision of design which unfortunately remains under-represented in design schools and the press. Rarely is the content discussed in this volume the subject of pedagogical focus or curricular attention.

 Producing edited books requires the support of many. There are those who assist financially and those who sustain in spirit. Thankfully, many do both. We are entirely grateful to Dean Frederick (Fritz) Steiner for having been an early supporter of PWID, for his continued and unwavering encouragement, and for having financially facilitated the production of B/M/T/S. We received grant support from The Trustees' Council of Penn Women and its Director, Theresa Welsh. Many individuals and organizations supported our authors, as recognized and acknowledged in their respective essays. The following organizations offered editors access to their graphic resources: the Blanton Museum of Art at the University of Texas at Austin, the Harvard Radcliffe Institute, the Zilkha Gallery, Wesleyan University, the Artists Rights Society, the National Archives [Goverment of the United Kingdom], San Francisco History Center at the San Francisco Public Library, New York Public Library, and Google Earth. Copy editing assistance was provided by UPenn PhD Candidate in Architecture, Rami Kanafani, and we are grateful to authors Behnaz Farahi, Patricia Olynyk, and Jacqueline Wu for having agreed to our use of their artwork in designing the cover. Lastly, we are thankful to have worked with commissioning editor Ricardo Devesa from ACTAR, who entertained this project from the very beginning and supported its creative goals throughout.

Most importantly, we acknowledge the efforts of our authors who ventured to walk with us on this journey of seeking, finding, and naming a new horizon for design wherein they have conceived and constructed new empathic, vibrant, and material relationships with and through sensible things and their extended worlds. Indeed, B/M/T/S has engendered an exceptional intellectual community whose 'design texts' encourage critical yet real affordances at the intra-section of biology, materials, and technology, seeking nothing less than their synthesis. This inspired gathering of makers tasked us, however, with recognizing the many ways that design is often complicit with the injustices experienced by humans and the more-than-human. Surely, surviving the next wave of environmental traumas – the result of our less than thoughtful deployment of material and energy resources – is of concern to many, including the women of B/M/T/S.

Most projects worth pursuing overcome challenges. When the most virulent of pandemics succeeded in upsetting the very definition of human resilience, existence, and survival, our authors channeled both anxieties and hope into their contributions. It was nothing less than a miracle when they answered our invitation to participate at the height of the contagion. Our ability to collaborate and 'make' by way of virtual tools was a welcome respite in a time of restricted human contact. The use of readily available digital platforms ensured the project thrived despite the cancellation of an in-person symposium dedicated to the topic. In keeping with the thesis of B/M/T/S, what a virus tried to separate, satellites, computers, software, servers, and fiber optics succeeded in reuniting. For this, the editors of B/M/T/S are entirely grateful.

Franca Trubiano, Philadelphia, August 2023

WHY BIO MATTER TECHNO SYNTHETICS?

FRANCA TRUBIANO

Bio/Matter/Techno/Synthetics: Design Futures for the More than Human (B/M/T/S) represents the design and intellectual work of women who create, make, and ideate at the intra-section of nature, artifice, and technology. Their desire to propagate, generate, calibrate, fabricate, interrogate, and animate all manner and matter of things motivates this collection of essays. The women of B/M/T/S have charted critical paths to design inspiration and material productivity from positions both embedded within and tangential to their disciplines. Future-ready visions and historically grounded critiques are offered across twenty-two texts authored by designers and thinkers who, living in a world of both abundance and trauma, take seriously their responsibility to make ethically. Operating in ever more complex aesthetic, ethical, environmental, and socio-political contexts and practicing at the intersection of art, architecture, landscape architecture, environmental design, material studies, emerging technologies, digital fabrication, media studies, robotics, and critical theory, the women of B/M/T/S have redefined the very origins, principles, and values of design.

B/M/T/S articulates the material and intellectual outlines of a rapidly shifting topography of ideas that challenges the hegemonic tendencies of parametricism, object-oriented ontology, parafictional realism, post-digital representations, and corporate environmentalism. Supported by the tenets of new materialism, post-humanism, and fourth-wave feminism, and by the difficult recognition that many forms of artifice invariably exacerbate life in the Anthropocene, the essays here collected contribute new resources, questions, practices, and projects for working and thinking differently in the space of design.

When was the last time we asked ourselves if a wearable device of trans-species agriculture is capable of mitigating late-capitalism's contamination of the soil that feeds us: how did we acquiesce to the erasure of a once thriving African American aquapelagic industry of oyster farming which sustained both human and non-human life in nineteenth-century New York City: how might we embrace biological design principles that celebrate material decay and decomposition: could we imagine a world built of seaweed: can we re-cycle vast quantities of glass waste as sand substitute: how might the internet of things help us visualize the thermodynamic space between human bodies and architecture: can designers leverage the spatiality of digital codes to represent hidden structural forces and patterns of movement: how might trans-disciplinary and trans-individual design principles activate complex data sets, modes of scientific analysis, and fabrication methods for responsive material structures: in what way can eco-feminism decolonize history and the harmful extractive practices upon which so much of our global supply chains are pred-icated: might we finally achieve a truly wholistic definition of life that includes zoë in biodesign principles? If any of these questions are of interest to you, the authors of B/M/T/S address these and other queries with carefully crafted texts that serve to synthesize an under-standing of life co-determined by bio, matter, and technology. How is it, therefore, that this sensibility towards practice remains at times marginalized amongst purveyors of design theory, art curation, and the awarding of projects? This was the question at the origins of B/M/T/S.

The texts, images, fabrications, installations, and conversations collected in this volume are instances of material creativity that seek to transcend the interminable influence of simple dualisms. They reveal and actualize the outlines of a counter-tradition wherein making is embedded, animated, and cultivated in and through the stuff of things. Enzymes, seaweed, oysters, mycelium, glass waste, green walls, robots, sentient signals, software, video projections, photogrammetric, nanoparticles, heat pumps, thermostats, digital weaving, chemical droplets, urban parks, and even words embody a shared horizon of meaning. B/M/T/S represents the critical practices of those who strive for synthesis in natural systems, material ecologies, and advanced technologies, as well as in the cultural constructions and intellec-tual speculations that each makes possible. Super-charging the still under-represented voice of women who make B/M/T/S underscores the interests of those who investigate design in all manner of organ-isms, be they human, more-than-human, or machines. Indeed, design in B/M/T/S embraces—without contradiction—ecology, and technology in its call for variety, diversity, multiplicity, and unity.

Embracing alternate forms of practice, especially those that cultivate the tenets of design and research as stewardship, has been central to the artists, architects, landscape architects, material scientists, builders, roboticists, coders, historians, and theorists of B/M/T/S. In a redirection of human hubris, they've cared for, tended to, and collaborated with their subjects. They've also engendered practices of resistance, much in need given our contemporary world of uneven powers.

FROM MUSE TO MAKER TO MATTER:

ARCHITECTURE AND THE MORE-THAN-HUMAN

FRANCA TRUBIANO

In the Shadow of the Muse

In writing this essay, I remain fully aware that women have not always been advantaged with access to the means of making. The now near-commonplace reference to them as designers, artists, architects, landscape architects, material scientists, builders, coders, historians, and theorists belie the fact that this is an extraordinarily recent phenomenon. Measured relative to the history of Western thought, for example, with Hellenistic Greece as a point of origin, women have sat at the intellectual 'table' of makers as it were for two percent (2%) of this history.[1] A hundred years ago, women were still considered muses, sources of inspiration, and objects of interest for those more gifted with the talent to write, draw, paint, compose, build, and lead. Even at the end of the 1960s, in Europe and the Americas, their work was hardly visible in publications, academia, the press, works of art, and public office. Indeed, during the second wave of feminism, women still struggled to achieve an independent voice and the freedom to create.

According to British art historian, author, and broadcaster Katy Hessel, even in the twenty-first century, the impulse to associate women artists with their male counterparts remains commonplace in historiography. In her essay for *The Guardian,* "Why do we still define female artists as wives, friends, and muses?" Hessel notes how even contemporary exhibits entirely dedicated to the work of women artists insist on identifying their personal relationships with the men in their lives.[2] Reporting on the spring 2023 Whitechapel Gallery show in London, *Action, Gesture, Paint: Women Artists and Global Abstraction 1940-1970,* Hessel observed that five of the eighty-six artist labels describing artwork dedicated portions of fairly limited fifty-word statements to husbands and other male artists who presumably influenced the work of these women. Hessel asked if this was characteristic of labels used to describe the work of men. Her careful review of labels in collections at The National Gallery, London, and Tate Britain suggested this was not the case. As Hessel observed, labels for Frans Hals (1582-1666) at The National Gallery speak not of Judith Leyster (1609-1660), and those for Piet Mondrian (1872-1944) at Tate Britain make no mention of Marlow Moss (1889-1958). But just as she suspected, Moss's label was named Mondrian. When validating a woman's intellectual and creative abilities, referencing the professional influence of her male peers and partners is presumably a requirement.

A similar observation was made by then-writer for *The Paris Review,* Cody Delistrary, during his 2018 review of the retrospective solo exhibit of expressionist painter Gabriele Münter (1877-1962) at the Louisiana Museum of Modern Art outside of Copenhagen, Denmark. He, too, remarked how, in reviewing Münter's prolific career—exhibited in one hundred and thirty works of art—"social conditioning dictates that you look first at the shadow of her long-term lover, the better-known Wassily Kandinsky," rather than strengths or shortcomings in the work itself.[3] Even art critics and scholars remain habituated by a form of unconscious bias in their reflex to assume that the work of women artists is derivative of that of their male colleagues. And yet, once more, the reverse is hardly ever the case.

In fact, what is vastly more common when narrating the work of male artists is citing the women who served as their companions, wives, and sources of inspiration. Indeed, as Delistrary reminds us in his essay, "When Female Artists Stop Being Seen as Muses," women historically acquired a place in the annals of art practice primarily in their role as creative motivators. This is the intellectual *pas de deux* that upholds the often repeated and rarely challenged dualism of 'genius' and 'muse,' wherein the former is characteristically associated with the gifted, inflamed spirit of creativity that emanates from the minds of men—the originator of all ideas, culture, laws, and designs—while the latter is allied with what inspires them, their women. Visionary and near-divine prodigies seek guidance in the musings of their companions, be they real or allegorical, corporeal or spiritual. How often has this narrative been played out: Pablo Picasso and Dora Maar, Diego Rivera and Frida Kahlo, Man Ray and Lee Miller, Andrew Wyeth and Helga Testorf.

Architecture—as the art of design and the stuff of buildings—has participated in its own way in propagating a gender-based narrative of creativity and inspiration. We need only recall the frontispiece of Marc-Antoine Laugier's (1713-1769) second edition of *Essai sur l'archi- tecture* (1755), whereupon the light of creativity is represented atop the cherub's forehead, gifted as he is with the flame of genius. This messenger conveys the 'invention' that is the Vitruvian primitive hut and his flame emblematic of the power to enlighten, undoubtedly possessed by the work's author and its engraver Charles-Dominique- Joseph Eisen (1720-1778).[4] The 'other' figure central to the engraving is that of Architecture who, allegorized as a woman, directs our attention to the dual origins of artifice. She points to 'nature's' temple with her right arm while resting atop the 'symbols' of architecture with her left. In this modern version of 'contrapposto', Eisen represented the disci- pline of Architecture as a seated female figure who gazes at structural principles while physically surrounded by fragments of an architectural language. In her role as muse, she binds the ideality of composition and abstraction (in the freestanding columns and pedimented rafters taking form out of the wooded wilds of the forest) with the corporeal- ity of ornamental fragments and building details (in the entablature, Ionic column capital, and remnants of fluted shafts scattered in the landscape). In and across her body, Architecture joins theory to prac- tice, idea to ornament, and figure to fragment. As the cherub's mind is gifted with the flames of genius, Architecture's body binds nature to artifice. (Figure 1)

Various versions of this narrative have appeared across the history of Western thought. Beginning in the Italian Renaissance, allegories of the arts and sciences were routinely characterized using the female figure. Sandro Botticelli's *Primavera* from 1482, Jacob Toorenvliet's *Allegory of Painting* from 1675-79, Pompeo Girolamo Batoni's *Allegory of the Arts* from 1740, and François Boucher's 1765 *Allegory of Painting* are but a few examples. Even Artemisia Gentileschi (1593–1656) painted a version of the allegorical female figure as an embodiment of the art of painting, except in her case, she pictured herself in the dual starring role of painter and muse (*Self-Portrait as the Allegory of Painting*, 1639–1639). (Figure 2) Amongst the representational arts, women and their bodies have repeatedly been the preferred sight for the intellectual projection of artistic and cultural ideas. In most cases, this innocuous use of allegory as a visual shorthand for the purposes of communicating an idea has been just that, harmless.

Figure 1. Marc-Antoine's Laugier's 1755 Frontispiece in the *Essai sur l'architecture,* Courtesy of the Perkins Rare Book Library of the Fisher Fine Arts Library, of the University of Pennsylvania.

When Science Deceives

Less innocent, however, were academic claims during the twentieth century that women's so-called receptivity and proclivity to inspire were near-biological truths and verifiable psychological profiles. Häns Jürgen Eysenck (1916-1997), for example, the Berlin-born British psychologist who sought to associate 'genius' with identifiable psychological characteristics, identified it with one gender. Unbelievably prolific as a Professor at the Institute of Psychiatry, King's College, London, for nearly thirty years, Dr. Eysenck had unparalleled influence over the field of psychiatry.[5] According to his *New York Times* obituary from 1997, he "published some 80 books and 1,600 journal articles."[6] And yet, it appears he "managed to offend a great many people."[7] The reasons for this are not entirely surprising. As an academic, intellectual, and psychologist, Eysenck insisted on using the scientific method (analysis, causation, and proof) to assert bio-genetic origins for intelligence and creativity. In his 1995 publication *Genius, The Natural History of Creativity*, Eysenck was categorical in his claim that genius was biologically male. Citing without irony the work of male historians and scientists as proof, he noted:

> Creativity, particularly at the highest level, is closely related to gender; almost without exception, genius is found only in males (for whatever reason!). Illustrations abound. In the list of geniuses studied by Cox, (1926) there are no women. There are no women among Roe's (1951a, b, 1953) eminent scientists, and very few in American Men of Science, or among members of the Royal Society; none in a list of the leading mathematicians (Bell, 1965), and none would be found among the 100 best known sculptors, painters, or dramatists. Simonton (1992) found no women in a list of the most famous 120 composers, from the Renaissance to the twentieth century, and hardly any women among his scientists.[8]

In this, and in other works of psychology circulating as late as 1995, the profession and its researchers normalized claims of male intellectual superiority, using science as their tool. Repeatedly, Eysenck placed his faith in the ability of diagrams and data to 'prove' his scientific claims. In only one example, he sought to measure the effect of solar exposure on "socially destructive" or "culturally positive behavior" for the purposes of evaluating the literary productivity of "European, Persian, Osmanic, Arabic, Chinese, and Japanese language groups."[9] Citing the work of minor German psychologist and self-proclaimed astrologer Suitbert (S.E.) Ertel (1932-2017) Eysenck suggested that periods of great artistic and scientific consequence were the result of measurable reductions in sun-spot activities on the earth: less cosmic sun, more human brilliance. While we can now appreciate that this so-called science was but the projection of prejudice, Eysenck's assertions regarding gender were systematically repeated. His feverish collection of 'data' had one stated goal: to "paint a rough portrait of the budding genius ... [as] clearly ... male, of middle or upper-middle parentage, ... [and] preferably born in

February."[10] Notwithstanding what is clearly a near-hysterical form of reasoning, Eysenck went on to cite from a 1993 work, *Intellectual Talent: Psychometric and Social Issues* that "mathematical reasoning" is a "discipline in which gender differences are particularly large (in favor of males)."[11] Its authors Benbow and Lubinski advanced that, "theoretical values, which are characteristic of physical scientists, are more characteristic of gifted males than gifted females," because "social values, which are negatively correlated with interests in physical science, are more characteristic of gifted females than males."[12]

And all these years I wondered what was at the origins of so much gender stereotyping and misinformation regarding the ability

Figure 2. *Self-Portrait as the Allegory of Painting (La Pittura)*, c.1638–9, Painting by Artemisia Gentileschi. Courtesy of The Royal Collection Trust.

of women to think and create. Fortunately, much of Eysenck's thesis of a scientifically verifiable correlation between creativity and gender, race, and solar sunspots has been debunked.[13] Alongside his erroneous claims about intelligence and race, the idea that white men who stay out of the sun are 'naturally' pre-disposed to invention remains nothing short of academic deception. This often repeated and until recently sanctioned account of human creativity has thankfully been discredited, even if for centuries it underwrote the erasure, dismissal, absence, denial, and rejection of women as creators, makers, and thinkers.

Lucky for us, Katy Hessel's 2022 award-winning book, *The Story of Art Without Men,* offers verifiable evidence of ingenuity among women who create. Her contributions not only disprove vacuous attempts to the contrary by unsubstantiated social scientists and psychologists but also actualize the important task of excavation of women who make from the buried annals of historiography. We now know of Plautilla Nelli (1524-1588), Lavinia Fontana (1552-1614), Carmen Herrera (1915-2022), Harriet Powers (1837-1910), and the Guerrilla Girls (founded 1985). But we almost did not. As architectural historian Despina Stratigakos reminds us in her piece, "Unforgetting Women Architects: A Confrontation with History and Wikipedia," women have been and continue to be excluded from history in countless ways, sometimes by neglect for not having kept archives, sometimes because the products of their practice escape the canonic understanding of monographic historiography, and sometimes by sheer prejudice.[14] As Stratigakos's example from 2013 makes clear—wherein a male Wikipedia editor, identified as Der Krommodore, unilaterally sought to remove an in-progress posting on the German architect Thekla Schild because 'he' was not aware of her—keeping creative women in the sightliness of scholarship is constant work.

Women who make, despite

Indeed, efforts to keep women from taking their rightful place in the history of artistic and cultural production are nothing new. In the early 1990s, academic discourse in the arts, steeped in the era's political conservationism and sanctioned by notions of sexual difference, promoted once more the prejudicial idea that genius was born of the male gaze. Camille Paglia's controversial academic best-seller from 1990, *Sexual Personae: Art and Decadence from Nefertiti to Emily Dickson,* used binary definitions of gender to justify behavioral and psychological stereotypes traceable in the figural and literary arts.[15] Tireless efforts by early feminist scholars—Andrea Dworkin (1946-2005), Julia Kristeva (b.1941), Elisabeth Grosz (b.1952), Luce Irigaray (b.1930)—judiciously and strategically de-constructed the idea that men make and women inspire.[16] We recall Grosz's pivotal reading of Irigaray's definition of 'phallocentrism', which Irigaray exposed as a totalizing force in Western philosophy and political thought in "[leaving] no space or form of representation for women's autonomy."[17] Highlighting Irigaray's critique of Jacques Lacan's (1901-1981) psychoanalytic models, Grosz reminded us that

> men—philosophers, psychoanalysts, scientists, writers—have spoken for women for too long. Women remain the objects of speculation, the source of metaphors and images necessary for the production of discourse, ... while they are denied access to positions as producing subjects.[18]

Grosz's own re-framing of creativity identified how Western religion—and Christianity, more particularly—participated in destroying women's access to making as a form of cultural expression. As Grosz discussed, men invented, gave voice to, and articulated the origins of religion, art, and culture in their "debt" to women for being born of their bodies. She noted:

> A whole history of philosophy seems intent on rationalizing this debt away by providing men with a series of images of self-creation culminating in the idea of God as the paternal 'mother', creator of the universe in place of women/mothers. Man's self-reflecting Other, God, functions to obliterate the positive fecundity and creativity of women. Born of woman, man devises religion, theory, and culture as an attempt to disavow this foundational, unspeakable debt.[19]

Men created gods (religion, art, culture) because they could not create life. Hence, men create while women pro-create: a fiction which, when repeated, whether shouted or whispered, continues to cost us all. For what is the cost of not inviting half of the world to the making of the world? Nothing less than trauma: political, social, religious, environmental, and material trauma. The arbitrary division of art from nature has contributed to the environmental precipice that we now face, to the orchestrated deterioration of women's rights by zealots, to the global crisis of human migration, and to coordinated threats to modern democracy and the rule of law.

And yet, if this essay assumes a moment of willful, self-imposed blindness when asserting that an alternate, previously marginalized definition of creative making—directed at improving, sustaining, cultivating, and caring for the destiny of our built and natural environments—is in fact possible, I believe it has earned it. If it calls for a reckoning of gender inequalities at the intra-section of artifice and nature, it has earned that, too. Because women make, despite everything, doing so in intellectual and material conditions that are far different than those offered by Western metaphysics and the history of aesthetics. This essay seeks an alternate 'way of making' that edifies while championing an expanded field of becoming, one wherein women challenge, subvert, ignore, and overturn—if they care to—the less-than productive binary of maker and muse. One that recognizes that the drive to create is not the purview of a single gender or of gender *tout court*. Indeed, momentary blindness is needed to offset continued preconceptions that stall the empowerment of women within design fields. Even with hard-won battles, women continue to contend with the silencing, misrepresentation, and misinterpretation of their work. Indeed, we have yet to reckon, in any sustained way, with whether the increased presence of women in design has motivated any real changes to its practice and reception.

Matter, again

I've asked this question of my own discipline, architecture. Have significantly larger numbers of women occasioned real changes in its teaching, practice, and leadership? Yes is the answer when recognizing the impact that biodesign, circular construction, ecological thinking, restorative design, regenerative materials, digital fabrications, smart computing, physical prototyping, and holistic thinking have had on the discipline. Architecture has successfully identified a new locus for the generation of its ideas, processes, and values. Doing so, however, comes at a price: ideas, observations, and understandings that once guided my work are now tenuous and insubstantial. Principles that previously were significant intellectual signposts no longer support my conversations, exchanges, and debates. Many books, references, and go-to-scholars offer little respite in the face of new, critical questions made possible by the making practices of women. Ideas I once had about materials—materials that detail our designs, materials that soar in gothic cathedrals, materials inscribed on the facades of Renaissance *palazzi*, materials we dissimulate, hide, and obfuscate, materials we desire and must have, and even materials sold in hardware stores—are henceforth no longer valid. Indeed, *tout court,* substance is no longer what I once thought it to be, as I no longer subscribe to the same definitions of matter, materials, and materiality. What changed? What now gives me cause to pause?

Perhaps for far too long I had believed that stone, clay, wood, metals, concrete, glass, and polymers were but substances waiting patiently, in reserve, for my manipulation, my handling as an architect. Surely, building materials were destined to serve my goals of architectural expression. Whether heavy or light, opaque or transparent, extracted from the ground or manufactured in a laboratory, material substances were inert, and it was the task of design to bring them to form. In this world, all categories of architectural matter, be they organic or inorganic, anticipate contact with the human imagination to acquire their purpose. The human will to form is what gives matter shape, profile, contour, thickness, and hence, meaning. Without the life force and motility of form, the matter has no perceivable figuration, no representational status, and no point of entry into the rarefied world of language. Only when transitioned out of nature by the sheer potency of human artifice does matter enter the cultural horizon of intentionality. In this way, architecture is defined as the use of materials transitioned from the wilds of brute matter to the materiality of civilization. Really?

Unthinking and Obstinate Vessels

This interpretation of architectural substances I had inculcated for decades via texts, both ancient and contemporary. Plato, for example, wrote during the fifth century BCE in his *Timaeus* of the origin of all things, describing the structure of matter as qualities. Fire, air, water, earth, and any combination thereof were for Plato vessels, receptacles "that receive all things" and that exist as "a kind of neutral plastic material on which changing impressions are stamped by the things which enter it."[20] For figuration and form to appear, "anything that is to receive in itself every kind of character must be devoid of all character."[21] Offering an easy to understand example, Plato affirmed how "we may indeed use the metaphor of birth and compare the receptacle to the mother, the model to the father, and what they produce between them their offspring."[22] So entirely 'receptive' were Plato's qualities that they were not in themselves "visible and sensible things."[23] Rather, as vessels they were "invisible and formless, ... possessed in a most puzzling way of intelligibility, yet very hard to grasp."[24] For Plato, matter, she, mother, lay in wait for the power of form, he, father to impress, mark, characterize, define, and bring her into appearance.

Similarly, and no less disquieting for the future of architectural substances, seventeen-century philosopher René Descartes (1596-1650) identified in his *Discourse on Method* and *Meditations* (1637) clear and categorical differences between things and thoughts as between things and self. He postulated that men had the power to name, designate, locate, and measure corporeal bodies with "magnitude, extension, position, substance, duration, [and] number."[25] So, Descartes not only projected a quasi-fiction regarding the status of knowledge, but he also invented a self-imposed philosophical gap between humans and all others.[26] Ready to acknowledge that both he and stones were substances, Descartes found it necessary to discriminate, saying that while he was "a thinking and non-extended thing," stone, on the other hand, was an "extended non-thinking thing." [27] So stating, he introduced not one but two disruptive propositions: the first, that humans are primarily "thinking things" for which the body is of little consequence; the second, that the rest of nature's substances (be they stone, sand, or stalagmites) have no access to thought and hence no prospect of anima, or animation. In one and the same philosophy, and to much acclaim since many continue to believe this even today, Descartes postulated that men were all-thinking and matter dead on arrival.

These and similar ideas about matter have continued to influence our understanding of physical substances, raw materials, and building practices in architecture. Most recently, historian of architecture and engineering Antoine Picon, in his *The Materiality of Architecture*, notes that despite being,

> the subject of endless philosophical speculation, the matter always remains somewhat enigmatic. It appears as fundamentally indifferent to human goals... Deprived of human intentionality, matter actually often resists those who want to use it, beginning with

makers and builders. Bending wood and carving stones are never easy; they require strength and skill. These operations may be interpreted as the result of cooperation...but they can also be seen as tense encounters, as fights against a stubborn material reality. ...Because of its indifference and resistance, matter may seem mute and even inscrutable or opaque, an impression reinforced by its general lack of immediate expressivity.[28]

In language and words uttered in the past or at present, matter is less than responsive, obstinate, and uncommunicative. A simple but effective narrative for inculcating the rarefied gap between mind and matter, as between architectural intention and the substance of the building. In this view of the world, matter is impenetrable, lacking in expression, and with no life force of its own. Indeed, when architects create, they are tested by their indifference and resistance. Only when one succeeds in making matter, speak does the architect initiate it into the near sacred space of language. According to Picon,

> The entire paradox of the architectural discipline consists in endeavoring to render matter expressive, often to the point of overlooking its stubborn resistance to language as such—even though a significant portion of its expressive power lies precisely in this resistance.[29]

Reluctant to participate in the project that is 'language,' architects are triumphant when they make matter communicate, even if against their will. This, it appears, is the definition of architectural language at the center of architectural theory. As Picon notes,

> In the final analysis, to situate architecture, or to think about it, necessarily implies confronting the constant oscillation between the silent obstinacy of matter, its resistance to human endeavors, and the desire to animate it so as to make it communicate.[30]

Hence, to be an architect is to be in a constant state of opposition with matter, wrenching it from silence and insisting it speak, perform, and service the discipline.

Snap

Something shattered, something changed. To be sure, it was not a sudden recognition or an entirely new thought but a gentle apparition, a coming into focus of an outline that had lived in my imagination for many a year. When, for example, an object surfaces from the depths of the sea, piercing the top of the water and floating with no intention of returning from where it once submerged. The first glint of light on the object's surface reveals what lay obfuscated for years: I no longer shared in this definition of architectural substance, even if for a very long time I had intellectually and academically accepted that matter was inanimate, passive, unintelligible, recessive, devoid of will, and obstinate. I could no longer support the idea that matter only spoke and hence participated in society when it accepted the will of form.

As a woman who makes assertions that matter is dialectically different from thought and devoid of form, I find this to be a source of much intellectual dissonance. Rock, soil, sand, dirt, and trash offer no less intellectual depth than the contemplation of geometrical theorems. Matter is given to thought, and thought is material. Human subjects are not the privileged bearers of life nor of intellect. All manner of things, including the more-than-human, offer configurable life forms. Movement, energy, and perpetual metamorphosis in stone as in sand are recognizable to all who empathize with their position, predicament, and inner power. Matter's undeniable potency is terrifyingly palpable to skilled artists who venture to leave the slightest lasting impression of substances. Camille Claudel's (1864-1943) *Sakuntala* (or *Vertumnus et Pomona*) is a case in point: qualitative differences in the terracotta, marble, and bronze versions of this composition were surely not silent to its artist; they were the very basis for three distinct versions of the work.

Awakening to this 'other' understanding of matter has been as if walking for the very first time in a landscape of new ideas, sensibilities, and possibilities. Transformative publications by contemporary authors Jane Bennett, Stacy Alaimo, and others have redefined my relationship with architecture. They have made it possible to seek alternative correspondences between things and thoughts. Their words have supported many an attempt at re-corporealizing matter and at fostering an intellectual practice that includes both 'substanc-iating' the architectural body and substantiating she who makes. (Figure 3)

Indeed, my own recent scholarship has focused on reinstating a sense of value to the very body of architecture that is building.[31] Addressing the long-standing disparagement of material substances at the hands of architectural theory has sought to understand why the 'thingness' of architecture has occasioned but the slightest of interest on the part of those who 'think' about architecture. Whether by neglect, *oublié,* or design, the last half century of architectural theory, in true Descartes fashion, has rejected its own corporeality. To understand why, I turn to literary scholar, political theorist, and philosopher Jane Bennett, who furthers our understanding of things by drawing us back to their very power as sensible matter. According to Bennett, language, more broadly, has inadvertently forgotten or strategically neglected the corpus of our material world. On the one hand, material things are ubiquitous, overt, easily consumable, and transactional, embedded in daily life to the point of being intellectually invisible; on the other, the age-old privileging of Western narratives, colonial histories, and gendered thought has sidelined the generative value of things. In response, Bennett's 2009 book *Vibrant Matter: A Political Ecology of Things* invites the reader to re-engage with the "positive, productive power" of material things, even if doing so requires celebrating—without irony—the very force and agency of their "negative power or recalcitrance."[32] And if, as previously argued, matter and bodies, more generally, are thought of as 'resistant', it is precisely this power to oppose that Bennett uses in recentering the agency of things and matter.

Things, for Bennett, have power. In the fact that they exist as bodies, they have the power to "affect other bodies".[33] Things are not objects. They transcend objectification, for things are "vivid entities not entirely reducible to the contexts in which (human) subjects set them, never entirely exhausted by their semiotics."[34] In this, things are never fully given to language. They possess their own agency and ability to create and, as such, escape the limits of both morphology and semantics. Even urban waste dumps, as she describes them, are teeming with material processes that are chemical, biological, and topographical. Trash, Petri dishes, silicon chips, limestone beds, forests, and oceans actively 'remake' themselves and all 'things' collected therein.

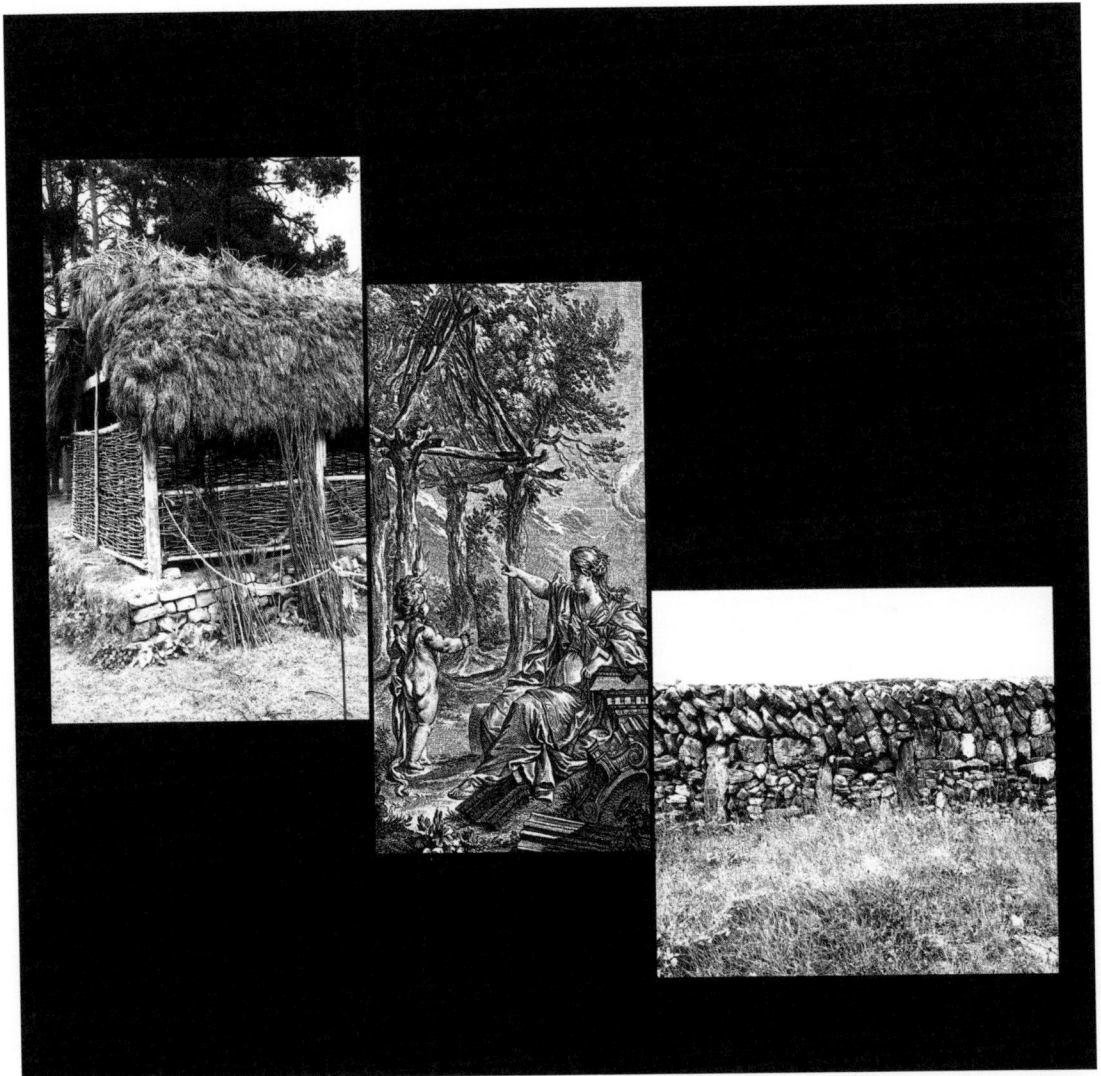

Bennett's ambition in *Vibrant Matter* is "to give voice to a vitality intrinsic to materiality, [and] in the process absolving matter from its long history of attachment to automatism or mechanism."[35] She seeks its freedom from a host of intellectual hierarchies and politically divisive dualisms exacerbated by Descartes's epistemology. Bennett achieves this by identifying an alternate understanding of things in the writings of European polymath and philosopher Baruch Spinoza (1632-1677). Near contemporary of Descartes, Spinoza "claim[ed] that all things are "animate, albeit in different degrees"."[36] Even the most stable, inert, and apparently unchanging of material substances possessed motility, drive, and capacity for self-transformation. According to Spinoza, "each thing [*res*], as far as it can by its own power, strives [*conatur*] to persevere in its own being."[37]

Might we not infer, therefore, that in this world, stone, clay, wood, metals, concrete, and all manner of polymers are alive and communicative? Might we not surmise that rather than waiting in standing reserve for architects to give them shape and meaning, all substances identifiable along the continuum that is 'matter to things' are always already involved in their own becoming? Although we've succeeded at transforming fossil fuels into building plastics by way of chemical synthesis—with near-apocalyptic consequences, I might add—the ground is no less capable of self-synthesis in support of its own life forms today than when first generated millions of years ago. Indeed, much that is more-than-human is perpetually animate, and according to Bennett, this is the "vitality of materiality" possessed by the family of "nonhuman bodies" whom we would do best to recognize "as actants rather than as objects."[38]

"Thing-power" occasions a humbling of human intent and of design.[39] The force that lives within all material things—regardless of its stage in the continuum of transformation from matter to materiality—invites us to "readjust the status of human actants: not by denying humanity's awesome, awful powers, but by presenting these powers as evidence of our own constitution as vital materiality."[40] As makers of material cultures, time and again, humans have demonstrated their overwhelming powers of creation. Yet, we've rarely acknowledged that we, too, are part of the 'nature' we manipulate. Having the foresight to do so, we might more empathetically participate in acts of making. In fact, according to Bennett, our ability "to *experience* the relationship between persons and other materialities more horizontally is to take a step toward a more ecological sensibility."[41] Doing so, however, requires a significant realignment between things, thoughts, and us. For as Bennett notes, it is easy

Figure 3. Primitive Hut redux, Architectural tryptic between thatch and field stone (Newtonmore, Scotland, Frontispiece of 1755 *Essai sur l'architecture*, and Inishmore, Oghil, Aran Island, Ireland). Digital composition by Franca Trubiano.

> to acknowledge that humans are composed of various material parts (the minerality of our bones, or the metal of our blood, or the electricity of our neurons). But it is more challenging to conceive of these materials as lively and self-organizing, rather than as passive or mechanical means under the direction of something non-material, that is, an active soul or mind.[42]

Following Descartes's formulation of how we know what we know, it is near impossible to imagine that our bodies have an independence of agency that is, at times, inarticulate to our minds: it is even more troubling to conceive that the materials with which we build have the same. But this we must, should we hope to build with greater empathy. Recognizing that the ground, the air, our waters, and the sources of energy we've learned to harness can be projective, cooperative, communicative, intelligible, expressive, and alive—as we are—offers an alternate path to re-corporealizing matter and to 'substanc-iating' the body that is architecture.

The Matter of the Muse Who Makes

Lastly, in seeking to substantiate she who makes critical theorist Stacy Alaimo, writing at the intersection of environmental humanities, animal studies, new materialism, and gender theory, reacquaints feminist scholarship with an expanded territory of the body. In her 2008 article "Trans-Corporeal Feminisms and The Ethical Space of Nature," Alaimo invites her readers to resist the intellectual silencing of all matters biological and generative. She calls for a discursive re-engagement with human corporeality for the purposes of restoring our relationship with 'nature.' It is hardly surprising, as she reminds us that given the general denigration of the body in the history of Western metaphysics, earlier feminists dismissed the intellectual and theoretical importance of materiality. And why wouldn't they, after all, they had been taught that to even think of matter and material things was beneath thought. Not to mention that for centuries, their own bodies had been the representational vessels for the projection of ideas conceived by others. In fact, the very subject of the body, their body, had been the reason for their exclusion from the arts, the sciences, and critical thinking in the first place. Obviously, earlier feminists stayed clear of matter, bodies, and corporeality. (Figure 4)

The cost of doing so, however, was high. According to Alaimo, the outcome was to "deny the material existence of the body" and to inadvertently "cast the body as passive, plastic matter."[43] More to the point, silencing of the "biological body" resulted in its removal from "evolutionary, historical, and ongoing interconnections with the material world."[44] For Alaimo, a more "ethical, political, and theoretically desirable" position is that of "trans-corporeality," a re-situating of the body within a vastly more expansive definition of matter that offers an important "contact zone between human corporeality and more-than-human nature."[45] In this alternate conception of the body, "the corporeal substance of the human is … inseparable from "the environment."[46] Trans-corporeality facilitates a more robust ecological reframing of the generative body when it instantiates "material fleshiness, [as] inseparable from "nature" or "environment."[47] According to Alaimo, "trans-corporeality" affords us the opportunity to 'believe' once again in the union of body and environment and in their ability to "meet and mingle in productive ways."[48] Here, "movement" between humans and more-than-human natures invites and makes possible "complex modes of analysis" that actualize the "material and discursive, [the] natural and cultural, [the] biological and textual."[49] Nature and environment are the hardly "mere background for [our] exploits," rather, they are "always as close as one's own skin," and never "inert, empty space" .. or [mere] ... resource for human use."[50] In this way, I argue that all architectural materials, be they clay, biogenic materials, metal ores, minerals, or polymers, are always already embedded in the environment, in nature, and hence, a part of us.

No convenient boundary exists between body and environment, matter and thought formless and formed, nature and artifice. Alaimo articulates this best when stating:

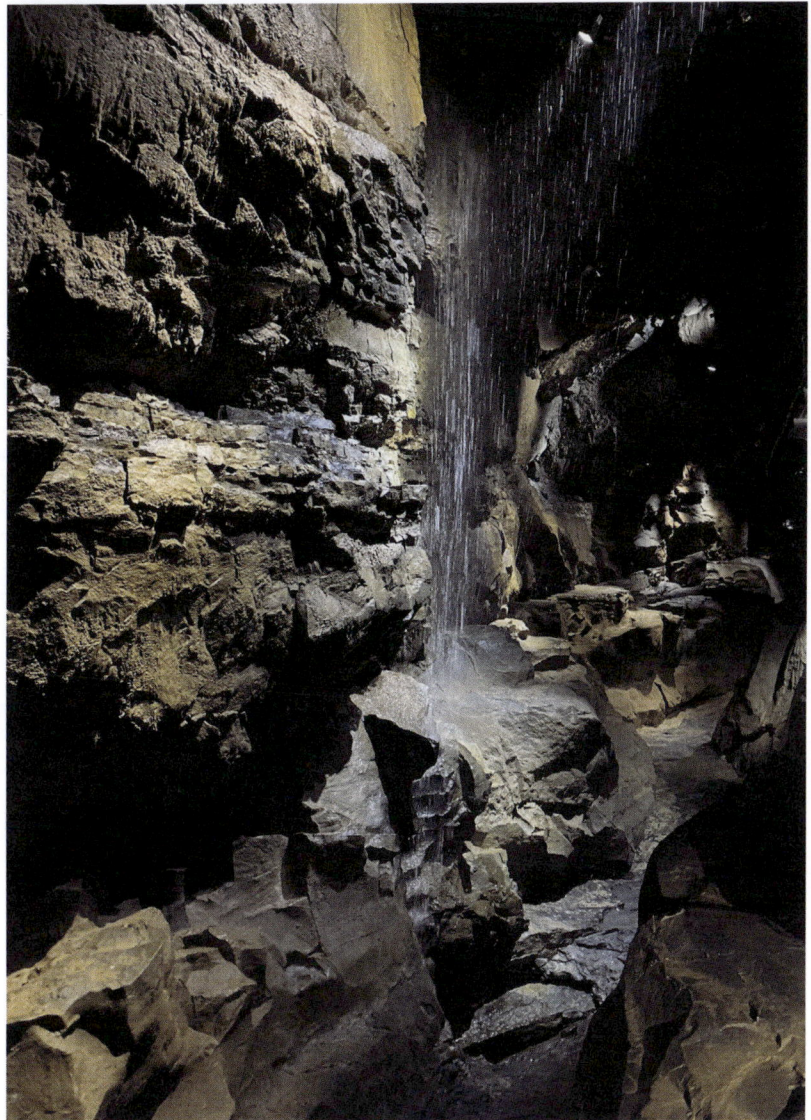

Figure 4. Interior of the Ailwee Cave, Baɬyvaughan, County Clare, Ireland. 2023. Photograph by Franca Trubiano.

Trans-corporeality opens up an epistemological "space" that acknowledges the often unpredictable and unwanted actions of human bodies, nonhuman creatures, ecological systems, chemical agents, and other actors. Emphasizing the material interconnections of human corporeality with the more-than-human world and at the same time, acknowledging that material agency necessitates more capacious epistemologies allows us to forge ethical and political positions that can contend with numerous late-twentieth-century/early-twenty-first-century realities in which "human" and "environment" can by no means be considered as separate: environmental health, environmental justice, the traffic in toxins, and genetic engineering, to name a few.[51]

The *prima materia* of architecture is as natural, alive, and corporeal as I am. There is no intellectually convenient divide separating human and more-than-human bodies, nor is there between stone, flesh, and mind. Trans-corporeality is the thread that weaves matter's environmental origins in the continuum that is nature, body, matter, artifice, and technology. Human intentionality is not the only force that animates the work of architecture. Materials speak, communicate, and self-organize. The question remains: are we willing to listen, engage, and follow? Indeed, are we willing to wait until the matter speaks first? If we do, we have a chance of learning its ecologically restorative powers; if we don't, we will continue to extract, manipulate, engineer, and endanger it into submission, forcing it to perform and speak in a voice not its own. For when matter speaks first, we find ourselves in the presence of what Alaimo calls a "fully vibrant, dangerously vibrant, matter, [where] ... not only is the difference between subjects and objects minimized, but the status of the shared materiality of all things is elevated."[52]

In a world where architecture—as the stuff of buildings and the art of design—might be less about imposing a language of signs than about our ability to act as mid-wives in matter's metamorphosis into materiality, ensuring we engage with life (both human and more-than-human) is the work of women who make.

Acknowledgements

The author graciously acknowledges the assistance of Ed Deegan, a Library Specialist who facilitated my access to Marc-Antoine's Laugier's 1755 Frontispiece in the *Essai sur l'architecture*, held in the Perkins Rare Book Library of the Fisher Fine Arts Library of the University of Pennsylvania (Figures 1 and 3), and Daniel Partridge, DAM and Digital Imaging Officer at The Royal Collection Trust for the permission to use the self portrait painted by Artemisia Gentileschi (Figure 2).

ENDNOTES

1. Assuming a critical mass of intellectual production by women since the early 1970s and designating the origin of recorded thought in sixth century BCE of the pre-Socratic philosophers, women have been active during fifty of the over two thousand six hundred years.

2. Katy Hessel, "Why do we still define female artists as wives, friends, and muses?" *The Guardian* (February 20, 2023), https://www.theguardian.com/artanddesign/2023/feb/20/why-do-we-still-define-female-artists-as-wives-friends-and-muses

3. Cody Delistray, "When Female Artists Stop Being Seen as Muses," *The Paris Review* (July 6th, 2018) https://www.theparisreview.org/blog/2018/07/06/when-female-artists-stop-being-seen-as-muses/

4. Marc-Antoine Laugier, *Essai sur l'architecture* (Paris: Chez Dechesne, 1755).

5. Willian H. Honan, "Hans J. Eysenck, 81, a Heretic in the Field of Psychotherapy," *The New York Times* (September 10, 1997).

6. Ibid

7. Ibid

8. Häns Jürgen Eysenck, *Genius: the natural history of creativity* (Cambridge; New York: Cambridge University Press), 127. I am indebted to the work of Helen Lewis whose BBC Podcast series "Great Wives" introduced me to the claims of Eysenck. https://www.bbc.co.uk/sounds/brand/m000z0cv

9. Eysenck, *Genius: the natural history of creativity*, 165.

10. Ibid., 169.

11. Ibid, 129

12. Ibid.

13. Sarah Bosley, "Work of renowned UK psychologist Hans Eysenck ruled 'unsafe'", *The Guardian* (October 11, 2019) https://www.theguardian.com/science/2019/oct/11/work-of-renowned-uk-psychologist-hans-eysenck-ruled-unsafe This new-story discusses challenges to his work that set out to identify a personality type most susceptible to cancer.

14. Despina Stratigakos, "Unforgetting Women Architects: A Confrontation with History and Wikipedia," *Women [Re]Build: Stories, Polemics, Futures* (Applied Research & Design, 2019).

15. Camille Paglia, *Sexual Personae, Art and Decadence from Nefertiti to Emily Dickinson* (New Haven: Yale University Press, 1990): https://doi-org.proxy.library.upenn.edu/10.12987/9780300182132

16. Luce Irigaray et al. 1993). *An ethics of sexual difference* Athlone Press. Retrieved from https://proxy.library.upenn.edu/login?url=https://www.proquest.com/books/ethics-sexual-difference/docview/839112306/se-2 ; Andrea Dworkin, *Letters from a war zone: writings 1976-1987* (London : Secker & Warburg, 1988); Elizabeth Grosz, *Volatile Bodies: Toward a Corporeal Feminism* (Bloomington : Indiana University Press, [1994])

17. Elizabeth Grosz, *Jacques Lacan, A feminist introduction* (London: Routledge, 1990 & 1998), 174.

18. Ibid., 177.

19. Ibid., 181.

20. Plato, *Timaeus and Critias* (Penguin Books, 1977), 69.

21. Ibid., 69-70.

22. Ibid., 70.

23. Ibid.

24. Ibid.

25. René Descartes, *Discourse on Method and the Meditations* (Penguin Books, 1968), 122.

26. Ibid., 123.

27. Ibid.

28. Antoine Picon, *The Materiality of Architecture* (University of Minnesota Press, 2020), 4.

29. Ibid., 7.

30. Ibid., 8.

31. Franca Trubiano, *Building Theories, Architecture as the Art of Building* (Routledge, 2023).

32. Jane Bennett, *Vibrant Matter: A Political Ecology of Things* (Duke University, 2010), 1.

33. Ibid., 3.

34. Ibid., 5.

35. Ibid., 3.

36. Ibid., 5.

37. Ibid., 2.

38. Ibid., 10.

39. Ibid., 16.

40. Ibid., 10.

41. Ibid.

42. Ibid.

43. Stacy Alaimo, "Trans-Corporeal Feminisms and The Ethical Space of Nature," in *Material Feminisms*, eds. Alaimo and Hekman (Indiana University Press, 2008), 217.

44. Ibid., 217.

45. Ibid.

46. Ibid., 218.

47. Ibid., 217.

48. Ibid.

49. Ibid., 218.

50. Ibid., 218.

51. Ibid.

52. Bennett, *Vibrant Matter*, 13.

BIO
MATTER
TECHNO
SYNTHETICS

ETHICAL AND FECUND DESIGN FUTURES

SUSAN KOLBER

The prefix 'bio' signals life and is an invitation for essential questions. Whose lives matter? What sustains life? This chapter explores how designers can consider these questions as they advance their own ethical objectives while navigating social, environmental, and technological provocations when working with diverse webs of life. The selected authors address uncertainties by exposing critical histories while proposing novel socio-ecological designs for multi-species thriving, radical landscapes, urban forms, and buildings. Bio features both realized projects and speculative designs across a range of scales. The authors document what it means to be at the forefront of resilient and sustainable design, to explore the spectrum of human and more than human agency, and to confront the challenges of resource scarcity in the face of survival.

Ayasha Guerin's writing challenges designers to rethink the meaning of 'resilience.' Guerin's work traces the historical, social, and ecological relationship between Black marine traders and oyster habitats in the nineteenth century. Guerin investigates their prosperous mutuality and the interdependent effects of technology, colonial capitalism, and racist policy along New York's shorelines. The essay's interspecies narrative invites designers to scrutinize their own projects' situated histories. By better understanding the connected social and environmental vulnerabilities inherent in a place, designers can create futures that foster more effective, just, and inclusive resilience.

In reflecting on three decades of shaping urban waterfront landscapes at OLIN, **Lucinda Reed Sanders** affirms the need for more nuanced considerations around the work of shoreline resiliency. Sanders defines the evolving concepts of 'sustainability,' 'resilience,' and 'adaptation' in projects and demonstrates how ideas and hypotheses can change and evolve from their initial concept to the completion of a project and beyond. The six case studies featured demonstrate the ever-expanding variables that designers must consider as they endeavor to spark true transformation and meaningful change at the threshold where cities meet the water.

Gundula Proksch envisions a speculative, aqueous Circular City that responds to challenges of water scarcity in the era of climate extremes. Through an innovative concept, Proksch introduces an interconnected architecture of water-based food web systems featuring aquaponics, microalgae cultivation, and ecological water treatment. This controlled urban ecology promotes both ethical living and resource efficiency. Proksch delves into the scaling and design of these systems, suggesting a technocentric approach that can support a city with integrated agricultural and life-sustaining practices.

Sonja Dümpelmann's contribution elevates the agency of plants, urging designers to seek an enhanced understanding of their full potential. Dümpelmann underscores the limitations of conventional landscape architecture by discussing the notions of 'duplicity of landscape' and 'plant blindness.' She examines two synthetic landscape materials—the extensive green wall constructed for the 1915 Panama Pacific International Exposition in San Francisco and 'Watson Pots,' a precursor to grass pavements developed during World War II to conceal British runways from German bombers. The 'duplicity of landscape' discusses the use of plants to mask and camouflage both spaces and political implications, rendering them invisible. 'Plant blindness' highlights the capacity of plants to go unnoticed despite their active presence and impact.

Aroussiak Gabrielian posits how speculative design can inspire novel, ethical trans-species relationality essential for collective survival. Gabrielian's "Posthuman Habitats" project reveals the intimate ecology between species and materials (humans, plants, bacteria, fertilizer, etc.) to sustain life. The project's site is the human body. Gabrielian imagines how a cloak, draped over the wearer like a second skin, can become a personal and mobile edible landscape. This act of wearing and caring for the second skin de-centers the human species, allowing for an expansive appreciation of multispecies agency.

Shifting scales to a sustainable facade for a building, **Anne Beim**'s work in Denmark utilizes abundant, locally sourced reed and clay, producing fire-safe thatch facades that minimize carbon emissions and building waste. These materials have been used in historic buildings for centuries. Beim's work reminds us of the importance of intuition and traditional construction alongside advancements in twenty-first century sustainable technology, offering designers a critical reframing of biogenic possibilities in contemporary building practices.

Seen holistically, the authors in the Bio section interrogate the issues of our interconnected survival, our ability to thrive with divergent webs of life creatively and responsibly, and the possibilities for a fecund future. Recognizing the indelible impact that humanity has left on our planet, these essays offer an empathetic approach, providing ground for more inclusive and life-affirming models of design.

OYSTERS AND THE BLACK STRUGGLE FOR FREEDOM:

AN INTERSPECIES HISTORY OF EXPLOITATION AND RESILIENCE IN NEW YORK'S AQUAPELAGO

AYASHA GUERIN

ABSTRACT

Before European colorization, New York was one of the most oyster-rich habitats in the world. These once productive reefs were exhausted, however, in just two centuries of settlement. A focus on Black life in the oyster trade highlights the ways in which Black work at sea was mediated by desires for freedom on land and how marine entanglements have assisted Black fugitivity, liberation, and community empowerment. This essay details how extractive practices in the aquapelago ultimately exploited both human and non-human life, reflecting inter-species interdependencies, endangerment, and habitat loss under colonial capitalist policies. Considering the intersection of environmental and social justice, this paper models the importance of historicizing the liminal space between land and sea for advancing ideas about race, nature, and value in considering future plans for 'resilience' along New York's waterfront.

+ oysters
+ African Americans
+ resilience
+ New York's
  Aquapelago
+ social justice

Figure 1. Detail of "Oyster Heap", Alice Austen Collection, Collection of Historic Richmond Town. Photo by Alice Austen, 1894.

Before European colonization in the early seventeenth century, the New York estuary was filled with millions of oysters. The waters, including Jamaica Bay, Raritan Bay, and the Jersey Flats, were believed to have held as many as half of the world's oysters, according to some environmental historians.[1] Oysters supplied Algonquin peoples, including Lower Manhattan's Lenape, with a large part of their diet, especially during winter when land life hibernated.[2] European settlers also adopted the oyster as a mainstay of their culinary culture, and the reefs would come to feed not only a growing New York population but also the southern and West Indies colonies connected to New York by company trade routes. This essay argues that New York's islands and waters have experienced extractive histories of colonial capitalism, which radically altered coastal ecologies and introduced social hierarchies on land. Oysters were pivotal to these histories and the processes by which these extractive narratives took place can be traced to former shorelines and reefs, many of which are urban flood zones today. This was the first land that enslaved people touched when they arrived in 1626 in New Amsterdam, the Dutch trading outpost where they were hired as day laborers by their enslavers during British Rule. During slavery's gradual abolition in the nineteenth century, Black people continued to learn sea skills in New York Harbor, which allowed them to find work as independent dock hands and mariners and to build networks of labor that offered spaces of autonomy and community solidarity.[3]

Harbor Histories

The context and consequences of marine trade activities in New York's Harbor are deeply entangled with the history of Native American and Black people's experiences.[4] Between 1740 and 1865, Black men had few alternatives to employment at sea, when they made up about one-fifth of the total maritime force on the eastern seaboard.[5] Maritime work not only influenced the Black community's coastal settlement but also the destinies of the aquatic species with which their lives were entangled at sea.

From the seventeenth century onwards, captains of oyster sloops relied on slave labor to help harvest reefs.[6] Being an oysterman in New York required mastery of sailing and the muscle to operate wind-powered boats. These were skills that many Black laborers developed after two centuries of enslaved labor in Lower Manhattan ports. Some possessed sea skills from West Africa and their sustainable indigenous knowledge for net-harvesting mollusks was learned and passed down through generations of enslaved people brought to the Americas. Many free Black people in New York chose to work as common oystermen, and in 1810, in the infancy of the oyster trade, more than half (sixteen of twenty-seven) of the oystermen listed in the city directory were free African Americans.[7] Unlike working a job on land, where competition with poor White laborers for employment could be dangerous, Black oystermen worked for themselves most of the year.

As the port city grew during the nineteenth century, struggles for Black liberation leveraged the knowledge of and access to sailing networks and to the hunting, gathering, and trade of sea animals. Some maritime work, like whaling, aided the escape of Black peoples because it took sailors out to sea for long periods, while steady wages from the oystering industry provided many Black longshoremen and deckhands seasonal jobs as shuckers in New York Harbor.[8] For Free Black people, to be a common oysterman meant one could support collective efforts in Black liberation by investing in local community services like Black schools and churches. It also meant they could participate in securing passage for enslaved Blacks by helping 'conduct' movements through stations of the Underground Railroad that were concentrated on the waterfront.

Coastal space served as an important place of Black fugitivity where countless escaped enslavement by finding work aboard the ships and barges that docked in the harbor. Free Black people actively led abolitionist activities on ships, docks, and in sailors' boarding houses in New York.[9] Income from their sea trades afforded them opportunities to buy kin out of slavery under gradual emancipation and to buy property, which was then a prerequisite for Black men to vote. For New York's Black residents, who were continually ghettoized in land-filled neighborhoods built on exploited reefs, this tenuous ground was often the first site of freedom. Despite insecure footings, these neighborhoods were hotbeds of progressive organizing, where interracial relationships and new sources of Black capital empowered political struggles for racial equality.[10] (Figure 2)

As New York's global position shifted from that of a European colony to its new status as the emergent manufacturing capital of North America, waste and pollution streamed into the harbor. The coastal flats of New York played a formative role in shaping and financing New York's urban expansion, but indigenous aquatic life suffered under this urban growth. Shellfish were poisoned, and reefs deteriorated due to habitat encroachment and eutrophication caused by the dumping of sewage.[11] These effects endangered life for human and nonhuman communities in nineteenth-century New York, and they continue to influence the risks of living in the flood zone, even today.

Aquapelagic Frameworks

In the field of Island Studies, a debate about conceptualizations of island space and the sociocultural life of islands can be traced to the emergence of the term 'aquapelago'. The term was proposed by coastal researcher Philip Hayward when referring to the integrated marine and terrestrial assemblages generated by human habitation and activity in particular island locales.[12] Hayward's 2015 essay "The Aquapelago and the Estuarine City: Reflections on Manhattan" describes New York's aquapelago and brings particular focus to the socioecological constructions on and between New York's islands'

MIDSUMMER IN THE FIVE POINTS.

waterfronts. It encourages a reading of the relationship between humans and their aquatic environments as "aqua pelagic assemblages," where human and nonhuman life change and develop each other.[13]

This work was influenced by that of political theorist Jane Bennett, whose 2010 book *Vibrant Matter* looks at the agency of nonhuman matter and engages the field of relational ontology with ecology, encouraging a new-materialist study of "the political ecology of things."[14] The question Bennett asks in her introduction—"How could political responses to public problems change were we to take seriously the vitality of nonhuman bodies?"— is layered with new meaning for this study when we consider the historical exclusion of Black bodies from humane treatment and the erasure of slavery's impact on New York's waterfront development.[15] Moreover, as Black Studies professor Christina Sharpe writes in her book *In the Wake: On Blackness and Being*, the "history of blackness is testament to the fact that objects can and do resist."[16]

Acknowledging the uneven hazards planned into contemporary flood geographies through racial and ecological exploitation, one might ask, "How might political responses to public problems change, were we to take seriously the vitality of Black bodies?" There are infinite ways to think materially about the influence of Black bodies on the ecologies of the New York waterfront and about that of aquatic "vibrant matter" on Black experience and geographies. I've previously argued that an aquapelagic framework might encourage us to study racially dissonant experiences of time and, thus, history.[17] This aquapelagic site reframes our analysis of urban socio-ecological relations as entanglements between land and oceanic space and helps us set this study of inter-species relations in New York's nineteenth-century harbor.

This essay also joins Katherine McKittrick and Clyde Adrian Woods' call to explore "how the Underground Railroad might be theorized as a complex, non-linear, diaspora geography."[18] I look to New York's maritime trades as a frame with which to study historical entanglements of sea life and Black life. A focus on oyster reefs, barges, and coastal community settlements which are in today's flood zone (formerly marshlands), highlights the interdependencies of precarious life lived in the space between land and sea, as well as the ways in which work at sea has been mediated by desires for freedom on land. In exploring the ways in which experiences of Black life have been bound up in the aquatic, I consider how activities of racial exploitation, coupled with species resource extraction, have produced and exacerbated the crisis in New York's urban waterfront flood zones, historically and presently.

Oysters and Oystermen

Oysters grew along the shores of Jersey City, Manhattan, Brooklyn, and the islands of Ellis and Bedloe.[19] Some of the richest oyster beds in the

Figure 2. "Midsummer in the Five Points", by C. A. Keetels (d.1898), The Miriam and Ira D. Wallach Division of Art, Prints and Photographs: Picture Collection, The New York Public Library. Permission to publish secured by Editors.

entire country were in the southern portion of the Lower Bay, known as Raritan Bay. Most of them were on shoals, under ten to twenty feet of water.[20] A huge bed, later known as the Great Beds, was located at the western end of the bay just beyond the mouths of the Raritan River and the Arthur Kill. Oysters grew along the Raritan River from its mouth to five miles upriver along the entire length of the Arthur Kill, and, to an extent, in the Kill Van Kull. Another natural bed known as the Chingarora Bed was located on the south of Raritan Bay.[21]

Oystermen saw their livelihoods threatened during the first oyster collapse, when most of the natural shellfish beds in Kill Van Kull, Arthur Kill, and Prince and Raritan Bays began to thin as early as the 1810s. To preserve their livelihood and meet increasing demand, they imported and transplanted to the New York aquapelago small "seed" or tiny, larval "set" oysters from fertile Virginia beds.[22] The immature oysters were left alone until they reached market size, which took from one to four years, according to how mature the oysters were, to begin with. Oystermen had to clean the empty shells and bottom trash off the beds where they were transplanted to spread them out as evenly as possible. Handled this way, oysters grew faster than they did on natural beds, were more uniform in size and shape, and were considered to have a better flavor, bringing higher premium prices.[23] By the end of the nineteenth century, there would be a second collapse, this time because waterfront industrialization and an extractive view of profit maximization of cultivated beds developed unsustainable practices without regard or care for ecosystem interdependencies. (Figure 3)

Oystermen sold their harvests to merchants, who controlled the marketing of the trade and oversaw it from brightly painted barges along the shoreline. The decorative barges were not seagoing vessels and resembled two-storied houseboats. As intermediary spaces, they provided a floating factory/market at the edge of the city. Extending the architecture and commerce of the city into the waves, oyster barges effectively blurred the boundary between land and sea, as one side of the barge was open to oystermen who arrived by boat, while the other side hosted a ramp to trade at the shore.[24] Merchants stood at balconies at both the front and back end of the barges to oversee business and watch for theft, while shuckers worked in the shucking and packing rooms on the water deck below. Because oysters were frequently sold by names according to their source (Malpeques, Cotuits, Robbins Islands, Blue Points, Rockaways, Sounds, Prince's Bays, Shrewsburys, Maurice River Coves, Bombay Hooks, Potomacs, Rappahannocks, and Chincoteagues), the barges educated passersby on the ecological web of the wider New England oyster trade and signaled their niche in a web of environmental relationships that extended beyond the docks of Manhattan.[25] The piles of oyster shells that accumulated at their wharf-side moorings, like those of the native Lenape middens before them, invited contemplation of the significance of the abundant marine environment that grew the diets and fortunes of New York City.

Figure 3. "Oyster Houses, South Street and Pike Slip, Manhattan," photo by Berenice Abbott. The Miriam and Ira D. Wallach Division of Art, Prints and Photographs: Photography Collection, The New York Public Library. Sourced from Wikimedia, Oyster_Houses,_South_Street_and_Pike_Slip,_Manhattan_(NYPL_b13668355-482643)

Figure 4. "Oyster Heap", Alice Austen Collection, Collection of Historic Richmond Town. Photo by Alice Austen, 1894.

OYSTERS AND THE BLACK STRUGGLE FOR FREEDOM

Sandy Ground, on the south shore of Staten Island, is one of the oldest surviving oystering communities established by freed enslaved people in North America and one of the region's most important. By 1865, at least four Black men captained their own sloops, and dozens more worked as common oystermen in the oyster trade. By the 1880s, Sandy Ground was "really quite a prosperous little place," where as many as forty African-American Sandy Grounders were common oystermen by 1900.[26] By the nineteenth century, the Sandy Ground community grew to include White settlers and oystermen as well, and the industry grew to host tens of thousands of full-time occupations for New Yorkers. (Figure 4)

Relationships developed in Sandy Ground influenced the larger socio-ecological landscape of the harbor. Oyster work in the aquapelago offered one of the few contexts where Black and White people did business as equals. The interracial nature of New York's oyster trade was unique, where Black oystermen sold seed to White oystermen. It also relied on oak baskets by the thousands, and Black Sandy Grounders utilized their basket-making skills by cutting white-oak saplings, splitting them into strips to soak in water, and weaving them into bushel baskets.[27] Common oystermen were hired by captains for part of the season, but they otherwise worked to cultivate the public, natural growth beds, where it served their best interest to act as ecosystem stewards. Everyone sold to the merchants on Manhattan barges. Their harvest was only regulated through a series of traditional customs that promoted the long-term sustainability of the beds, as they were seen as a resource that was common to all local oystermen.

Traditionally, common oystermen were permitted to glean the oysters left on the captain planters' holdings after the harvest. Underwater, private property was an unsettled concept, as the ownership of a planted bed was considered forfeited if the planter did not return to harvest annually. Time served as an important mediator in the trade because it determined public access to private beds. On the Lower Manhattan waterfront, it was urgent that the bulk commodity be handled efficiently so that the tens of thousands of oysters that arrived in vessels to the barges each day did not spoil before they were redistributed around the city by cartmen. One of the most prominent jobs (on land) for African Americans in the nineteenth century was that of cartmen who transported goods from the port to businesses throughout the city. Their importance is acknowledged in the relatively high wages oyster loads paid. While cartmen rates were fixed by law and set according to the commodity transported, oysters were one of the better-paid loads at thirty-one cents per load.[28] (Figure 5)

Cooperative and intimate relationships between oyster, oysterman, captain, cartman, and merchant supported the growth of the industry and of Black wealth in New York. Black people worked at all levels of the trade, from sailing and planting on the water to shucking, carting, and managing oyster saloons on land. Moreover, oyster cellars catered

Figure 5. "Men Shucking Oysters", Alice Austen Collection, Collection of Historic Richmond Town.

to different clientele depending on their neighborhood, but in the first half of the nineteenth century, it was widely accepted that oyster cellars were run by Black people.[29] One such person was Thomas Downing, who listed himself in the 1823 city directory as an oysterman and who rose to prominence as the trade's wealthiest businessman. Like the oyster spats that were brought from fertile Virginia beds to grow in New York, Downing was born in the Chesapeake Bay area and made his way north at age twenty-one, following troops during the War of 1812.[30] He began tonging for oysters with a small skiff and he sailed across the Hudson to the New Jersey Reefs (known as Communipaw and Harismus coves). By 1830, he had opened Downing's Oyster House, located on the corner of Broad and Wall Street, a fine-dining restaurant with plush rugs and chandeliers that catered to New York's elite. His was the most celebrated oyster cellar. Fine dining was, however, segregated, and with the money he made from his elite White patrons (who often had investments in business that relied on slavery themselves), Downing founded several Black schools, a church, and the all-Black United Anti-Slavery Society of the City of New York.[31] In fact, Downing's extended family was engaged in abolitionist organizing and activism. In a 1910 biography of his son George Thomas Downing, it is noted that "Downing's public career began when he was but a youth as an agent of the Underground Railroad."[32]

The Downing family story highlights Black New Yorkers' relationship with the oyster as an important example of interspecies history; it not only allowed for plentiful diets but powered the financing of liberatory Black politics. Selling large quantities of oysters daily, Downing expanded his business in 1835, renting the basements of two neighboring buildings on Broad Street, which today face the New York Stock Exchange.[33] Conveniently located a short walk to the water's edge, one might surmise that Downing's family likely used these cellars to hide Black fugitives on the run. With damp floors and walls, it was an ideal environment for oyster storage, but the underground hiding spot would have offered difficult conditions for people.

Extractive Zone

The shift from subsistence gathering to seeding and growing of the commons and to the industrial extraction era of oystering coincided with the increased privatization of underwater commons. New legislation after 1900 imposed a costly licensing procedure, requiring the registration of beds and taxation on oyster habitats by area, which allowed for merchants to cut out the oyster trade middlemen and to invest in thousands of acres of harbor waters themselves.[34] It also refused common oystermen the right to work planted beds after the end of the season, a right that had allowed a large sector of the industry to grow. The merchants behind these changes called themselves "oyster capitalists," who turned the trade into integrated corporations that employed dredge technologies on powered boats.[35]

In this way, a subsistence relationship with oyster reefs was changed into a colonial extractive enterprise that sought to maximize profits without regard for ecosystem dynamics and needs. In addition, the overgrowth of the city's population at the end of the nineteenth century overwhelmed the harbor with sewage contaminants, causing severe oxygen drops in regional waters and occasional fish kills.[36] Not only was New York's artisanal production of oysters industrialized with the help of new technologies like the dredge, spoil from the slips at the city's littoral was hauled to areas with oyster reefs. In this way, factories refining sugar and petroleum along Brooklyn and New Jersey waterfronts added industrial pollution to the problem. Runoff from John D. Rockefeller's Standard Oil, for example, released three million gallons of crude oil per week into the New York estuary, poisoning shellfish.[37] All of these factors accelerated the industry's collapse at the turn of the twentieth century.[38]

Not only did the dredge era of oystering reorganize archipelagic relations underwater, but it also changed race relations in the industry. Unlike the diverse groups of actors who had understood the fragile tipping points of harbor ecosystems and acted as stewards in an artisan-like culture of oystering, oyster capitalists viewed beds as short-term resources for their own fast profit. The final component of the attack on the customary culture of oystering was the mobilization of Jim Crow racism, which legally separated and excluded African American oystermen from spaces and work opportunities. Many Black oystermen found themselves unemployed as they faced discrimination by the new oyster companies during harvests. These companies would not buy seed or mature stock from them, as captains once had. When they were hired, Black oystermen were only allowed to work as deck and shore hands.[39] The final nail in the coffin for the oyster trade in New York came in 1916, when industrial pollution from the Brooklyn, Queens, and New Jersey waterfronts forced the closure of oyster beds, with many oystermen abandoning their trade for land labor.[40]

What happened to Black communities who relied on the oyster trade? Anthropologist William Askins, who studied race relations from the nineteenth to the twentieth century in the Black oystering community of Sandy Ground, suggests that the degree of equality experienced in the trade served as a significant model for social relations between non-oystering Blacks and Whites of the region, such that the oystering culture of respect and conviviality was extended and accepted by the residential community at large.[41] This meant that the interracial society responded to attacks by oyster capitalists by calling upon and reinforcing rituals of community solidarity. The ripple effect of equal race relations in Sandy Ground's oyster culture had implications for the community on land, where they rejected racist practices and symbols used to divide the harbor workforce.

However, by 1956, when a profile on Sandy Ground was published in *The New Yorker*, the community consisting of "forty or fifty

Southern-looking frame dwellings and a frame church," was described as a ghost town with a disproportionate number of old people.[42] Mr. Hunter, an elderly man who had moved to Sandy Ground as a child, was raised by an emancipated Black mother and a Black oysterman father. Hunter explained that after the collapse of the Oyster industry, "the men in Sandy Ground had to scratch around and look for something else to do, and it wasn't easy."[43] Abandoning their sea skills for undesirable jobs on land as janitors, garbage men, and porters caused "a lot of life to go out of the settlement."[44] The Church, the center of community life, was especially hard hit. As Mr. Hunter told it:

> Many of the young men and women moved away, several whole families, and the membership went down. The men who owned oyster sloops had been the main support of the Church, and they began to give dimes where they used to give dollars.[45]

Traditional maritime skills could not compete with new industrialized waterfront uses that were less mindful of the limits of the natural world. But it was the rise of Jim Crow era racism which excluded Black people from the industry they helped to foster, and that hit Black enclaves like Sandy Ground especially hard. The stories of Thomas Downing and the Sandy Ground oystermen offer historical lessons for navigating anti-Blackness that might be incorporated into acknowledgments of racial disparities in conversations about social and ecological coastal resilience today.

New York's Flood Zones Today

Flood zones are often thought of in the future tense because they reference an insurance term associated with flood risk. In ecology and sociology, "resilience" refers to a view of life made up of systems and their ability to survive disturbance. More practically, author Ashley Dawson identifies that the National Academy of Sciences defines resilience as "the ability to prepare and plan for, absorb, recover from, and more successfully adapt to adverse events," a term increasingly deployed to connote "an ability to withstand the various, unpredictable shocks of the catastrophic convergence of urbanization and climate change."[46] In the twenty-first century, New York's flood zone has emerged as a space of future-oriented resilience planning and policy, yet few of these efforts take seriously the aquapelago's past. New York City's waterfront—the site which witnessed the discrimination of Black oystermen—is at the frontline of the socio-ecological crisis and continues to host precarious constructions of urban living today. Yet, the future orientation of resilience discourse often overlooks the importance of historical and political conditions that contributed to the environmental and social crisis in the first place.

To understand how climate change is affecting the world of aquapelagos, we might view their waterfronts as interrelated, mutually constituted, and co-constructed places, where coalition building and interspecies interdependencies are vital for robust waterfront life. Providing food, storm protection, and water filtration makes reefs absorb the energy of storm surges, reduce their impact inland, and are the best ways to protect coastal communities from threats like hurricanes and sea level rise.[47] Rather than understanding islands as isolated, territorially bounded political spaces, as processes of colonization and urban jurisdiction have been assigned, it is imperative that we also trace how constructions of racial hierarchies have driven speculative and exploitive treatments of life and land across coastal ecologies.

Ocean acidification, waste-dumping, dredging, and other extractive activities continue to drive estuarine and coastal ecosystems into systematic ecological decline, endangering coastal communities that are now at the frontline of flooding and storm surges. While plenty of architectural and landscape designs propose to tackle the question of climate adaptation, it is less clear how future waterfront plans might address the social conditions that exacerbate risks associated with displacement, including plans that tackle the question of social resilience within the flood zone.

Since European colonization, the reefs and marshlands between New York's islands have been reclaimed for human settlement. But the water and species that settlers encroached upon, as well as the chemicals and bacteria that streamed into these spaces, have also been agents and co-creators of the city and its political conditions. There is more to learn from the trans-species intersectionality of exploitation in historical littoral relationships for modern questions of coastal resilience. The ongoing challenges facing vulnerable communities in New York require ways of thinking that are responsive and responsible to the historical conditions that continue to produce risky assemblages of being.

More broadly, there is a need for greater marginalized perspectives in New Materialist studies so as not to overlook the importance of racialization and environmental racism as determining factors shaping the ecological entanglements of the aquapelago at any given time and across time. More directly, the largest drawback of the aquapelago framework within Island Studies, Post-humanism, and New Materialist viewpoints is the lack of attending to how historical, colonial relationships have shaped the socio-ecological assemblages between species—both animate and inanimate matter—differently across racialized geographies. To advance discussions of "the political ecology of things," as in the work of Bennett, Hayward, and others from Island and Ocean Studies, we must acknowledge how historical constructions of racial difference and space have influenced human and nonhuman relationships.

New York's reefs and waters were heavily exploited and polluted within just two and a half centuries of colonial settlement, and parallel histories of human and environmental subjugation at the waterfront demonstrate the mutual embeddedness of environmental and social justice challenges. Navigating anti-blackness was communal practice for New Yorkers who worked at sea and who lived and worked in the flood zone. Collective bed care fostered the sustainability of reef life and social life among oystermen, with the relationship between oysterman and oyster a mutually supportive one. Both the livelihoods of Black mariners and oysters were challenged by the privatization of the coastline, the encroachment of a growing city on its reefs, and the polluting of the aquapelago by industrial runoff. This suggests that contemporary social and environmental justice movements can also have mutually constitutive goals and benefit from interspecies considerations. Moreover, a close study of what resilience has looked like for Black communities in the history of New York has meant looking at systems of inter-species relation, care, and cultivation that supported their mutual survival. The importance Black oystermen had in nineteenth-century cities as planters and stewards of harbor ecology provided them with economic freedoms withheld from them inland trades. Their stories offer the future of New York's social and ecological resilience planning, as well as important warnings that are sure to inform climate justice efforts in the flood zone today.

Resilience Planning Projects and Social Justice

A refreshing treatment of resilience thinking, which considers the important role of oysters and reef stewards in New York, is SCAPE's *Living Breakwaters* project, a winner of New York's Rebuild by Design Competition launched by the United States Department of Housing and Urban Development (HUD). The competition solicited landscape architecture designs for urban adaptation to climate change after Hurricane Sandy caused an estimated nineteenth billion dollars in damages to the city in 2012. Because channel dredging and the diminishment of natural and farmed oyster reefs have left Staten Island's south shore increasingly exposed to storm surges, *Living Breakwaters* proposed a 'necklace' of offshore breakwaters to "reduce risk, revive ecologies, and connect residents and educators to Staten Island's southeast shoreline."[48] The proposal called for collaboration with *The Billion Oyster Project*, a program that works with public high school students at the Harbor School on Governors Island to restore oysters to seven restoration sites throughout New York Harbor.[49] The *Living Breakwaters* project is expected to be completed in 2024.[50]

Further out on Long Island, NY, the *Shinnecock Indian Nation Climate Vulnerability Assessment and Action Plan* (first drafted in 2013 and again in 2019), describes the risk of land loss along the shoreline. The plan reports that the southern portion of the tribal land already experiences routine flooding which will become more pronounced with sea level rise. The Shinnecock response has been the implementation of a seven-component coastal habitat restoration project to restore three thousand feet of shoreline to Shinnecock Bay. This will add acres of habitat for marine life such as oysters, crabs, fluke, and flounder and halt the erosion of the land in that area.[51] Leadership by the Shinnecock

Nation to protect tribal land and community, human and non-human, should serve as a model for how we approach resilience plans on New York City's flood zone. (Figure 6)

Wetland and oyster restoration projects matter because had New York City's oyster beds been intact the damage from Hurricane Sandy, which flooded New York's subway systems and inflicted billions of dollars in damage to the flood zone, would have been attenuated by thirty to two hundred percent.[52] Furthermore, they matter in the context of environmental and racial justice. The Mohawk artist Alan Michaelson makes work that engages the destruction and transformation of Indigenous environments by colonialism. He is interested in exploring these effects "from the tiniest plant to the largest nonhumans around."[53] Michaelson's work in the MoMA PS1 2021 *Greater New York* show was an installation titled *Midden*, referring to the mounds of oyster shells Dutch colonial explorers found when they settled Mannahattan. Video captures of industrially polluted New York waterways, which were once flourishing oyster habitats, are projected onto a bed of oyster shells (on loan from the *Billion Oyster Project*). The viewer watches the images shift and change while the audio of the Delaware Skin Dance group sounds like a traditional call-and-response song with drums. *Midden* provokes attention to the materiality of socio-ecological loss in New York, eliciting a spiritual response to the violence of extraction and exploitation. It reminds us of the importance of connecting Native American and Black histories of dispossession, displacement, and environmental justice and, more broadly, of addressing the impacts of settler colonialism and anti-blackness together.

The problems faced by vulnerable coastal communities in New York require plans and policies that foster relations of care and reparation. Designing for the New York aquapelago today necessitates research across social and environmental histories to identify both forms of resistance and companion relationships across racial and species differences. The Black struggle for freedom in the harbors of the Algonquin people reveals the importance of attending to socio-ecological relations at the shoreline. Designers are challenged to connect the effects of colonialism and racial subjugation towards ways of thinking, creating, and caring that are responsive to the historical conditions that continue to contribute to risky assemblages of being.

Acknowledgments

This is an abbreviated and edited version of an article that originally appeared in *Shima* 13 n.2 in 2019.

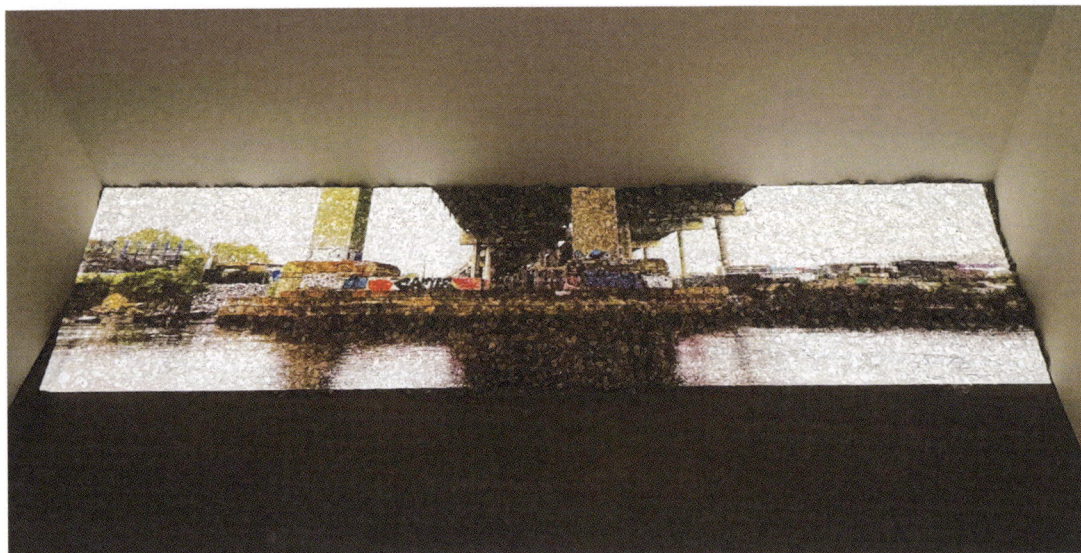

Figure 6. Installation view of Alan Michelson, *Midden* (2021) in *Greater New York* 2021 on view at MoMA PS1 from October 7, 2021, to April 18, 2022. Image courtesy MoMA PS1. Photo by Noel Woodford.

ENDNOTES

1. Mark Kurlansky, *The Big Oyster* (New York: Random House, 2006), 35.

2. Evan T. Pritchard, *Native New Yorkers: The Legacy of the Algonquin People of New York* (San Francisco: Council Oak Books, 2007), 39. See also, Eric W. Sanderson and Markley Boyer, *Mannahatta: A Natural History of New York* (New York: Abrams, 2009), 106-07.

3. New York's strong trading relationship with the West Indies resulted in more than three thousand slaves being imported into New York in the first four decades of the 18th Century. Men, women, and children continued to be bought and sold throughout the city until 1799, when New York passed a Gradual Emancipation act that freed slave children born after 4th July 1799 but indentured them until they were young adults. In 1817 a new law passed that would free slaves born before 1799 but not until 1827. See John L. Rury, "Philanthropy, self-help and social control: The New York Manumission Society and free Blacks, 1785-1810," *Phylon* 46 no.3 (1985): 232; Edgar J. McManus, *A History of Negro Slavery in New York.* (Syracuse University Press, 2001), 154, 159.

4. Ayasha Guerin, "Shared Routes of Mammalian Kinship: Race and Migration in Long Island's Whaling Diasporas," Island Studies Journal 16 no. 1 (2021): 43-61.

5. W. Jeffrey Bolster, *Black Jacks: African American Seamen in the Age of Sail* (Cambridge: Harvard University Press, 2007),159. See also, Michael Sokolow, *Charles Benson: Mariner of Color in the Age of Sail* (Amherst: University of Massachusetts, 2009), 45-46.

6. Charles R. Foy. "Blacks on the New York Waterfront during the American Revolution." *Red Hook Waterstories.* https://redhookwaterstories.org/exhibits/show/blacks-on-thewaterfront/blacks-waterfront-american-rev.

7. John H. Hewitt, "Mr. Downing and His Oyster House: The Life and Good Works of an African-American Entrepreneur – 19th Century New York, New York Restaurateur, Thomas Downing," *New York History* 74, no. 3 (1993): 240.

8. Kurlansky, *The Big Oyster,* 181.

9. Mary Malloy, "African Americans in the Maritime Trades: A Guide to Resources in New England," *Kendall Whaling Museum Monograph* Series 6 (Canton: Blue Hill Press, 1990), 7.

See also, Martha Putney, *Black Sailors: Afro-American Merchant Seamen and Whalemen Prior to the Civil War*, (Westport: Greenword Press, 1987), 100.

10. Carla L. Peterson. *Black Gotham: A Family History of African Americans in Nineteenth-Century New York City.* (Yale University Press, 2011).

11. Peter J. Jacques, "The Origins of Coastal Ecological Decline and the Great Atlantic Oyster Collapse," *Political Geography* 60 (2017):154–64.

12. Philip Hayward, "Aquapelagos and Aquapelagic Assemblages," *Shima: The International Journal of Research into Island Cultures* 6 no. 1 (2012): 1–11.

13. Philip Hayward, "The Aquapelago and the Estuarine City: Reflections on Manhattan," *Urban Island Studies* 1 (2015): 81–95.

14. Jane Bennett, *Vibrant Matter: A Political Ecology of Things* (Durham: Duke University Press, 2010).

15. Ibid., vii.

16. Christina Sharpe, *In the Wake: on Blackness and Being*, (Durham: Duke University Press, 2016), 7.

17. Ayasha Guerin "Underground and at Sea: Oysters and Black Marine Entanglements in New York's Zone A," *Shima* 13, no.2, (2019).

18. Katherine McKittrick and Clyde Adrian Woods, "No One Knows the Mysteries at the Bottom of the Ocean," in *Black Geographies and the Politics of Place,* eds. Katherine McKittrick and Clyde Adrian Woods (Boston, South End Press: 2007), 9.

19. Clyde L. MacKenzie "A history of oystering in Raritan Bay, with environmental observations." in Pacheco, A (ed) *Raritan Bay: Its Multiple Uses and Abuses*, American Littoral Society, New Jersey Marine Sciences Consortium, National Marine Fisheries Service—Proceedings of the Walford Memorial Convocation, Sandy Hook Laboratory Technical Series Report n30: (1984), 38.

20. Joseph Mitchell, "Mr. Hunter's Grave," *The New Yorker* 22nd September 1956: www.newyorker.com/magazine/1956/09/22/mr-hunters-grave

21. MacKenzie "A history of oystering in Raritan Bay, with environmental observations," 38.

22. Edwin G. Burrows and Mike Wallace, *Gotham: A History of New York City to 1898*, (New York: Oxford University Press, 1999), 662.

23. Mitchell, "Mr. Hunter's Grave".

24. Michael J. Chiarappa, "New York City's Oyster Barges: Architecture's Threshold Role along the Urban Waterfront," *Buildings & Landscapes: Journal of the Vernacular Architecture Forum* 14, n. 1 (2007): 89.

25. Ibid., 91.

26. Ibid. See also, William, Askins, "Oysters and Equality: Nineteenth Century Cultural Resistance in Sandy Ground, Staten Island, New York," *Anthropology of Work Review* 12, n. 2 (June 1991): 8. https://doi.org/10.1525/awr.1991.12.2.7

27. Joseph Mitchell, "Mr. Hunter's Grave," *The New Yorker*, September 14, 1956. www.newyorker.com/magazine/1956/09/22/mr-hunters-grave

28. See Kurlansky, *The Big Oyster*, 176.

29. Ibid., 167.

30. John H. Hewitt, "Mr. Downing and His Oyster House: The Life and Good Works of an African-American Entrepreneur – 19th Century New York, New York Restaurateur, Thomas Downing," in *Protest and Progress* (Routledge, New York. 2000), 81.

31. Kurlansky, *The Big Oyster*, 46-47, 240-41.

32. S. A. M. Washington, *George Thomas Downing: sketch of his life and times.* (Newport, R.I.: Milne Printery, 1910), 7

33. Ibid, 5. See also Hewitt, "Mr. Downing and His Oyster House: 83.

34. Askins, "Oysters and Equality: Nineteenth Century Cultural Resistance in Sandy Ground, Staten Island, New York," 9.

35. Ibid, 9.

36. Eric Goldstein and Mark A. Izeman, *The New York Environment Book*, (Washington: Island Press, 1990), 50.

37. Andrew Hurley, "Creating Ecological Wastelands," Journal of Urban History 20, no. 3 (1994): 345-46. 340-64.

38. Ted Steinberg, *Gotham Unbound: The Ecological History of Greater New York*, (New York: Simon & Schuster, 2015),156.

39. Askins, "Oysters and Equality: Nineteenth Century Cultural Resistance in Sandy Ground, Staten Island, New York," 10.

40. Ibid., 9-11.

41. Ibid.

42. Mitchell, "Mr. Hunter's Grave."

43. Ibid.

44. Ibid.

45. Ibid.

46. Ashley Dawson, *Extreme Cities: The Peril and Promise of Urban Life in the Age of Climate Change* (New York: Verso, 2019), 156.

47. Katie Arkema, Greg Guannel, Gregory M. Verutes, "Coastal habitats shield people and property from sea-level rise and storms," *Nature Climate Change* 3, (October 2013): 913-18. https://doi.org/10.1038/nclimate1944

48. "Living Breakwaters Rebuild by Design Competition." SCAPE https://www.scapestudio.com/projects/living-breakwaters-competition/

49. The Billion Oyster Project (BOP) is leading the effort with funding by the New York State Department of Environmental Conservation. See: https://www.billionoysterproject.org/

50. Eric Klinenberg "The Seas are Rising. Could Oysters help?" *The New Yorker.* August 2, 2021. Online.

51. Shinnecock Indian Nation, *Climate Vulnerability Assessment and Action Plan* (2019), 7.

52. Christine M. Brandon et al. "Evidence for elevated coastal vulnerability following large-scale historical oyster bed harvesting." *Earth Surface Processes and Landforms* 41 no. 8. (2016), 1136-43.

53. Patricia Leigh Brown, "Oyster Shoreline at 'Greater New York' Has a Pearl of a Message." *New York Times*, Oct. 4, 2021. https://www.nytimes.com/2021/10/04/arts/design/alan-michelson-oysters-moma-ps1.html

AT THE EDGE AND ON THE GROUND:

INTERPRETING SUSTAINABLE, RESILIENT, AND ADAPTIVE DESIGN

LUCINDA REED SANDERS

ABSTRACT

It is incumbent on practicing landscape architects, planners, and urban designers to interpret the more theoretical principles of sustainability, resilience, and adaptation by translating them into realizable ideas that become the basis of the built environment. Landscape architects believe that professional practice is naturally inclined to implement ecological design principles. However, over the last two decades, even landscape architects have had to make a case for the benefits of ecosystem services that use empirical tools such as 'SITES' and the 'Landscape Performance Series.' In the face of climate change and shifting consciousness surrounding equity and human displacement, the profession has had to embrace increased ecological, hydrological, social, institutional, and economic considerations and complexities at a meteoric pace. Yet the profession's business model does not lend itself to independent research or post-occupancy evaluations of completed projects. By drawing upon six OLIN-designed case studies spanning three decades, I argue that most research on sustainability, resilience, and adaptation in the realm of professional practice is conducted through the process of designing a project, underscoring the intersectionality of design and science. Purpose-driven clients help to set the stage for advancing our research aspirations, and increasingly, collaboration with a wide breadth of specialized consultants helps us accomplish this work. All six projects are located along urban waterfronts, and the challenges include flooding and human displacement resulting from climate change.

+ urban design
+ measurement
+ sustainability
+ resiliency
+ adaptive design

Figure 1. The wetlands of Pier 26 at Hudson River Park. 2015-2020 Park Opened, Ongoing Playscape in Construction. Partnerships: Hudson River Park Trust, OLIN, Biohabitats, Pentagram, Mueser Rutledge Consulting Engineers, Silman, Tillett Lighting Design Associates, Wesler-Cohen, Northern Designs, Inc., GSESP, Inc., MONSTRUM, Gilbane. Image credit, Barrett Doherty.

Part I: Context

During my forty-plus-year career in landscape architecture and urban design, my colleagues at OLIN and I have had the opportunity to plan and design significant waterfront projects in urban settlements. Many proposed designs are actively being implemented, while those that have been constructed have been absorbed into the memory of their communities. What have these decades of designing urban environments at the 'edge' taught us about sustainable, resilient, and adaptive designs? In answering this question, OLIN has a rich yet sometimes inconclusive set of project outcomes to examine.

Because cities have the greatest concentrations of humanity, it is provocative to contemplate the intersectionality of our institutional, social, and economic systems with our human-constructed physical environment in the context of dynamic biotic and abiotic worlds. Historically, these considerations have not always been relevant to urban design and landscape architecture. Cities have been places where the predictability of controlled environments is preferred over-reliance on natural systems, where consumption and transaction are prioritized over outdoor activity, and where spectacle matters more than spirituality. Across four decades of practice, the world and our consciousness about the world have changed dramatically. We now understand that natural systems are far more complex and multivalent, and we urgently call upon ourselves to adopt new strategies that enhance our consciousness of sustainability, resilience, and adaptation as we continue to shape our cities. This is most critical for those who live along vulnerable waterfronts. Today, many of these aqueous-adjacent settlements find themselves in precarious situations as new realities take hold. People who reside within inherited settlement patterns face looming threats, the urgency of which is not uniformly felt or valued. Escalating pressures from climate change and rising seas are increasingly putting billions of dollars of investment in cities at risk. We have an urgent need to rethink outdated urban plans that have left the most vulnerable in peril while bureaucratic institutions that could establish much-needed policies work at a glacial pace. Those who are being attentive observe human displacement alongside a dearth of equitable and affordable housing. All this is unfolding in an atmosphere of super-charged politics.

As practicing landscape architects and urban designers, what do we claim to have done to build more sustainable, resilient, and adaptive environments? Seeking to posit an answer to this question, in this essay, I offer a brief overview of the terminology of 'sustainable, resilient, and adaptive' and then draw on six OLIN-designed urban waterfront projects as each yields a slightly different and likely incomplete response to the urgent needs we are facing. Many landscape architects, especially those educated under the mantle of Ian McHarg (1920-2001), a Scottish-born landscape architect who taught at the University of Pennsylvania for nearly five decades, believe that our work is inherently environmentally responsible.[1] Because it is a discipline steeped in ecological heritage, it is all too easy to claim our work as sustainable. Even today, landscape architects look to assert their prescient 'Olmstedian' roots, seeking to prove that even the earliest designs stand up to today's more rigorous assessment of resilience while acknowledging their interests in activism when actively advocating for the adoption of sustainable, resilient, and adaptive strategies.[2]

When I joined Hanna/Olin in the early 1980s, it seemed that many enlightened landscape architects aspired to be environmentalists, working diligently to incorporate, wherever they could, ecological principles into designs. In those early years of practice, Hanna/Olin distinguished itself with a parallel emphasis on cultural tradition. We were regarded as early integrators of cultural *and* ecological design, a significant claim during a time when the nascent field of ecology reluctantly accepted humans as a part of the ecological equation.[3] This dual frame shaped the later firm OLIN's approach to design when aligning its emerging theory of sustainability with social and economic considerations. Long before it was popular, we intuited that it was impossible to isolate natural systems from human systems.

By the end of the twentieth century, theories of sustainability and its affiliated principle, resiliency, were taking shape. Our allied professionals, engineers, and architects were busily establishing criteria for designing a more sustainable and resilient world. The well-known LEED accreditation system for buildings, which had its early roots in the 1990s, followed by LEED-ND for neighborhoods, served as building blocks for the development of the 'City Resiliency Index' (CRI) developed by ARUP in 2013.[4] The CRI is far more than a tool for designers, as it also helps cities measure resilience and guide proactive intervention in the most vulnerable systems. At the turn of the century, the profession of landscape architecture was under increasing pressure to not only claim its longstanding heritage of ecological design but to empirically substantiate its assertions. As a result, landscape architects began formulating, testing, and implementing strategies often informed by their more measured colleagues, engineers. Professional organizations of landscape architects, spurred on largely by academics and supported by practitioners, developed their version of guidelines and performance metrics for landscapes known as the 'Sustainable SITES Initiative' (SITES).[5] In 2004, Fritz Steiner, then Dean of the School of Architecture at The University of Texas, Austin, convened allied professionals with the intention of developing a rating system for landscapes that was credible, rooted in science, and measurable. Steiner wrote: "Very early on, we decided to use the ecosystem services concept to guide the Sustainable SITES Initiative, thus grounding the system design in science."[6]

'The Landscape Performance Series' (LPS) developed by the Landscape Architecture Foundation (LAF) soon followed SITES.[7] LPS arose out of discomforts that I, and other practicing professionals voiced: If SITES was a rating system intended to measure sustainability in projects as they were being designed, how would we be assured that these projects were performing as intended? LPS, instead, was developed as a platform for assessing claims of sustainability and performance post-construction. To achieve this, LAF organized a program in which faculty-led student research teams, working alongside designers and clients, assess the performance of installed projects and document the benefits of those landscapes. Since 2010, this research program has produced over one hundred and fifty Case Study Briefs, available on their website. LAF has aimed to answer an additional question: What does it mean to claim a project is sustainable when the benchmarks of sustainability continuously change? The Landscape Performance Series platform is a place to gain insights into best practices as new knowledge is generated. While I laud and support SITES and LPS, it must also be recognized that measuring sustainability in designs and assessing post-occupancy performance are challenges faced by the profession, as clients rarely support post-occupancy research.

As we climb this challenging slope, designers have been called to deepen their understanding of sustainability by incorporating resilient designs. Landscape educator Jack Ahern argues that resilience is the fourth dimension of sustainability—following the familiar tri-partite dimensions of economic, social, and environmental sustainability—in a context of non-equilibrium.[8] Resilience is described by some as the elastic function of a system to revert to its pre-stimulus state.[9] The National Academy of Sciences (NAS) defines resilience as "the ability to prepare and plan for, absorb, recover from, and more successfully adapt to adverse events. The common features [of resilience] include critical functions (services), thresholds, cross-scale (both space and time) interactions, and memory and adaptive management conceptualizations of resilience."[10] Notably, in the 2021 NAS definition of resilience, concepts of adaptation entered the lexicon. In some instances, resilience is defined as an insufficient strategy, with landscape architects now required to consider who and what is adapting and what kind of social and political activism is needed to ensure adaptation. As Keenan notes: "[A]daptation is about the capacity to transform to an alternative domain of operation. Resilience has a threshold, and beyond that threshold, one either adapts or fails. Today in the US [United States], much of the discourse in design and planning is oriented toward the notion of resilience without fully contemplating the nature of adaptation."[11] Considering impending climate change, authors Nate Kauffman and Kristina Hill also insist that our sights need to be focused squarely on adaptation.[12]

Regardless of when projects are designed and built, the world of professional practice has been left to interpret theories of sustainability, resilience, and adaptation and to bring meaning to them. While this quickly changing landscape presents its share of challenges to do 'the right thing', it hasn't prevented our profession from forging ahead, striving to create our best interpretation in the context of our human-induced shifting biotic and abiotic world. Our goal is for the multitude of species, including homo sapiens, to not merely survive but also adapt and thrive within these shifting contexts. As landscape architects and urban designers, we must seek to be prescient moving forward.

The projects discussed in the second part of this essay traverse forty-five years of practice. They are a sampling of case studies, all of which are located along a waterfront environment and for which concepts of sustainability, resilience, and adaptation were operative. This body of work represents some of what I and my partners at OLIN have been able to accomplish with our clients and co-collaborators. It represents our collective understanding of these ideas at the time of the projects' design and implementation. When useful, the approach I have taken in reviewing them is one of critical examination because what seemed right at the time of the project's conception may not be when viewed through our contemporary lens. They have been arranged chronologically with the intention of underscoring the ever-increasing need to incorporate new insights and complexities into how we design the landscape.

Part II: Case Studies
Robert F. Wagner Jr. Park, New York, New York (1991-1996)

Wagner Park in Battery Park City, located across from the Statue of Liberty, is currently being demolished to guard against rising seas.[13] The OLIN / Machado-Silvetti designed park, which opened in 1996, will be replaced by yet another project. Wagner Park, as built thirty-two years ago, has many of the hallmarks of resilient design before people used that language in the context of the built environment. As a testament to its foresightedness, the park survived Hurricane Sandy with minimal damage.[14] Its prescient design included the siting of infrastructure above the hundred-year flood line, incorporating sandy soil mixes to ensure rapid drainage in the event of floods, and using rugged plants to withstand salt-laden spray. Wagner Park will be the first among many parks in Battery Park City to fall to the threat of rising sea levels and future Sandy-like storms, but not for the reasons one thinks.

In the late 1970s, Hanna/Olin joined the team of Cooper-Eckstut as landscape architects and urban designers to develop a master plan for ninety acres of land.[15] While the plan represented an important historical marker for the fields of urban design and landscape architecture, it called for the creation of new land by filling the Hudson River to the ends of the former shipping piers, an activity that fifty years ago may have seemed benign relative to the former industrial operations that took place there.[16] Now, with hindsight, we can see that this action compromised the ecological health of the Hudson River and disrupted hydrological flows. And yet, while the plan was conceived well before the aspirations of sustainable and resilient design were developed, there are two significant design strategies employed that most would agree are integral to the creation of socially resilient environments. The plan established connections from the waterfront to the existing fabric of the city, and most importantly, it declared the waterfront one hundred percent publicly accessible. While these principles, I would argue, are relevant and essential markers of resilient places, today, they are clearly insufficient in the context of sea-level rise and climate change.

Unfortunately, Manhattan's lack of resilience in the face of rising seas is tied to the relentlessly hard infrastructure found at the water's edge; it is now impossible to rely on purely ecological systems to match the scale of the threat. To protect the vast resources of New York City, a proposed system of mostly hard subterranean and above-ground infrastructure is being implemented, occasioning the destruction of Wagner Park The project known as 'The BIG U' put forth during the 2013 post-Sandy resiliency competition, 'Rebuild by Design', calls for "ten continuous miles of protection".[17] Unlike prior generations of single-use infrastructure, the intention of The BIG U is not only to protect significant assets but to incorporate multi-functionality, offering social resiliency and an enriched terrestrial ecology. However, in the context of a politically mandated wholesale redesign of lower Manhattan—that includes an already resilient Wagner Park—a new park perched fourteen feet in the air, disconnected from the urban context, will result.

The end of Wagner Park raises questions about sustainability, resilience, and adaptation: Who gets to decide when projects end, and how transparent are these decisions; is the Environmental Impact Review (EIR) process an objective tool; how can the public know if fear is being used to drive an outcome; what is the role of memory and artistry in the discussion on climate change; do we have an obligation to accomplish resiliency with minimal carbon outputs? Few dispute the reality of climate change and the need to protect built environment assets. And yet, the questions I pose above invite clients and designers to rethink their methods when choosing to rebuild existing environments.

Mill River Park and Greenway, Stamford, Connecticut (2005)

In 2005, OLIN was commissioned by the Mill River Collaborative (MRC) to prepare a master plan for the thirteen-and-a-half-acre park along the Mill River corridor, which wends its way through the downtown of Stamford, Connecticut. The United States Army Corps of Engineers (ACE) was mobilized when flooding of the downtown persisted due to increased runoff from upstream development and the impoundment of water by the remnant mill dam. The impoundment area was surrounded by tall concrete walls that effectively perched people above the debris-laden river, creating a physical and social barrier between the downtown and adjacent neighborhoods. To open access to the park, the OLIN team designed porous and accessible edges inviting people to the river as well as the landscape infrastructure for multiple community-based programs including hiking and biking trails, playgrounds, gardens, skating rinks, and amphitheaters. To accomplish this, the dam was removed and the river rebuilt, both of which were bold moves that would not have been feasible without a thorough understanding of hydrological and ecological sciences.[18] (Figures 2 and 3)

This segment of the river flows through a compressed urban setting, at the very end of the watershed. In partnership with the Army Corps of Engineers and our ecologists, we collectively built and fine-tuned

Figure 2. Mill River Park in low water flow. Mill River Park Collaborative, OLIN, Nitsch Engineering, Stantec, Habitat by Design, GZA, Army Corps of Engineers, Tillett Lighting Design Associates, Northern Designs. Image credit, OLIN/ Sahar Coston-Hardy.

Figure 3. The river in a bank full condition. Mill River Park Collaborative, OLIN, Nitsch Engineering, Stantec, Habitat by Design, GZA, Army Corps of Engineers, Tillett Lighting Design Associates, Northern Designs. Image credit, Courtesy of Mill River Park Collaborative.

hydrological and ecological models to ensure there was room for the river during flood events. The new outlines of the river were carefully designed in plan and section, paying close attention to the behavior of naturally flowing waters. As 'natural' as this river appears, it requires the most sophisticated engineering and construction techniques to maximize hydrological and ecological performance. The sectional contouring of the river was designed to respond to the most frequent floods that occur within a precisely defined bank full datum; rock veins were used to direct and slow the water, creating deeper pools for fish. The hydrologic work was foundational to establishing the rich ecology for which this project is known. Because plants follow the contour of the water, and likewise, the fish, mammals, reptiles, and birds follow those same contours, we were able to use science to develop a unique ecology that now runs through the city within the water and along the embankments and uplands of the river.[19]

Hydrologic and ecological sustainability and resilience are present in this project. But what about social sustainability and resilience? As designers, we assumed that science could be interpreted into form and texture by making spaces for people to gather and by selecting plants and materials. Taken together, we believed these decisions contributed to poetry, elevated science, and, ideally, the hearts and consciousness of people. We firmly held, then and now, that alongside fostering social sustainability and resilience, these are reasons good design matters. Yet good design alone is insufficient. Social resilience is also realized through the tireless work of the Mill River Collaborative who is responsible for the oversight of the park. Hosting a myriad of programs and providing educational opportunities they too contribute forms of adaptation.[20]

11th Street Bridge Park, Washington, DC (2014–present)

As in the case of Hanna/Olin's design for Bryant Park from the early 1980's, well-designed and managed parks—think Jane Jacobs and William H. Whyte—have been hailed as models of city plans that spur urban regeneration.[21] With such models replicated across the United States, today, communities are raising cautionary flags as the effects of 'Greenlining' and its displacement are increasingly seen as threats. Clients and designers understand the importance of getting ahead of the potential unintended consequences associated with the design and construction of new parks. The 11th Street Bridge Park in Washington, DC, offers an excellent case study on this topic.[22] (Figures 4 and 5)

Won through competition in 2014 with the architecture practice OMA, the 11th Street Bridge project seeks to close the chasm and heal the wounds of inequity through the design of the public realm. The social and contextual themes of the project are familiar to us now. Urban renewal swept across the United States in the mid 1900s and in its wake, federally funded programs displaced hundreds of thousands of people, inordinately affecting people of color. In 1949, twenty-three thousand people, mostly African Americans, were displaced

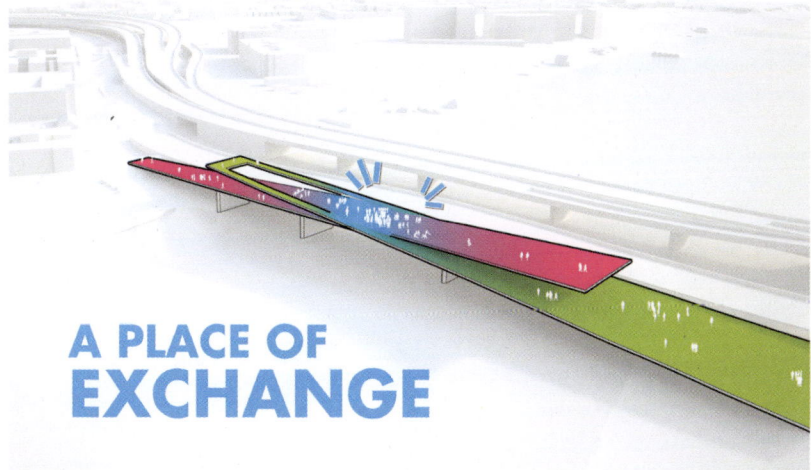

Figure 4. Conceptual diagram of the 11th Street Bridge Park. The District Department of Transportation (DDOT), Building Bridges Across the River, OMA, OLIN, Whitman, Requardt and Associates, LLP, Setty & Associates, Delon Hampton & Associates, MCLA, Inc., Lynch & Associates, Ltd., Anacostia Watershed Society, WDP & Associates Consulting Engineers, Inc., Dharam Consulting, Forecast Public Art. Image credit, OMA + OLIN.

A PLACE OF EXCHANGE

from the west side of the Anacostia River in the name of 'urban renewal'.[23] Redlining policies further sealed the fate of those who were displaced. Financial disparities are well documented, but a cascade of problems followed those policy decisions including a lack of access to clean air and water, green space, housing, and healthcare, all impacting the quality and longevity of life.

In 2014, through many forms of community engagement, goals were established for the project that included re-engagement with the river, connecting communities, improving health disparities, and providing inclusive economic opportunities. The overarching challenge for the project was to design a public realm that accomplished these goals. A bridge in the form of a symbolic "X marks the spot" was devised as the place for civic exchange. The artful armature was conceived to knit the two sides of the river together with community-generated programs equitably distributed throughout the new park.[24] In the words of sociologist Elijah Anderson, it is intended to become a "cosmopolitan canopy," a place where differences slide away and our common humanity is embraced and celebrated.[25]

This project aims to bring social sustainability and resilience deeper into the communities, and in response, an equitable development task force was formed that included the design team. The resulting 2018 Equitable Development Plan works to prevent the displacement of current residents in this economically challenged area of Washington, DC, which is already experiencing escalating real estate pressures. The plan states: "Recognizing that signature parks can increase surrounding property values, the 11th Street Bridge Park is committed to working with partners and stakeholders to ensure that existing residents surrounding the Bridge Park can continue to afford to live in their neighborhood once the park is built, and that affordable homeownership and rental opportunities exist nearby."[26] The plan is ambitious and far-reaching, aspiring to support local residents by improving their economic futures through job training

An Equitable Development Plan was prepared with local, and Federal agencies to proactively prevent displacement in the face of current and future real estate pressures.

Figure 10

Housing Cost by Neighborhood Group

Ward 6

Ward 7

Ward 8

Gross rent as a percentage of household income (GRAPI) 35% and more:

48.4 - 59.6%
43.3 - 48.4%
39 - 43.3%
33.7 - 39 %
29.4 - 33.7%
19 - 29.4%
Data Suppressed

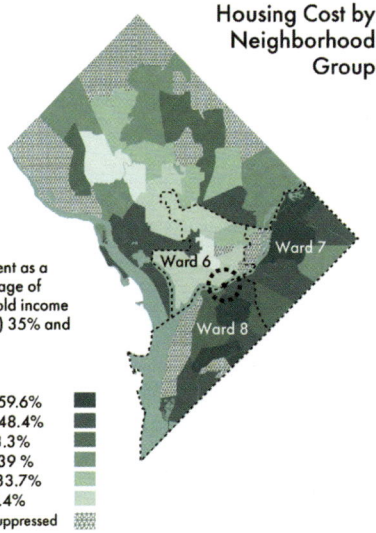

HOUSING AND EMPLOYMENT INEQUITIES

East of the River   surrounding 11th Street Bridge Park

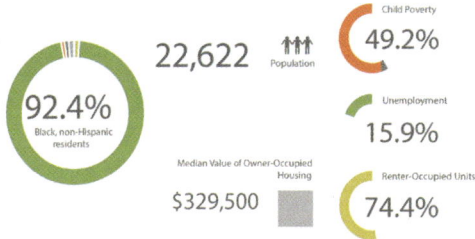

92.4%
Black, non-Hispanic residents

22,622
Population

Median Value of Owner-Occupied Housing
$329,500

Child Poverty
49.2%

Unemployment
15.9%

Renter-Occupied Units
74.4%

West of the River   surrounding 11th Street Bridge Park

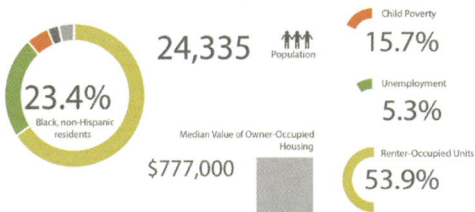

23.4%
Black, non-Hispanic residents

24,335
Population

Median Value of Owner-Occupied Housing
$777,000

Child Poverty
15.7%

Unemployment
5.3%

Renter-Occupied Units
53.9%

2017 Median Sales Price
Single Family Units

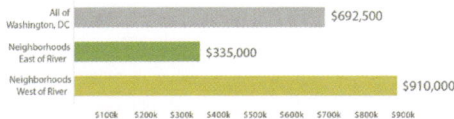

All of Washington, DC	$692,500
Neighborhoods East of River	$335,000
Neighborhoods West of River	$910,000

$100k $200k $300k $400k $500k $600k $700k $800k $900k

Figure 5. Historic analysis of the housing and employment inequities on both sides of the Anacostia River. The District Department of Transportation (DDOT), Building Bridges Across the River, OMA, OLIN, Whitman, Requardt and Associates, LLP, Setty & Associates, Delon Hampton & Associates, MCLA, Inc., Lynch & Associates, Ltd., Anacostia Watershed Society, WDP & Associates Consulting Engineers, Inc., Dharam Consulting, Forecast Public Art. Image credit, OMA + OLIN.

programs, local community land trusts, small business enterprises, and other leadership development programs.

A 2022 *New York Times* article posed the question: "Can Anacostia Build a Bridge Without Displacing Its People?" The author suggests that the 11th Street Bridge "could be a model for how to create public space while lessening the effects of gentrification."[27] The project and its varied initiatives clearly strive beyond the traditional benchmarks of social sustainability and resilience by having elevated and enriched how we think about their principles within the design community.

Pier 26 at Hudson River Park New York, New York (2015)

OLIN was commissioned in 2015 to design Pier 26 immediately north of Battery Park City.[28] The pier is a part of the larger five-hundred-and-fifty-acre Hudson River Park system and is overseen by the public benefit corporation, Hudson River Park Trust (HRPT).[29] That body is empowered to design, build, operate, and maintain the park and estuarine sanctuary in and along the three miles of the Manhattan shoreline. OLIN's directive was to design the pier as a vehicle for educating about the Hudson River estuary and promoting long-term environmental stewardship. Our engagement work confirmed that the public yearned for the very thing that a century-plus of industry had stripped away. (Figure 6)

Our canvas for design was a pier that sits at the southern edge of the estuary. Eager to learn everything we could about the Hudson River estuary, which reaches far into Upstate New York, we researched the indigenous ecology of the river, discovering the existence of diverse biomes: upland woodland, grasslands, scrub-shrub, and inter-tidal zones.[30] Because hydrologic profiles are linked directly to terrestrial profiles, the former vegetal ecology existed in a predictable rhythm from higher elevations to lower elevations. Moreover, this vegetal and hydrologic cross-section supports a diverse array of wildlife. These findings inspired us to organize the pier in a similar manner, using the ecological construct we had uncovered in our research.

The two-acre pier had high ecological ambitions for its rich terrestrial and riverine habitat. Yet, given the scale of the project, as designers, we were under no illusion that we could significantly impact the ecology of the Hudson River estuary. The primary questions we needed to answer were: Could we design a landscape that was meant to teach, and if so, how? Merely installing a re-creation of an ecological cross-section of the estuary was not the answer. This is where we, as designers, believed that project ideas inclusive of beauty, craft, and poetry must be part of the process and significantly participate where design and science intersect.

We answered the challenge by inserting an eight-hundred-foot-long folded piece of furniture into the ecological framework, piercing through the topographically arrayed biomes. This furnishing became an armature to connect the variety of project programs. The design was focused on heightening the senses while immersing users in the interpretation of indigenous ecologies. Experiences are created while walking through the upland woodland, idling in the grassland, swinging in the scrub-shrub, hovering above the tide deck, or playing in the Sturgeon-inspired playground. Heightening the design of this ecological cross-section was intended to offer pleasure and inspiration while answering the didactic mission of education. To this end, the Hudson River Park Trust hosts robust educational programs for all ages whose lessons are directed at fostering the next generation of environmental stewards. There is, however, the other, less overtly didactic side of how we as humans absorb knowledge through experience outside of the classroom. It is here, I argue, that beauty has a role in sustainability, resilience, and adaptation. By all measures, the pier is beautiful, uplifting, and inspiring. Beauty, shaped by craft, artistry, and poetry, makes this a place people want to be. American landscape architect Elizabeth K. Meyer states, "Design matters because it alters the ethos of people who use the spaces."[31] This offers us hope that users of beautifully designed ecological spaces may, indeed, be influenced to become stewards of the earth.

Origin Park, Clarksville, Indiana (2018-present)

In 2018, OLIN was retained by the River Heritage Conservancy to design a master plan for a new four-hundred-and-fifty-acre park on the Indiana shores of the Ohio River, across from the city of Louisville, Kentucky.[32] The site sits entirely outside of the levee built in response to the 1937 flood that devastated the region. As a result, the site succumbed to flooding and was forgotten and abused over the ensuing eighty-five years. Gravel was quarried, and soils were excavated from the site to build nearby infrastructure projects. Illegal landfills and dumping produced skyward mounds as contaminants from polluting industries leached into the ground. In addition, the Army Corps of Engineers (ACE) reconfigured and dammed the Ohio River to enable barge passage around the shallow Devonian Era fossil beds. However, the dams caused erosion of the Indiana shoreline at a shockingly rapid pace. Some might say this is a landscape with a tragic environmental condition. (Figures 7a, 7b, and 7c)

Buried under the debris is another vitally important historic past. Most visitors to this landscape know the stories of George Rogers Clark (for whom Clarksville, Indiana is named) and of his brother, William Clark, of 'Lewis and Clark' fame. There is documented evidence of indigenous settlements having occupied this territory and of Bison using it to cross the Ohio River, as noted in fossil beds. Moreover, it is likely that the underground railroad ran proximate to the site. These stories, now mostly invisible, will be brought to the surface with compelling narratives, which in turn are inseparably tethered to the hydrology and ecology of the site.

Inspired by its proximity to the mighty Ohio River and its emergent woodlands, the OLIN team coined the term 'raw awe' to describe the site. The landscape is poised to become an extraordinarily significant

Ecological Communities

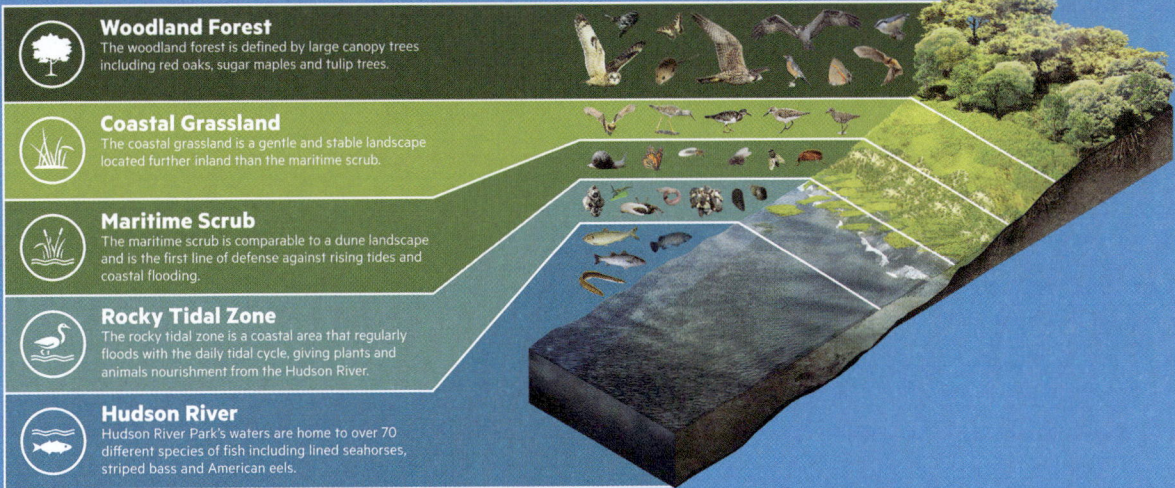

Woodland Forest
The woodland forest is defined by large canopy trees including red oaks, sugar maples and tulip trees.

Coastal Grassland
The coastal grassland is a gentle and stable landscape located further inland than the maritime scrub.

Maritime Scrub
The maritime scrub is comparable to a dune landscape and is the first line of defense against rising tides and coastal flooding.

Rocky Tidal Zone
The rocky tidal zone is a coastal area that regularly floods with the daily tidal cycle, giving plants and animals nourishment from the Hudson River.

Hudson River
Hudson River Park's waters are home to over 70 different species of fish including lined seahorses, striped bass and American eels.

Figure 6. The five ecological communities present in the park today, Pier 26 at Hudson River Park. Hudson River Park Trust, OLIN, Biohabitats, Pentagram, Mueser Rutledge Consulting Engineers, Silman, Tillett Lighting Design Associates, Wesler-Cohen, Northern Designs, Inc., GSESP, Inc., MONSTRUM, Gilbane Building Company, Trevcon Construction, Steven Dubner Landscaping, E-J Electric Installation Co. Image credit, OLIN / Pentagram.

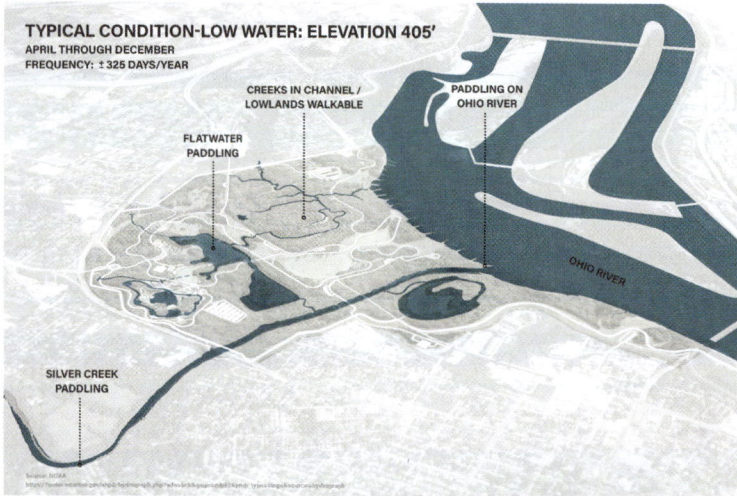

TYPICAL CONDITION-LOW WATER: ELEVATION 405'
APRIL THROUGH DECEMBER
FREQUENCY: ± 325 DAYS/YEAR

CREEKS IN CHANNEL /
LOWLANDS WALKABLE

FLATWATER
PADDLING

PADDLING ON
OHIO RIVER

SILVER CREEK
PADDLING

OHIO RIVER

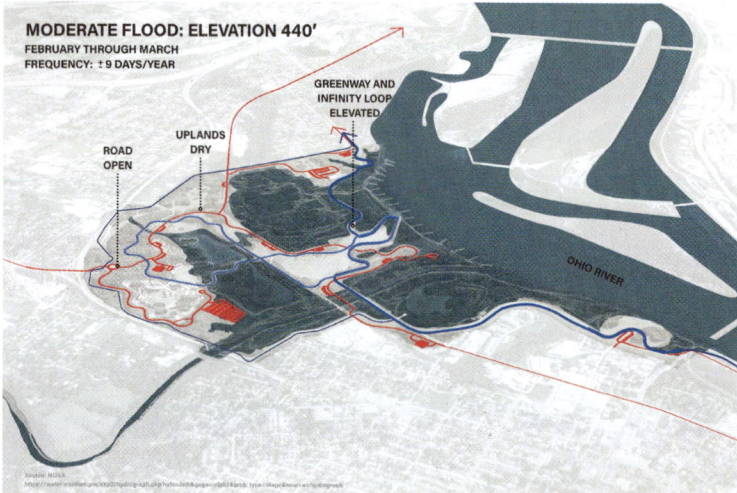

MODERATE FLOOD: ELEVATION 440'
FEBRUARY THROUGH MARCH
FREQUENCY: ± 9 DAYS/YEAR

GREENWAY AND
INFINITY LOOP
ELEVATED

ROAD
OPEN

UPLANDS
DRY

OHIO RIVER

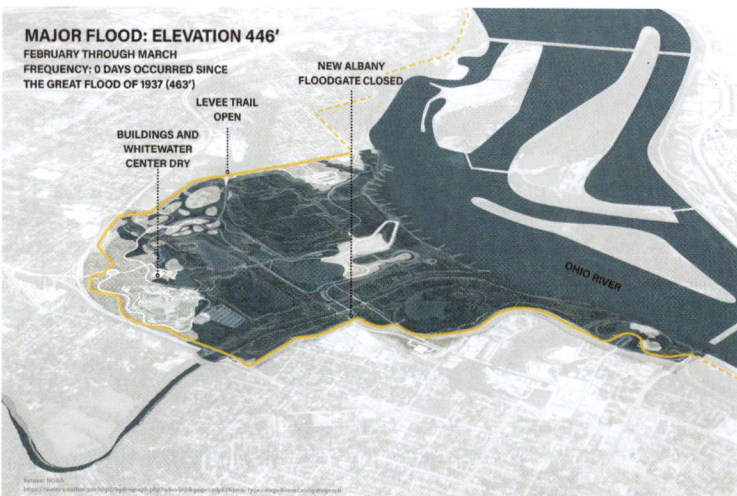

MAJOR FLOOD: ELEVATION 446'
FEBRUARY THROUGH MARCH
FREQUENCY: 0 DAYS OCCURRED SINCE
THE GREAT FLOOD OF 1937 (463')

NEW ALBANY
FLOODGATE CLOSED

LEVEE TRAIL
OPEN

BUILDINGS AND
WHITEWATER
CENTER DRY

OHIO RIVER

Figure 7a. Typical low water, Origin Park, Clarksville IN. River Heritage Conservancy, OLIN, RES, Clark Dietz, Geosyntec Consultants, Jacobi, Tooms, and Lanz, QK4 Engineers, MRCE, S2O, SME, Dharam Consulting, Endres Studio, Tillett Lighting Design Associates, Cultural Resource Analysts, Joseph and Joseph and Bravura Architects. Image credit, OLIN.

Figure 7b. Moderate flood, Origin Park, Clarksville IN. Image credit, OLIN.

Figure 7c. Major flood, Origin Park, Clarksville IN. Image credit, OLIN.

and unique public park in the region, linking to the network of important Olmstedian parks across the Ohio River.

But what do terms such as sustainability, resilience, and adaptation mean in this highly manipulated context? In response, we decided to work with the site, not against it. We coordinated our design with the rhythms of flooding to welcome people into the park during different flood stages. Hydrological engineers collaborated with us to model the site, establishing the elevation of roads and a hierarchy of paths accessible under different flood regimes. In low water, the entirety of the park will be opened; it will be possible to paddle on flatwater in the quarry ponds and to hike through the lowlands, defined by a creek restoration project. In moderate floods, the roads, regional trails, and uplands will be open, serving the intensely civic platforms. And hence, by utilizing careful hydrologic modeling, we have designed a park that embraces the flood and celebrates the 'raw awe'.

Likewise, at the outset, we made a commitment to sustain the emerging ecology. We will selectively intervene in the existing ecologies only where we are able to enhance biodiversity. In consultation with our team of ecologists and environmental scientists, one hundred and fifty acres of meadows will be installed, and seventy-five thousand new trees will be planted, creating two hundred and fifty acres of emergent and proposed upland and lowland forests. Science has informed the design of this park, yet intuitively, we understand that what we are proposing will make a more environmentally sustainable and resilient landscape. It is an adaptation of consciousness around what a park can be, and it is an adaptation of our relationship with nature.[33]

Caño Martín Peña, San Juan, Puerto Rico (2020-2022)

Caño Martín Peña demonstrates an evolution in OLIN's ability to work with communities to build a more resilient and adaptable future. The project dares to embrace a breadth of complexity that expands well beyond most traditional design practices.[34] There is one theme that underlies this expansion: social capital. OLIN partners Richard Roark and Marni Burns assert that the eight communities of the Caño Martín Peña District, built piece-by-piece by their residents, recognize that their greatest asset in the face of a changing climate is their social capital. Over the past decade, Roark has been deeply influenced by the work of Eric Klinenberg, whose book Heat Wave: A Social Autopsy in Chicago is one of the most impactful studies on the very topic of human survivability in the throes of climate change. The conclusions of Klinenberg's research on the devastating 1995 heat wave point to enhanced survivability of individuals who were socially connected, while those who perished were isolated.[35] The Caño Martín Peña Comprehensive Infrastructure Master Plan (CIMP) embraced this research by protecting, first and foremost, the social bonds amongst inhabitants. The plan was developed to serve the communities' needs by improving their collective health and quality of life, incorporating nature-based strategies and climate change risk analysis to develop holistic solutions while safeguarding the communities' deep social bonds.

Grounding the plan in social and environmental frameworks was only possible through extensive collaboration with the Group of Eight (G8), a series of close-knit communities that emerged informally around a channel of the San Juan Bay Estuary. Emphasizing the community agency of these 12,000 residents beset with risks resulting from frequent flooding and a lack of infrastructure ensured that the G8 was instrumental in shaping proposals that reflected their needs for a safe and thriving future. The plan underscores community agencies as they confront the impacts of climate change. Only then is it possible for a co-created plan to be firmly embedded with environmental and sustainability principles that incorporate nature-based solutions and risk analyses based on local sea level rise and the Intergovernmental Panel on Climate Changes AR6 report. The Caño Martín Peña Comprehensive Infrastructure Master Plan targets safe drinking water, sanitary and stormwater collection, flood protection, access to safe housing, open space, and economic opportunities. It also accounts for the financial and political realities of implementing infrastructure and housing projects.

If the plan is realized, the future it will create will likely be sustainable and resilient, addressing adaptation on multiple levels. That is, innovative approaches to stormwater management, flood protection, and housing development are designed to be scalable and adaptable, allowing for continued improvement in response to changing environmental conditions. The communities can adapt to the new realities because they are able to remain intact, protect their social capital, and be empowered with tools that ensure environmental adaptation. In this way, as designers, we are also adapting to new ways of working, expanding the breadth of considerations and fine-tuning them to the places we work.

ENDNOTES

1. Richard Weller, Meghan Talarowski, *Transects: 100 Years of Landscape Architecture and Regional Planning at the School of Design of the University of Pennsylvania* (Novato, CA: Applied Research and Design 2014).

2. See Fadi Masoud, Elspeth Holland, "Landscape Architecture is Resilient Design: Enduring Strategies and Frameworks Adapted from the Olmsted Office", *Journal of Landscape Architecture*, 16:3 (2021), 50–65. DOI: 10.1080/18626033.2021.2046769 ; Rolf Diamant, Ethan Carr, E., *Olmsted and Yosemite: Civil War, Abolition, and the National Park Idea* (Amherst, MA: Library of American Landscape History 2022).

3. Alf Hornborg, Carole L Crumley, Eds, *The World System and the Earth System: Global Socioenvironmental Change and Sustainability Since the Neolithic* (New York, NY: Routledge 2006).

4. US Green Building Council, "LEED v4 for Building Design and Construction" (July 25, 2019) usgbc.org/resources/leed-v4-building-design-and-construction-current-version. US Green Building Council, "LEED for Neighborhood Development" (2014) https://www.usgbc.org/leed/rating-systems/neighborhood-development. ARUP, The Rockefeller Foundation, "City Resilience Index: Understanding and Measuring City Resilience" (2012) https://www.arup.com/perspectives/city-resilience-index

5. The Sustainable SITES Initiative, https://sustainablesites.org/

6. Fritz Steiner, The Sustainable SITES Initiative, "The Emergence of SITES: A Retrospective", (October 2020) https://sustainablesites.org/emergence-sites-retrospective

7. The Landscape Architecture Foundation, Landscape Performance Series, https://www.lafoundation.org/what-we-do/research/landscape-performance

8. Jack Ahern, "Urban Landscape Sustainability and Resilience: The Promise and Challenges of Integrating Ecology with Urban Planning and Design. *Landscape Ecology* 28, (2013), 1203–1212, https://doi.org/10.1007/s10980-012-9799-z

9. Jesse M. Keenan, "The Resilience Problem: Part 1" in *Climates: Architecture and the Planetary Imaginary,* ed. James Graham (New York, NY / Zurich: Columbia University Press / Lars Muller Publishers 2016).

10. Elizabeth B Connelly, Craig R Allen, Kirk Hatfield, Jose M Palma-Oliveira, David D Woods, Igor Linkov (2017) "Features of resilience" *Environment Systems and Decisions, 37,* 46–50 (2017). https://doi.org/10.1007/s10669-017-9634-9

11. Keenan, "The Resilience Problem: Part 1", 161.

12. Nate Kauffman, Kristina Hill, "Climate Change, Adaptation Planning and Institutional Integration: A Literature Review and Framework" *Sustainability* 13 (19) (2021) 10708. https://doi.org/10.3390/su131910708.

13. More information on the demise of Wagner Park can be found in the following sources: Winnie Hu, Anne Barnard, "A Plan to Save a Beloved Park from Flooding has Angered its Biggest Fans" *The New York Times* (October 21, 2022). https://www.nytimes.com/2022/10/21/nyregion/wagner-park-manhattan.html; Audrey Wachs, "Architect's Aren't Happy with Plans to Remodel This Park" *The Architects Newspaper* (May 16, 2017), https://www.archpaper.com/2017/05/wagner-park-remodel/; "The Cultural Landscape Foundation's Landslide Program identifies threatened works of landscape architecture," https://www.tclf.org/landscapes/robert-f-wagner-jr-park; "New York City Lower Manhattan Coastal Resiliency", https://www.nyc.gov/site/lmcr/progress/battery-park-city-resilience-projects.page

14. Paul Goldberger, "A Small Park Proves That Size Isn't Everything" *The New York Times*, (November 24, 1996). https://www.nytimes.com/1996/11/24/arts/a-small-park-proves-that-size-isn-t-everything.html . The project description can be found on the OLIN website: https://www.theolinstudio.com/robert-f-wag-ner-jr-park. See also, Samantha Maldonado, "In Battery Park City, Another Plan to Destroy a Green Space in Order to Save It" (May 16, 2022), https://www.thecity.nyc/2022/5/16/23070850/battery-park-plan-destroy-green-space-resiliency.

15. The project description can be found on the OLIN website, https://www.theolinstudio.com/battery-park-city.

16. For examples of this assertion see Anna Doud, "Battery Park City, Reimagining Lower Manhattan" SSRN Elsevier (April 25, 2011), http://dx.doi.org/10.2139/ssrn.2322718 and Carter Wiseman "The Next Great Place: The Triumph of Battery Park City" *New York Magazine 119* (24) (1986) 34–41.

17. For a description of this project see, https://rebuildbydesign.org/work/funded-projects/the-big-u/

18. William S Saunders, "Change the Channel" *Landscape Architecture Magazine* (December 2015) 94–112.

19. The project description can be found on the OLIN website, https://www.theolinstudio.com/mill-river-park-and-greenway.

20. For information on the Mill River Collaborative see, https://millriverpark.org/

21. Jane Jacobs, *The Death and Life of Great American Cities* (New York, NY: Vintage Books 1961); William H. Whyte, *The Social Life of Small Urban Spaces* (Washington DC: The Conservation Foundation 1980).

22. The project description can be found on the OLIN website, https://www.theolinstudio.com/11th-street-bridge-park. For information on the park and on Building Bridges Across the River (BBAR) see, https://bbardc.org/project/11th-street-bridge-park/

23. Francesca Russello Ammon, "Southwest Washington, Urban Renewal Area HABS DC-856," *Historic American Buildings Survey,* National Park Service U.S. Department of the Interior, (Summer 2004), 2.

24. Hallie Boyce, "Bridging a Divide Through Park Design" *Parks & Recreation Magazine* (2018). https://www.nrpa.org/parks-recreation-magazine/2018/october/bridging-a-divide-through-park-design/. See also Lazo, Luz, "DC's First Elevated Park Will Link Neighborhoods Divided by River" *The Washington Post* (2022). https://www.washingtonpost.com/transportation/2022/08/04/dc-anacostia-river-park-bridge/

25. Elijah Anderson, *The Cosmopolitan Canopy* (New York, NY: W.W. & Co, 2012).

26. Building Bridges Across the River, "11th St. Bridge Park's Equitable Development Plan" (2018). https://cdn2.assets-servd.host/material-civet/production/images/documents/Equitable-Development-Plan_09.04.18.pdf.

27 "Can Anacostia Build a Bridge Without Displacing Its People?" *New York Times:* Headway (2022). https://www.nytimes.com/interactive/2022/08/09/headway/anacostia-bridge.html

28. The project description can be found on the OLIN website, https://www.theolinstudio.com/pier-26

29. Additional project information can be found on the Hudson River Park Trust website, https://hudsonriverpark.org/locations/pier-26/

30. Eric Sanderson, *Mannahatta: A Natural History of New York City* (New York: NY Harry N. Abrams, 2009).

31. Jared Green, "Beth Meyer: Natural Beauty Has a Ripple Effect" The Dirt (11/2009)"

32. Project information can be found on the River Heritage Conservancy website, https://www.originpark.org/

33. The project description can be found on the OLIN website, https://www.theolinstudio.com/origin-park

34. The project description can be found on the OLIN website, https://www.theolinstudio.com/cano-martin-pena

35. Eric Klinenberg, Heat Wave: A Social Autopsy of Disaster in Chicago (Chicago: University of Chicago Press 2003).

CITYFOOD:

INTEGRATED, PRODUCTIVE WATER-BASED SYSTEMS FOR NOURISHING THE CIRCULAR CITY

GUNDULA PROKSCH

ABSTRACT

Climate change has a significant impact on water and food security, primarily due to reduced availability of water for sustenance. Clean freshwater has become a finite asset, though it is still extracted, used, discharged, and polluted in many places as if it were a ubiquitous commodity. It is critical, therefore, to include water and food within our resource loops when seeking to foster circular sustainability strategies. What if cities and buildings could capture and reuse water while producing food and biomass to serve their inhabitants? This essay investigates how three water-based living systems—aquaponics, microalgae cultivation, and anaerobic digestion—do just this in closed-loop resource cycles. It reviews how these biological systems connect across three scales in the built environment—technology, building, and city—to grow, power, and feed sustainability. It also discusses how these water-based systems are linked according to inputs and outputs so that one system's byproduct becomes a resource for the next. The physical enclosures they require to thrive can be integrated with buildings and urban infrastructures to mutually improve their performance. The integration strategies discussed herein follow the trans disciplinary Food-Water-Energy (FWE) Nexus and circular city models. While the goals and principles of circular cities are well understood, the generative processes that help us establish how, where, and at what scale to design and implement them are not. This essay invites architects, landscape architects, planners, and designers to define new economies for implementing integrated, productive water-based systems for nourishing the circular city.

+ aquaponics
+ microalgae cultivation
+ food-water-energy nexus
+ circular city
+ controlled environment agriculture (CEA)

Figure 1. Detail of Algae Cultivation Center. Photo credit: Andreas Heddergott/ TUM.

Water scarcity and food insecurity are intrinsically connected and pose a direct threat to human and environmental health.[1] Climate change and its impact on equity, economics, and geopolitics compound these challenges even further.[2] Despite water's vital role as a resource for all life on Earth, humans use seventy percent of all available freshwater for food production and, unfortunately, in many cases, by using wasteful irrigation methods.[3] Cities, industries, and buildings do not manage water carefully; they discharge it after a single use without sufficient efforts made to recover, recycle, and reuse it. To make matters worse, industrial agriculture creates water waste run-off that results in eutrophication. Nutrients travel to the wrong places, where they must be removed using costly water treatment operations.[4]

Alternatively, circular bio-economies can leverage opportunities for collecting diverse water sources and practicing controlled urban agriculture that capitalizes on the self-cleaning potential of water. What if buildings and neighborhoods could capture, clean, and reuse precious water resources while producing food and biomass to serve their inhabitants? What if cities could turn into productive networks capable of exchanging byproducts at the most efficient, environmental, and economically beneficial scale?

The international interdisciplinary research consortium CITYFOOD (supported by the Belmont Forum, the Joint Programming Initiative Urban Europe, and the National Science Foundation) started its work in 2018 with six partners in Germany, Sweden, Norway, the Netherlands, Brazil, and the United States. The consortium is tasked with investigating urban integration and scaling up environmentally sustainable, economically feasible, and socially beneficial food production in cities.[5] Its members have backgrounds in marine biology, aquaculture, engineering, urban planning, and architecture, and their research is focused on the water-based, circular growing system of aquaponics. In combining aquaculture and hydroponic cultivation to create emergent strategies that sustainability provide wholesome food, aquaponics has a positive impact on the urban environment and its resource cycles.[6]

Technology Supported Food Production

Over the past decades, urbanites have increasingly grown interested in reconnecting with the sources of their food. This has accelerated the development of multi-layered urban agriculture, whose operations include the growing of food on and in buildings.[7] One rapidly expanding facet of this movement is the integration of natural growing systems with high-tech infrastructure, in which living systems depend on tightly controlled, technical environments that provide them with optimal growing conditions. This approach, known as Controlled Environment Agriculture (CEA), has generated various typologies of urban farms such as rooftop greenhouses, vertically integrated greenhouses, and vertical indoor farms.[8]

Initial architectural responses to this technology-based approach to urban food production emerged with the publication in 2010 of *The Vertical Farm, Feeding the World in the 21st Century* by Dr. Dickson Despommier. This first book, with photographs of hydroponic growing systems and depictions of vertical farms, featured projects using glossy utopian renderings that explored the design of high-rise towers entirely dedicated to food production. These speculative images garnered broad interest in the idea of urban farming by the design community, even if its projects were never realized.[9]

Driven, however, to connect natural and technical systems, the current generation of urban CEA enthusiasts focus on the implementation of actual urban farms using technological innovations and creative solutions for achieving economic profitability. They are led by entrepreneurial initiatives, governmental incentives (to support environmental sustainability, resource recovery, food security, and business development), and industry suppliers that contribute their expertise in controls and in the automation of growing systems. Many of these new hydroponic technologies combine the precision of growing infrastructures and greenhouse enclosures with the sprouting vitality of the crops grown. Gotham Greens, for example, claims that its high-tech CEA greenhouses produce thirty-five times more heads of lettuce per acre than conventional field production while using ninety-five percent less water.[10] However, the construction of CEA urban farms is a great deal more capital-intensive and whose inputs, such as synthetic and mineral fertilizer, have a larger environmental footprint. While conclusive life cycle assessments (LCA) are not yet available, research in more sustainable and circular versions of CEA is increasingly taking place.[11]

Frameworks Towards Circularity

The development of more sustainable models for urban food production and operations requires interdisciplinary approaches. Professionals with expertise in water management, horticulture, CEA, engineering, sustainability science, urban planning, and policy writing need to collaborate across their disciplines.[12] A transdisciplinary framework that supports such a complex task in sustainable development is the Food-Water-Energy (FWE) Nexus.[13] Delivering food, water, and energy sustainable to all remains one of the most significant challenges for the near future.[14] The FWE Nexus is an academic sub-field in sustainability science that has garnered a great deal of research activity in the last ten years. Nexus thinking, which models the interdependencies, connections, synergies, and trade-offs across these three sectors (food/ water/ energy), helps to secure these resources more equitably.[15] This policy-based approach reduces the risk that any one of these three sectors is managed and regulated separately without ecosystem awareness and without regard for the interdependencies of their supply chains and resource cycles.[16] Naturally, a comprehensive, concurrent assessment of the three sectors leads to more sustainable solutions.

Notwithstanding research on the FWE Nexus has flourished, solution-based and implementation-focused studies are still rare, especially at the urban scale.[17] The Nexus synergistically connects resource streams, but it does not inherently produce circular solutions. It is essential that new approaches be framed within concepts of circular economies and, more specifically, within circular bio-economies. Both concepts follow the same core principles (reducing waste and pollution through design, using and reusing materials if possible, and supporting natural regenerative systems) and aim to resolve environmental, economic, and social goals.[18] However, circular bio-economies focus explicitly on delivering renewable biological resources.[19] A related concept, that of the circular city, applies circular strategies more holistically at the scale of cities, moving beyond a purely economic focus to a sustainable urban development framework.[20] As discussed by Professor Jo Williams from the Bartlett School in her *Circular Cities: A Revolution in Urban Sustainability,* circular economies focus on more resource-efficient production and circular city frameworks are based on efficiencies in urban and spatially defined systems that include infrastructures, the delivery of goods and services, and social benefits.[21] Collective resource management and careful consideration of the urban metabolism made possible by the FWE Nexus connect all actors in the ecosystem of the circular city by closing resource loops beyond the goals of a circular economy.[22] This paper describes three such loops at three different scales of engagement.

Scale One: Productive Water-Based Systems for Cultivating Food

At the center of CITYFOOD's investigation is aquaponics, a growing system that combines fish rearing and hydroponic plant cultivation. (Figures 2a and 2b) This closed-loop technology mimics a natural ecosystem with three symbiotic partners—fish, microorganisms, and plants—connected by the water exchanged between each of the growing environments. Aquaponics can produce, recover, and reuse nutrients within one system. Its ability to clean and recycle fish water makes it a more sustainable version of the recirculating aquaculture system (RAS).[23] In addition, the use of closed-loop hydroponic distribution reduces the water needed for plant production, a main advantage of this technology. Yet, the infrastructure required for aquaponics is intensive in that it needs two parallel systems: aquaculture tanks and the hydroponic growing mechanism contained in high-tech enclosures.[24] Operationally, however, it is advantageous to combine fresh produce (commonly leafy greens or tomatoes) with freshwater fish (a source of protein for urban food markets).[25]

Figure 2a. Aquaculture with tilapia swarming in fish tank. Photo credit: Daracha Thiammueang, Shutterstock.

Figure 2b. Hydroponic growing system in a rooftop greenhouse. Photo credit: Lufa Farms.

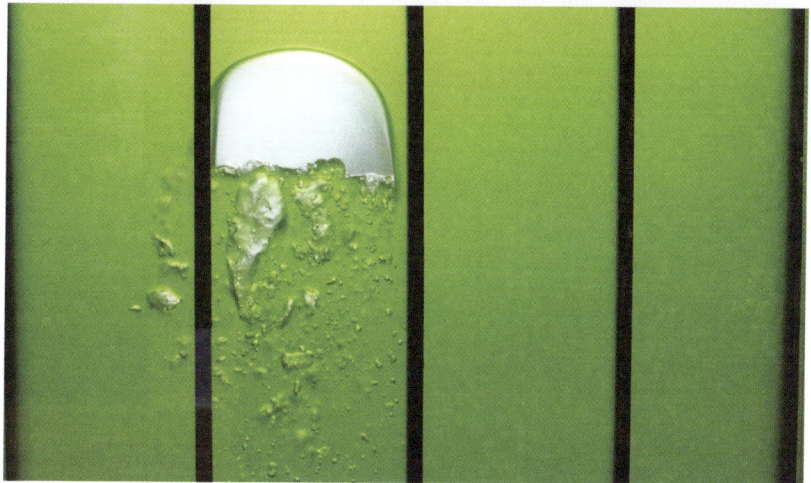

Figure 2c. SolarLeaf algae bioenergy façade panel installed in the BIQ House, IBA Hamburg, Germany. Photo credit: Arup, Colt International, SSC GmbH.

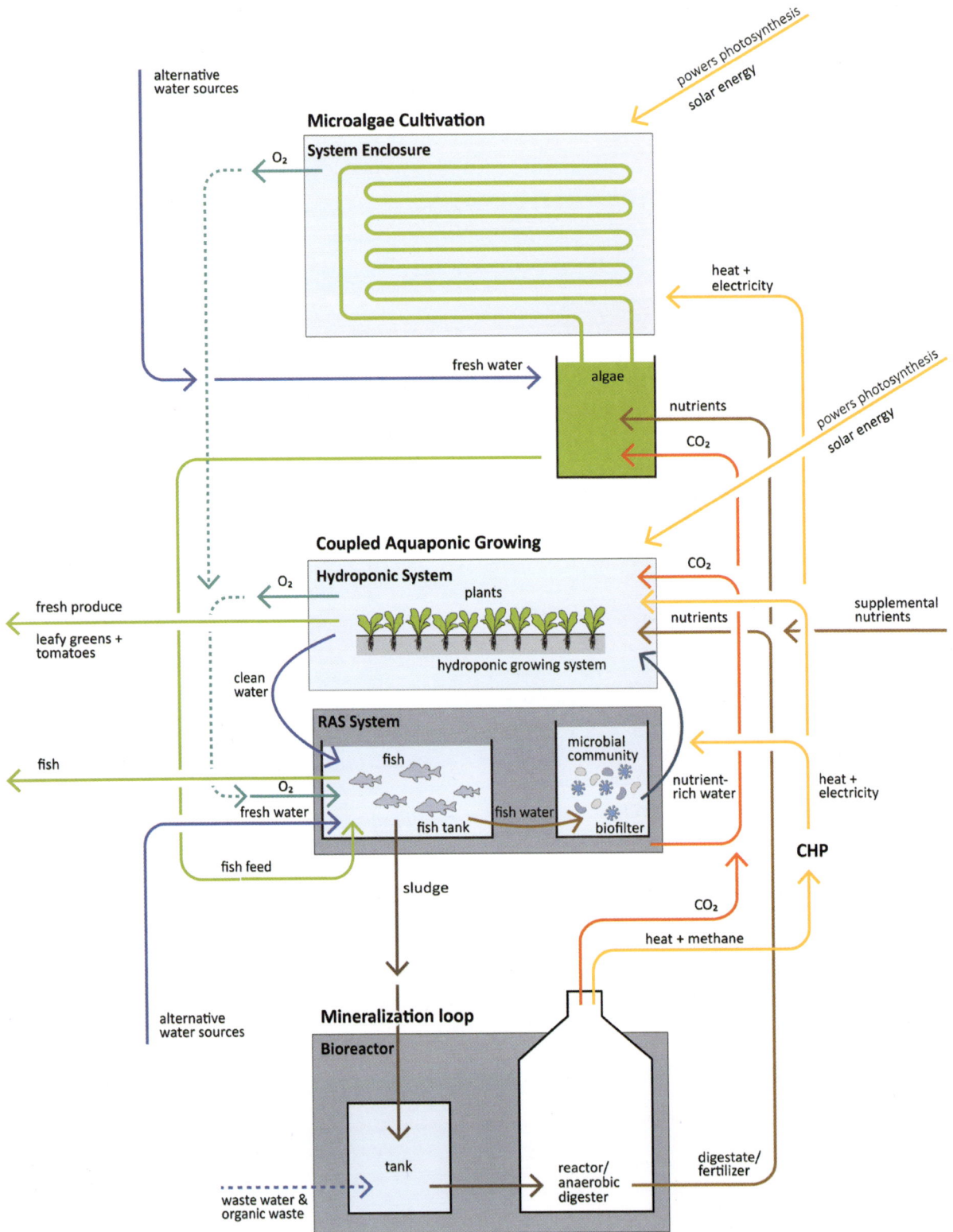

Microalgae Cultivation

System Enclosure

alternative water sources

O₂

powers photosynthesis
solar energy

fresh water

algae

nutrients

CO₂

powers photosynthesis
solar energy

heat + electricity

Coupled Aquaponic Growing

Hydroponic System

CO₂

plants

O₂

fresh produce
leafy greens + tomatoes

nutrients

supplemental nutrients

hydroponic growing system

clean water

RAS System

microbial community

fish

fish

O₂
fresh water

fish water

nutrient-rich water

heat + electricity

fish tank

biofilter

fish feed

CHP

sludge

CO₂

heat + methane

alternative water sources

Mineralization loop

Bioreactor

tank

reactor/ anaerobic digester

digestate/ fertilizer

waste water & organic waste

Aquaponic systems are as highly productive as hydroponic CEA systems, able to produce thirty times more food per acre than traditional farms.[26] However, while resource-efficient in terms of water use and nutrient recovery, their primary resource input—fish feed—has a significant environmental footprint. Currently, fish feed production relies on fishmeal derived from forage fish. This practice leads to overfishing, which is not sustainable, especially as the demand for farmed fish is expected to double in the next ten years.[27] Alternatively, many microalgae species contain valuable proteins, lipids, and bioactives, which make them a viable supplement or alternative for producing fish-free feed.[28] Photosynthetic microalgae effectively sequester carbon dioxide while generating oxygen (over fifty percent of global atmospheric oxygen) in a rapidly growing biomass.[29] (Figure 2c) Its ability to convert sunlight efficiently is five times higher than that of terrestrial plants.[30] Algae cultivation has garnered much research interest in the context of biofuel production and eco-services, such as wastewater treatment. More recently, attention has been focused on the production of higher-value products, such as human food (supplements), valuable raw materials for bio-based innovations, and (as proposed here) fish feed.[31] Producing at least part of the fish feed within the system is one of the highest aspirations of aquaponics, and it is entirely possible that microalgae production connected to an aquaponics system could supplement or replace the conventional fish feed.[32] Given the resources shared by aquaponics and microalgae cultivation, this essay connects these two water-based systems to assess possible symbiotic exchanges when using an expanded FWE Nexus lens for linking food, water, nutrients, and energy cycles. (Figure 3)

The primary purpose of the combined system is food production; aquaponics provides fish and produce for human consumption, and the microalgae feed the fish.[33] Water is the system's primary growing medium, needed for the metabolism of all organisms.[34] Aquaponics and algae photobioreactors are both recirculating systems with a relatively low need for freshwater input compared to other growing methods. Water can be replenished using alternative sources such as rainwater, air conditioning condensate, and groundwater adjacent to building foundations.[35] Water is the growing environment for the fish and algae, transporting dissolved nutrients to plants, which perform the system's self-cleaning functions while benefiting from the nutrients. Additional nutrients are contained in the solid waste of the fish, which settles as sludge at the bottom of the fish tank and is recoverable in a mineralization loop.[36] This third operation utilizes an anaerobic digester to contribute to several resource loops. Besides recovered nutrients that supplement the hydroponic loop and algae cultivation, it creates methane, heat, and carbon dioxide (CO_2).

Both autotroph systems—hydroponically grown plants and algae—require sunlight or photosynthetic active radiation (PAR) together with CO_2, water, and nutrients to perform photosynthesis. CO_2 is an essential building block for plant matter, and closed CEA environments require CO_2 fertilization to replenish and increase the

Figure 3. Schematic diagram of aquaponics system integrated with algae photobioreactor and anaerobic digester. Diagram by Gundula Prosch.

CO_2 concentration available in the air for photosynthesis. Doubling or tripling the ambient CO_2 level (400ppm) generates the best results in commercial greenhouses.[37] The CO_2 generated by fish, anaerobic digesters, and potentially other built environment sources of CO_2 could be used for this purpose. The heat and methane produced in the anaerobic digester contribute to the systems' energy needs for heating, pumping, and supplemental lighting. Whether additional energy is needed depends on the system size, the quantity of organic matter available, and the local climate. Anaerobic digesters typically work best at the infrastructure scale, though current research advances the benefits of smaller containerized versions.[38]

Scale Two: Building Integration

In temperate climates, hydroponic, aquaculture, and microalgae growing systems require enclosures that provide optimal growing conditions for year-round operation, similar to other forms of controlled-environment agriculture (CEA). Transparent enclosures needed for photosynthetic plants and organisms, as well as well-insulated spaces for aquaculture, offer architectural design opportunities for building integration. When production greenhouses are located on top of or within buildings, this is referred to as building-integrated agriculture (BIA). Social entrepreneur Ted Caplow and his team at NY Sun Works coined this term while developing the *Science Barge*—a floating greenhouse that in 2006 was an early prototype for rooftop greenhouses.[39] This form of innovative thinking led to the development of the first large-scale commercial rooftop greenhouses by urban farming companies Gotham Greens and Lufa Farms in 2011.[40] Both companies operate four rooftop greenhouses each, ranging in scale from 15,000 to 164,000 square feet.[41] Lufa Farms' rooftop greenhouses, for example, save up to fifty percent of the energy required for their heating compared to conventional on-grade greenhouses because they are situated on top of heated warehouses. In return, they insulate the roofs of their host buildings, which leads to additional energy savings in the cold climate of Montreal.[42] Gotham Greens's rooftop greenhouses, some of which are in Brooklyn, New York, offer similar benefits, even as the eight rooftop greenhouses do not share heating, ventilation, and air handling systems with their host buildings. However, research conducted on the building-integrated rooftop greenhouse of the *ICTA-ICP Research Center* in Cerdanyola des Vallès, Barcelona, Spain, indicates that connecting air and water handling systems can make the greenhouse and its host building more sustainable.[43] Similar investigations on the reuse of waste heat and greywater in innovative urban farming are currently being conducted at the *Altmarktgarten Oberhausen* rooftop greenhouse of the Fraunhofer Institute for Environmental Safety and Energy Technology (UMSICHT) research facility.[44] This research lab is housed in one of the most remarkable examples of architectural integration of rooftop greenhouses to date. The architects Kuehn Malvazzi created an elegant, seamless union between a state-of-the-art greenhouse and the five-story administrative building in the historic center of Oberhausen, Germany.[45] This project leads the way in terms of technical, aesthetic, and programmatic integration. (Figure 4)

Figure 4. Altmarktgarten Oberhausen, rooftop greenhouse integrated with administrative building in Oberhausen, Germany. Architect: Kuehn Malvezzi. Photo credit: OGM Oberhausener Gebäudemanagement GmbH.

Photobioreactors have captured the interest and imagination of architects and designers in several speculative design projects and prototypes, even if commercially, the building integration of algae cultivation has not yet advanced.[46] Only one project, the 2013 bio-reactive façade system, *SolarLeaf* developed by Arup, Colt International, and Strategic Science Consult and installed in the BIQ housing development at the International Building Exhibition (IBA) in Hamburg, Germany, has successfully tested its technical feasibility and constructability at the building scale.[47] Most research projects that focus on the advancement of algae production utilize photobioreactors in freestanding greenhouses, such as the Algae Cultivation Center of the Technical University Munich in Munich, Germany.[48] (Figure 5)

Once the growing system is integrated, the next step is to connect it with other productive food systems. Hybrid programming fosters the exchange of resources and byproducts, such as wastewater, heat, and biomass. One such zero-waste approach was pioneered by the food business incubator *The Plant*, located in the Back of the Yards neighborhood of Chicago's South Side.[49] Here, the adaptive reuse of a 1920s meatpacking facility provides space for artesian food production, research, and aquaponics. Over the last ten years, the 94,000-square-foot building and the twenty small businesses it hosts have followed circular city principles by recovering, recycling, and reusing building materials and resources.[50]

Figure 5. Algae Cultivation Center, greenhouse with algae bioreactor at the Technical University Munich (TUM) in Munich, Germany. Photo credit: Andreas Heddergott/ TUM.

The aquaponics farm *Ferme Abattoir,* built and operated by Building Integrated Greenhouses (BIGH), is another example of advanced building integration and hybrid programming, which redefines the intersection of built environments and food systems. This commercial urban rooftop farm is constructed atop a new mixed-use food market in Anderlecht's Abattoir neighborhood, a historical slaughterhouse district in the urban center of Brussels. *Ferme Abattoir* generates a multitude of positive impacts through its synergies with the host building and the urban context: it helps build the local economy, recover resources, and revitalize an urban district. The city offers access to a consumer base, workforce, and supportive community. At the same time, the farm provides year-round locally produced fresh food to urban residents and employment opportunities: exchanges that strengthen the local economy. This urban farm has access to utilities and environmental resources. It benefits from the city's higher ambient temperature (about four to six degrees Fahrenheit higher than the surrounding rural areas), reducing operational costs.[51] The farm can recover from the site and its host building resources such as rainwater, well water, stormwater, greywater, waste heat from cooling systems, electricity from photovoltaic (PV) panels, and CO_2 from the building's systems and tenants. The use of the rooftop as a greenhouse site provides additional rental income for owners and reduces the operational cost of the host building.

Urban integration of farms also contributes to urban revitalization. Cities provide many underutilized spaces, such as rooftops and former industrial sites, for this purpose. The presence of an urban farm in these places increases real estate values and offers numerous social benefits for residents.[52] The *Abattoir* site operates this way; the aquaponics *Ferme* in joining forces with the food market is the first project completed in the *Abattoir 2020* master plan. Devised by the architecture firm ORG Permanent Modernity, the plan aims to redevelop the industrial district, renew its civic importance, and offer economic opportunities to members of the vibrant but underserved neighborhood. Urban warehouses will eventually frame the outdoor market with the historic market hall at its center. This almost fifteen-acre-large public space will remain the heart of the neighborhood for social-cultural events and market activities.[53] This example connecting rooftop farms, host buildings, and urban sites demonstrates how the different symbiotic scales work together to fulfill the goals of the circular city. (Figure 6)

Scale Three: Circular City

The third scale of interest is that of the circular city, whose framework aims to generate interconnected resource flows in a network of industries, producers, and users. While circular city goals and principles are well articulated—as we discussed in the work of Jo Williams—the generative processes of designing and implementing such approaches require more detailed analysis, particularly in what concerns the water-based and food-related economy of the city. The question here is more specific: How, where, and at what scale should aquaponic urban farms and hybrid operations (that combine multiple industries) be implemented to become part of the circular city?

Two projects currently under development address this question by integrating large-scale, ground-up initiatives that follow circular city principles. The *Integrated Bioeconomy Project* by the Center of Solar Biotechnology at the University of Queensland in Brisbane, Australia, will connect hydroponic, algae, and aquaculture systems in a controlled biosphere.[54] Its extensive greenhouse will utilize various advanced technological controls, energy production, and nutrient recovery systems to house infrastructure for crop and algae production. This multifunctional greenhouse is scalable such that at its largest projected scale of multiple acres, it will undoubtedly boost the local economy. The primary goal of this project is to test new technologies at an economy of scale. Effekt Architects in Copenhagen, Denmark, is taking a different approach to integrating circular city principles. Their *ReGen* project, an ecovillage proposed in Almere in the Netherlands, is conceived as a small circular city whose construction will include twenty-five single-family houses and all the infrastructure needed to close the water, food, energy, and waste cycles; with the latter located on-site for ensuring this small community is self-sufficient. (Figures 7a and 7b)

Figure 6. Abattoir 2020: Masterplan for the redevelopment of Abattoir neighborhood in Brussels by ORG Permanent Modernity. The Ferme Abattoir (BIGH) atop the FoodMet market (ORG) is located to the left of the historic market hall in the center. Photo credit: ORG Permanent Modernity.

The most challenging question, however, is how to strategically integrate urban farms and hybrid food production within existing cities. Planning and design processes require reconciling circular city principles with the physical requirements needed for such complex operations, all the while identifying a suitable location. Urban locations should ideally offer opportunities for the recovery and re-use of underutilized resources.[55] In general, urban neighborhoods zoned for commercial and light industrial use are best suited for the physical requirements of size, building type, and infrastructure of potential host buildings. These urban areas are also more likely to offer potential partner industries for hybrid programming and the exchange of byproducts. So far, built environment professionals have insufficiently addressed how to generate a network of resource exchanges between productive nodes in the circular city. Currently, there is no consolidated urban dataset that lists such resources. An information model for crowdsourcing and GIS-based compilations of different public urban datasets has been proposed.[56] This map of potential alternative resources would be an essential tool for identifying possible connections between existing industries and sites for new construction. One could adopt material flow analysis to determine available resources generated by existing partner industries and to estimate the quantity of resources needed in one's operations.[57] This process could be assisted by the hybrid programming approach in which possible couplings of symbiotic industry relations are identified.

The primary needs for operating an aquaponics site are energy, heat, water, and CO_2. To source heat and energy, one could pair the greenhouse with an industry that generates exhaust heat, either from cooling (see *Ferme Abattoir*) or high-temperature processes, such as baking or brewing.[58] Additionally, other food industries generate large amounts of organic waste, which anaerobic digesters can transform into energy.[59] The most suitable alternative source of water for aquaponics is rainwater or well water, which are site-specific resources. A unique option for the coupling of water sources is that of aquaculture farms and hatcheries, which provide nutrient-rich water and can be paired directly with a hydroponic greenhouse. Such a combination would be considered a decoupled aquaponics system. CO_2 is produced in many industrial processes and commercial buildings (exhaust air from office spaces contains CO_2 levels of up to 2500 ppm).[60] Challenges include potential pollutants and the lack of well-established exchange systems for gas transfer.

Besides closing resource loops, aquaponics and water-based food production should consider other interdependent factors when following circular city principles. Depending on the business type—production facility (based on sales of produce and fish alone) or mixed-income operation (offering additional services and products)—their operations will likely require the selection of different sites. Production facilities tend to be located on larger, peri-urban sites at lower cost, while mixed-income businesses benefit from urban sites

Figure 7a. ReGen Ecovillage proposed for Almere in the Netherlands by Effect Architects. Rendering of aquaponics greenhouse with vertical growing structures. Photo credit: EFFEKTArchitects.

Figure 7b. ReGen Ecovillage proposed for Almere in the Netherlands by Effect Architects. Circular city material flow diagram. Photo credit: EFFEKTArchitects.

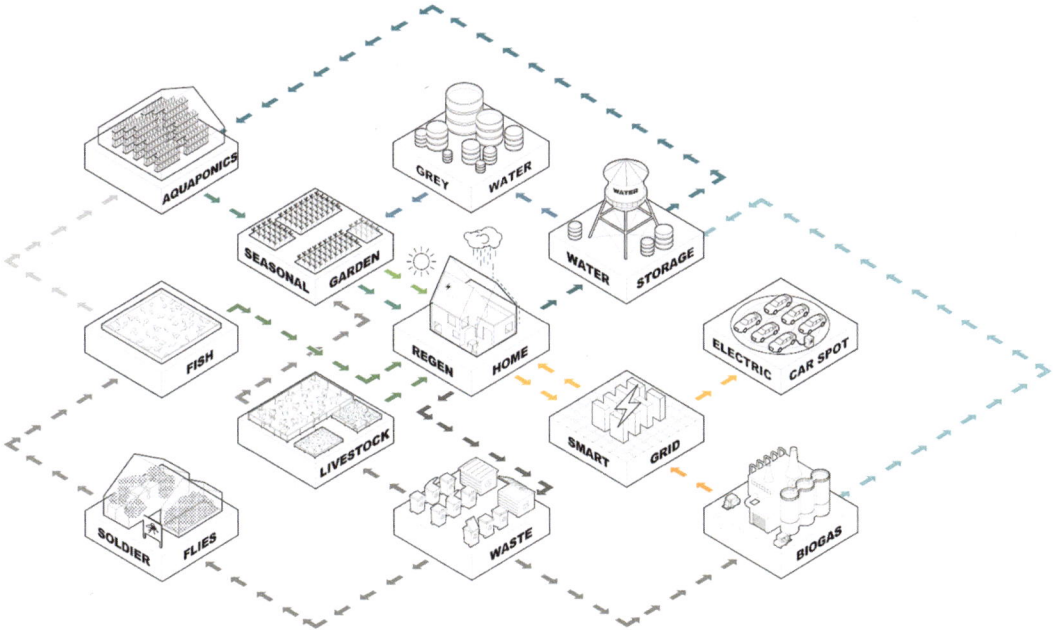

REGEN SYSTEM

closer to their customer base.[61] The latter often utilize smaller sites, rooftops, and host buildings with a symbiotic program, an excellent example of which is the new aquaponics greenhouse atop the recently opened "Market of the Future" in Wiesbaden, Germany.[62] The market, which is constructed of stacked timber, is further integrated architecturally by a large skylight over the produce section. (Figure 8) A special viewing area at roof level allows shoppers to observe greenhouse and farm operations. These spatial connections are opportunities to inform customers and create awareness about food production and its environmental impacts.

Figure 8. Aquaponics rooftop greenhouse on the "Market of the future," the recently constructed REWE "Green Farming" supermarket in Wiesbaden, Germany by ACME Architects. Photo credit: REWE/ Jürgen Arlt.

Synergies in Water-Based Productive Food Networks

The integrated, water-based systems for nutrient generation and recovery proposed here and investigated as part of the CITYFOOD project are founded on ideas from the Food-Water-Energy Nexus and those supported by circular cities. Individual technologies, buildings, and urban sites support the coupling of productive biological systems and their technical infrastructure and innovations. The uniqueness of aquaponics is the result of inherent synergies between a hydroponic and aquaculture system. Other productive processes involving algae, bacteria, and microorganisms may also be included in resource loops, providing microalgae-based fish feed, methane, and heat from anaerobic digestion. Technical infrastructures support these interconnected water-based systems in capturing the photosynthesis, metabolism, and self-cleaning power of natural ecosystems located within these highly controlled environments. Growth parameters for the different types of organisms are optimized to improve their productivity using customized sensing, regulation, and automation. Host structures and enclosures offer the next level of technology integration at the building scale for recovering resources, reducing operational energy needs, and improving building performance. Hybrid programming between multiple businesses fosters the exchange of byproducts. At the urban scale, additional information exchanges are needed to bridge knowledge gaps and access data by built environment professionals seeking to develop circular cities.

Beyond advancing their industrial symbiosis, the circular city framework fosters more holistic social, cultural, and spatial dimensions for cities. The urban integration of CEA offers many benefits in this area. Aquaponics is an excellent teaching tool for students from kindergarten (pre-school) to grade twelve (high school) and for job training in ecosystems, food production, and system regulation. Outstanding examples in this area are The Greenhouse Project by NY Sun Works and The Farm on Ogden by Windy City Harvest. Urban operations that embrace hybrid programming and connect urban food production to everyday experiences such as grocery shopping, restaurant visits, and cultural events benefit the city's population. In addition to the food they produce, urban farms create food system awareness, generate supportive communities of customers, and communicate their integration with the FWE Nexus for environmental and human well-being.

This multiscale approach to integrating new, circular food production methods in urban environments reveals and conveys the importance of urban water cycles. Most resource and energy exchanges are based on water and are often closely connected to nutrient and food cycles. The implementation of circular city principles helps expand our awareness of water management and strengthens our physical responses to recovering, recycling, and reusing water in the built environment. Overall, the symbiotic integration of growing systems, building infrastructures, and urban networks reduces the environmental footprint of food production. It increases water accessibility and food security by replacing untenable practices with the power of natural systems, recirculating operations, and integrated bioeconomy strategies.

Acknowledgments

The author thanks the entire CITYFOOD research consortium and all members of the Circular City + Living Systems Lab (CCLS) for their inspirations and contributions to the project.

Funding

The CITYFOOD project is part of the Sustainable Urbanisation Global Initiative (SUGI) Food Water Energy Nexus, a Collaborative Research Action (CRA) jointly initiated by the Belmont Forum and JPI Urban Europe. It received funding from the U.S. National Science Foundation (Award # 1832213).

ENDNOTES

1. Jianguo Liu et al., "Nexus Approaches to Global Sustainable Development." *Nature Sustainability* 1, no. 9 (September 2018): 466. https://doi.org/10.1038/s41893-018-0135-8.

2. Hugh Turral, Jacob Burke, and Jean-Marc Faurès, *Climate Change, Water and Food Security: FAO Water Reports 36* (Rome: Food and Agriculture Organization of the United Nations, 2011), 40-41.

3. Gundula Proksch, *Creating Urban Agricultural Systems: An Integrated Approach to Design*, (New York: Routledge, 2016), 44. https://doi-org.offcampus.lib.washington.edu/10.4324/9781315796772.

4. Hannah Hislop, ed. *The Nutrient Cycle: Closing the Loop* (London: Green Alliance, 2007), 5.

5. Gundula Proksch and Daniela Baganz, "CITYFOOD: Research Design for an International, Transdisciplinary Collaboration," *Technology|Architecture + Design* 4, no. 1 (January 2, 2020): 35-36. https://doi.org/10.1080/24751448.2020.1705714.

6. Ibid.

7. Kathrin Specht et al., "Urban Agriculture of the Future: An Overview of Sustainability Aspects of Food Production in and on Buildings," *Agriculture and Human Values* 31, no. 1 (March 1, 2014): 34. https://doi.org/10.1007/s10460-013-9448-4.

8. Khadija Benis and Paulo Ferrão, "Commercial Farming within the Urban Built Environment – Taking Stock of an Evolving Field in Northern Countries," *Global Food Security* 17 (June 1, 2018): 31. https://doi.org/10.1016/j.gfs.2018.03.005.

9. Dickson Despommier, *The Vertical Farm: Feeding the World in the 21st Century* (New York: Thomas Dunne Books/ St.Martin's Press, 2010), 146-47, 178-79, 242-43.

10. Benjamin Wooddy, "Sustainable Farming with Gotham Greens," U.S. Department of Agriculture, https://www.usda.gov/media/blog/2021/04/22/sustainable-farming-gotham-greens.

11. Oliver Körner et al., "Environmental Impact Assessment of Local Decoupled Multi-Loop Aquaponics in an Urban Context," *Journal of Cleaner Production* 313 (September 1, 2021): 127735, 11. https://doi.org/10.1016/j.jclepro.2021.127735.-

12. Jianguo Liu et al., "Systems Integration for Global Sustainability," *Science* 347, no. 6225 (February 27, 2015): 1258832, 963. https://doi.org/10.1126/science.1258832.

13. Maryam Ghodsvali, Sukanya Krishnamurthy, and Bauke de Vries, "Review of Transdisciplinary Approaches to Food-Water-Energy Nexus: A Guide towards Sustainable Development," *Environmental Science & Policy* 101 (November 1, 2019), 267. https://doi.org/10.1016/j.envsci.2019.09.003.

14. Future Earth Knowledge Action Network, "Water-Energy-Food Nexus," Future Earth, https://futureearth.org/networks/knowledge-action-networks/water-energy-food-nexus/ (retrieved August 5, 2021).

15. Jianguo Liu, "Nexus Approaches," 466.

16. Tony Allan, Martin Keulertz, and Eckart Woertz, "The Water-Food-Energy Nexus: An Introduction to Nexus Concepts and some Conceptual and Operational Problems," *International Journal of Water Resources Development* 31, no. 3 (July 3, 2015): 302. https://doi.org/10.1080/07900627.2015.1029118.

17. Darin Wahl, Barry Ness, and Christine Wamsler, "Implementing the Urban Food–Water–Energy Nexus through Urban Laboratories: A Systematic Literature Review," *Sustainability Science* 16, no. 2 (March 2021), 663-64. https://doi.org/10.1007/s11625-020-00893-9.

18. D. D'Amato et al., "Green, Circular, Bio Economy: A Comparative Analysis of Sustainability Avenues," *Journal of Cleaner Production* 168 (December 1, 2017), 716-17. https://doi.org/10.1016/j.jclepro.2017.09.053.

19. Ellen MacArthur Foundation, "Cities and Circular Economy for Food," 2019.

20. Jo Williams, "Circular Cities," *Urban Studies* 56, no. 13 (October 1, 2019): 2755. https://doi.org/10.1177/0042098018806133.

21. Jo Williams, *Circular Cities: A Revolution in Urban Sustainability*. Routledge Studies in Sustainability. (Milton: Taylor and Francis, 2021), 11.

22. COST Action Circular City, "Implementing nature based solutions for creating a resourceful circular city," https://www.cost.eu/actions/CA17133/

23. Wilson Lennard and Simon Goddek, "Aquaponics: The Basics," in *Aquaponics Food Production Systems*, ed. Simon Goddek, Alyssa Joyce, Benz Kotzen, and Gavin M. Burnell (Cham: Springer International Publishing, 2019), 138. https://doi.org/10.1007/978-3-030-15943-6_5.

24. Gundula Proksch, Alex Ianchenko, and Benz Kotzen, "Aquaponics in the Built Environment," in *Aquaponics Food Production Systems*, ed. Simon Goddek, Alyssa Joyce, Benz Kotzen, and Gavin M. Burnell (Cham: Springer International Publishing, 2019), 527. https://doi.org/10.1007/978-3-030-15943-6_21.

25. Gundula Proksch, Erin Horn, George Lee, "Urban Integration of Aquaponics," in *Urban Agriculture and Regional Food Systems*, ed. Peter Droege, Elsevier, 2021 (forthcoming).

26. Superior Fresh, "Regenerative Agriculture Reimagined," Superior Fresh. https://www.superiorfresh.com/our-farm

27. Cliff White, "Algae-Based Aquafeed Firms Breaking down Barriers for Fish-Free Feeds," SeafoodSource, https://www.seafoodsource.com/news/aquaculture/algae-based-aquafeed-firms-breaking-down-barriers-for-fish-free-feeds

28. Jennifer Yarnold et al., "Microalgal Aquafeeds As Part of a Circular Bioeconomy," *Trends in Plant Science* 24, no. 10 (October 1, 2019): 959-60. https://doi.org/10.1016/j.tplants.2019.06.005.

29. Benz Kotzen et al., "Aquaponics: Alternative Types and Approaches," in *Aquaponics Food Production Systems*, ed. Simon Goddek, Alyssa Joyce, Benz Kotzen, and Gavin M. Burnell (Cham: Springer International Publishing, 2019), 304. https://doi.org/10.1007/978-3-030-15943-6_12.

30. Rosa Rosello Sastre, "Products from Microalgae: An Overview," in *Microalgal Biotechnology: Integration and Economy*, ed. Clemens Posten and Christian Walter (Berlin, Boston: De Gruyter, 2012), 82. https://doi.org/10.1515/9783110298321.

31. Ibid., 13-16.

32. Kotzen, "Aquaponics: Alternative Types," 310.

33. Robin Shields and Ingrid Lupatsch, "Algae for Aquaculture and Animal Feeds," in *Microalgal Biotechnology: Integration and Economy,* ed. Clemens Posten and Christian Walter (Berlin, Boston: De Gruyter, 2012): 82. https://doi.org/10.1515/9783110298321.

34. Steven Beckers, "Aquaponics: A Positive Impact Circular Economy Approach to Feeding Cities," *Field Actions Science Reports* 20 (September 1, 2019), 81.

35. Proksch, *Creating Urban Agricultural Systems,* 63-64.

36. Simon Goddek and Oliver Körner, "A Fully Integrated Simulation Model of Multi-Loop Aquaponics: A Case Study for System Sizing in Different Environments," *Agricultural Systems* 171 (May 1, 2019): 145–47. https://doi.org/10.1016/j.agsy.2019.01.010.

37. Ontario Ministry of Agriculture, Food and Rural Affairs, "Carbon Dioxide in Greenhouses." http://www.omafra.gov.on.ca/english/crops/facts/00-077.htm#suppl.

38. Impact Bioenergy, "Horse AD25 Series." https://impactbioenergy.com/horse-ad25/

39. Ted Caplow founded New York Sun Works in 2004 to develop the Science Barge and started to publish research on building-integrated rooftop greenhouses. From 2008-2011 he contributed to several sustainable greenhouse projects in the New York region with his design consultancy Bright Farms Systems. In 2011, he partnered with Paul Lightfoot to transform the consultancy into the full-service commercial farming company BrightFarms, which opened its first greenhouse in 2013. Ted Caplow and Jennifer Nelkin, "Building-integrated Greenhouse Systems for Low Energy Cooling," 2nd PALENC Conference and 28th AIVC Conference on Building Low Energy Cooling and Advanced Ventilation Technologies in the 21st Century, Crete Island, Greece (2007), 172-76.

40. Proksch, *Creating Urban Agricultural Systems,* 215-20.

41. Gundula Proksch and Alex Ianchenko, "Commercial Rooftop Greenhouses," in *Urban agriculture and regional food systems,* ed. Peter Droege, Elsevier, 2021 (forthcoming).

42. Lauren Rathmell, "Growing vegetables on Montreal's rooftops with Lufa Farms," YouTube video, 2:07 min, Nov. 21, 2019. https://www.youtube.com/watch?v=aORkDkMC-kE.

43. Ana Nadal et al., "Building-Integrated Rooftop Greenhouses: An Energy and Environmental Assessment in the Mediterranean Context," *Applied Energy* 187 (February 1, 2017): 348. https://doi.org/10.1016/j.apenergy.2016.11.051.

44. Frauenhofer IRB, "ALTMARKTgarten Oberhausen – Experimentierfeld für urbane Landwirtschaft," Bauen +. https://www.bauenplus.de/aktuelles/ALTMARKTgarten-Oberhausen-Experimentierfeld-fuer-urbane-Landwirtschaft/.

45. Kuehn Malvezzi Associates, "Administrative Building with Integrated Rooftop Greenhouse," Kuehn Malvezzi. http://kuehnmalvezzi.com/?context=project&oid=Project:33673

46. Valerio R.M. Lo Verso et al., "Photobioreactors as a Dynamic Shading System Conceived for an Outdoor Workspace," *Journal of Daylighting* 6, no. 2 (December 2019): 148–149. https://doi.org/10.15627/jd.2019.14.

47. Jan Wurm, "SolarLeaf, Hamburg - Worldwide first façade system to cultivate micro-algae to generate heat and biomass as renewable energy sources," Arup. https://www.arup.com/projects/solar-leaf. See also, IBA Hamburg, "Smart Material Houses: BIQ," International Building Exhibition IBA Hamburg 2013. https://www.internationale-bauausstellung-hamburg.de/en/themes-projects/the-building-exhibition-within-the-building-exhibition/smart-material-houses/biq/projekt/biq.html.

48. Technical University of Munich (TUM), "Algae cultivation center opened at the Ludwig Bölkow Campus," TUM. https://www.tum.de/en/about-tum/news/press-releases/details/32656/

49. The Plant, "About The Plant," The Plant. https://www.insidetheplant.com/about-the-plant .

50. Ibid.

51. Beckers, "Aquaponics," 80.

52. Ibid.

53. ORG, "Abattoir 2020," ORG Urbanism. https://orgpermod.com/urbanism/projects/abattoir-2020-brussels-meat-market-district.

54. Evan Stephens and Ben Hankamer, "Integrated Bioeconomy Project," Institute for Molecular Bioscience, The University of Queensland. https://imb.uq.edu.au/integrated-bioeconomy-project.

55. Gösta Baganz et al., "Site Resource Inventories – a Missing Link in the Circular City's Information Flow." *Advances in Geosciences 54* (October 1, 2020): 23. https://doi.org/10.5194/adgeo-54-23-2020.

56. Ibid., 27-28.

57. Erin Horn and Gundula Proksch, "Building an ecosystem: integrating rooftop aquaponics with a brewery to advance the circular economy," in Open: Proceedings of the 108th ACSA Annual Meeting, San Diego, CA, (Washington, DC: ACSA Press, 2020), 7.

58. Ibid., 7-9.

59. The Plant, "Research + Learning," https://www.insidetheplant.com/researchlearning (retrieved August 5, 2021).

60. US EPA, ORD, 2017, "Indoor Air Quality," Reports and Assessments, US EPA. November 2, 2017. https://www.epa.gov/report-environment/indoor-air-quality.

61. Proksch, "Urban Integration of Aquaponics."

62. REWE, a German supermarket chain, ECF Farm Systems, and the architecture office ACME developed this project in collaboration. ACME, "Market of the Future," ACME. https://www.acme.ac/acme-space/164-market-of-the-future/

PLANT(S) MATTER:

ON THE DICHOTOMY AND DUPLICITY OF GREEN WALLS AND GRASS PAVEMENTS

SONJA DÜMPELMANN

ABSTRACT

Landscape architecture is a synthetic practice that uses plants, soil, wood, bricks, and other materials to shape the built environment for the well-being of human and non-human nature. While most designed landscapes consist of both forms of matter, on occasion, landscape designers and horticulturalists have also developed synthetic materials that combine the two. The green wall built for the 1915 Panama-Pacific International Exposition in San Francisco and "Watson's Pots," invented in Britain for the camouflage of concrete runways during the Second World War, are examples of early synthetic vertical and horizontal building materials. They relied on the integration of live and inert materials as well as the agency of plants to create modular structures used at a variety of scales. As forerunners of today's elaborate green walls and mundane grass pavers, their histories exemplify the "duplicity of landscape" and "plant blindness" that landscape architecture continues to suffer from. Despite the relative invisibility of plants in Western thought and culture, and despite their ability to 'veil' both cultural artifacts and their (own) histories, plants have been forces and agents of history. The landscapes they have rendered duplicitous are worth exploring to further our understanding of the relationship between human and non-human nature.

+ plants
+ green wall
+ grass paver
+ duplicity of landscape
+ plant blindness

Figure 1. Detail of Assistant Landscape Engineer Donald McLaren inspecting *Mesembryanthemum* Hedge, Panama-Pacific International Exposition, San Francisco, ca. 1914, The San Francisco History Center, San Francisco Public Library.

As a synthetic practice that uses live and inert material to shape the built environment for the well-being of human and non-human nature, landscape architecture strives to achieve the synthesis of utility and beauty, which the ancient Roman poet Horace called *dulce utili*. The integration of live plant matter and inert materials, like stone and sand, can be observed at a variety of scales, from the scale of the individual plant to that of entire cities and landscapes. For example, gardens and parks are inserted within the urban fabric, and roads and other forms of infrastructure are integrated into 'more or less' productive vegetated lands. While early professional landscape architects conceived of parks, gardens, and what we today call urban green infrastructure as elements standing in contrast to otherwise 'lifeless' built structures, they were also consciously bridging the nature–culture dichotomy. Although "the beauty of the park should be the other," as Frederick Law Olmsted reminded his contemporaries in 1870, alluding to its vegetation and pastoral scenery, the park "should ... complement the town."[1] Synecdoches of nature, parks, gardens, and plants were integral components of cities and the many comprehensive urban plans drawn up for cities in the early twentieth century. Landscape architecture, as defined by its early professionals in the late nineteenth century, was a synthesis of agriculture, horticulture, and forestry, as well as engineering and architecture.[2]

While most designed landscapes consist of both live and inert matter, on occasion, landscape designers and horticulturalists have developed synthetic materials that intentionally combine the two. Whereas plants have colonized horizontal and vertical vegetal surfaces for millennia, the green wall built for the 1915 Panama-Pacific International Exposition in San Francisco and "Watson's Pots" invented in Britain for the camouflage of concrete runways during the Second World War are examples of early synthetic vertical and horizontal building materials. They relied on the integration of live and inert materials as well as on the agency of plants to create modular structures used at a variety of scales. As will become clear, although the various combinations of live and inert matter belong to landscape architecture's defining characteristics, they have also proved to be among its challenges.

Greening Walls

An extraordinary feature greeted visitors at the 1915 Panama-Pacific International Exposition held on 635 acres along the Bay in San Francisco between February and December. A twenty-foot high, eight-foot wide green "wall" of more than one thousand feet in length, referred to variously as "hedge" or tautologically as "living hedge," separated the exhibition grounds from the city in the south. Running along Chestnut Street, the hedge formed archways above turnstile entrances and was described as an "elaborate [Spanish] Mission [revival] design at the main entrance" on Scott Street.[3] It was constructed out of 7,500 wooden trays six feet long and two feet wide, containing a two-inch

Figure 2. Wooden frames filled with *Mesembryanthemum*, lying on the plaza in front of the Palace of Horticulture before installation, Panama-Pacific International Exposition, San Francisco, ca. 1914, The San Francisco History Center, San Francisco Public Library.

Figure 3. Installation of *Mesembryanthemum* hedge, Panama-Pacific International Exposition, San Francisco, ca. 1914, The San Francisco History Center, San Francisco Public Library.

Figure 4. Assistant Landscape Engineer Donald McLaren inspecting *Mesembryanthemum* hedge, Panama-Pacific International Exposition, San Francisco, ca. 1914, The San Francisco History Center, San Francisco Public Library.

layer of earth covered with wire netting, and planted with thousands of ice plant cuttings (*Mesembryanthemum spectabilis*). (Figure 2) These planted trays were assembled on a wooden scaffold to form the dividing "wall." (Figure 3) A one-inch perforated pipe installed along the top of the scaffold irrigated the plants.[4] The initial design idea was attributed to architect W.B. Faville of the San Francisco architecture firm Bliss & Faville who suggested building a fence that was to appear like an old English wall overgrown with moss and ivy.[5] The idea was further developed by the exposition's Chief of Landscape Gardening and superintendent of Golden Gate Park, John McLaren, and his son Donald. They were already experimenting with wooden trays planted with ice plant, maidenhair vine, and pink and red ivy geraniums to be applied to the walls and columns of Bernard Maybeck's Palace of Fine Arts to give them a rustic and weathered appearance.[6] The exhibition's Assistant Landscape Engineer, Donald McLaren, was credited not only with realizing the hedge but with executing all planting work on the grounds.[7] (Figure 4)

The hedge was a welcome opportunity for the McLarens to both experiment with and illustrate the use of *Mesembryanthemum*, a plant indigenous to South Africa that John had found not "used so freely as it ought to be" in California gardening.[8] As a hardy ornamental and good window box plant, ice plant grew abundantly without artificial watering and was, therefore, also a good "substitute for grass," "forming a carpet of rich green" even on dry shifting slopes which it stabilized.[9] Planted in the hedge's trays which were installed vertically, *Mesembryanthemum*'s succulent leaves formed a green backdrop, and "in certain lights the water trickling through [them] shimmered like gems."[10] In the summer, the plants' profusely blossoming flowers added a layer of pale purple. Testifying to the achievement of its intended effect, the exposition's *Official View Book* erroneously described the hedge as a "mass of flowery moss."[11] (Figure 5)

The hedge became part of the exhibition ground's planting scheme that, begun in 1912, was in keeping with the buildings' color palette created by artist Jules Guerin.[12] For this purpose, plants were grown in newly established nurseries, with seeds and bulbs imported from Japan, the Netherlands, Belgium, and England. The fairgrounds featured gardens representing various states and regions in the United States as well as monochromatic color gardens displaying masses of one particular plant variety.[13] Plants whose blossoming season had passed were replaced almost over-night by others.[14] Like the elaborate and striking mass flower displays, the hedge contributed to the fairground's "imperial aesthetic" that celebrated the opening of the Panama Canal, symbolized the United States' ascendancy as an imperial nation, and cast San Francisco as its emerging center.[15] Furthermore, as Elizabeth A. Logan has shown, the exposition's plants and plant displays not only expressed San Francisco's recovery and rebirth after the 1906 earthquake, they also portrayed the superiority of California's climate and soil by enabling the growth and acclimatization of plants from across the world, turning the state into the "garden of the earth."[16]

Figure 5. Man posing in front of *Mesembryanthemum* hedge with the Tower of Jewels in the background, Panama-Pacific International Exposition, San Francisco, ca. 1915, Edward A. Rogers collection of Cardinell-Vincent Company and Panama Pacific International Exposition photographs, BANC PIC 2015.013:04348-NEG, The Bancroft Library, University of California, Berkeley.

As a type of fence or wall, the *Mesembryanthemum* hedge symbolically laid claim to the land, the very subject of the nation's expansionist practices. It enclosed a miniature world whose architecture, spaces, and plantings gave expression to the taming of nature, closing of the frontier, and violent taking of Indigenous lands. The calming green hedge provided a unifying frame for this busy, sparkling new world that also sought to construct social cohesion and national identity amongst a diverse body of politics. As "a green hedge that looked a hundred years old," it legitimized the United States' imperial aspirations, providing it with a cultural gravitas that could compete with its older European peers.[17] The hedge was at once new and old, alive and inert, planted and built, changeable and static. It was both a horticultural and technological feat. Although it was dismantled after the exposition and lost from sight faster than the time required for its construction, its innovative potential was clear. "McLaren devised a new kind of hedge likely to be used the world over," remarked John D. Barry in 1915, one of the exposition's early chroniclers.[18] In 1921, another commentator, Frank Morton Todd, noted that, to his knowledge, w had never before been used in this way.[19] And, in the hyperbolic praise of yet another contemporary observer, the "wall" was "second only in wonder to the Great Wall of China."[20]

Grass Pavements

Whereas the hedge was installed during the early months of the First World War, another innovative synthetic construction material involving plants and inert matter was developed during the Second World War; this time, not for peacetime purposes but for use in military camouflage. Britain's military airfields were ill-prepared for the war. Lacking any inherent form of concealment, many of the sites selected were conspicuous and easy targets for German bombers. Fast remedies became essential. In the early 1940s, British horticulturalist Leslie Watson, while working for the British military's Camouflage and Decoy Section, devised "a method of achieving camouflage for concrete runways, aprons, aerodrome roadways, paths, and hard standings by the growth of grass through interstices in the surface."[21] The task was challenging. Whereas runways had to be high-duty and practicable for the Royal Air Force, they had to be hidden or rendered invisible to the enemy's aerial view. What came to be called "Watson's Pots" after their inventor described a system of vegetated holes that penetrated the full thickness of concrete runways. The holes were cast in the concrete with the help of sheet-metal cones, then filled with soil and planted with tufts of grass. The plants gave the finished surface an appearance indistinguishable from adjacent grassland, especially at low angles of aerial vision, for example, from a height of one thousand feet and at distances of a quarter mile and beyond. Watson's Pots also provided a hard landing surface that drained easily and did not dust.[22]

1. Aerial photo taken 9.7.41 of experimental strip of Watson's Pots, 100 yds x 5 yds, laid at Dodwell Farm, Stratford-on-Avon in November 1940.

2. Aerial photo' of Watson's Pots at Dodwell Farm, taken April, 1942.

Figures 6a. and 6b. Photographs showing aerial views of test plots for various grasses used in "Watson's Pots," Dodwell farm, Stratford-upon-Avon, 1940–1942, HO217/7/RS7, The National Archives, Kew, UK.

3. Photo' taken at Fowlmere,
August, 1942.

4. Photo' taken at Dodwell Farm, 17.7.41.

Figures 6c. and 6d.
Photographs showing
ground views of test plots
for various grasses used in
"Watson's Pots," Dodwell
farm, Stratford-upon-Avon,
1940–1942, HO217/7/RS7, The
National Archives, Kew, UK.

5. Photo' taken at Dodwell Farm, 17.7.41

6. Photo' taken at Dodwell Farm, 17.7.41

Figures 6e. and 6f.
Photographs showing
ground views of test plots
for various grasses used in
"Watson's Pots," Dodwell
farm, Stratford-upon-Avon,
1940–1942, HO217/7/RS7, The
National Archives, Kew, UK.

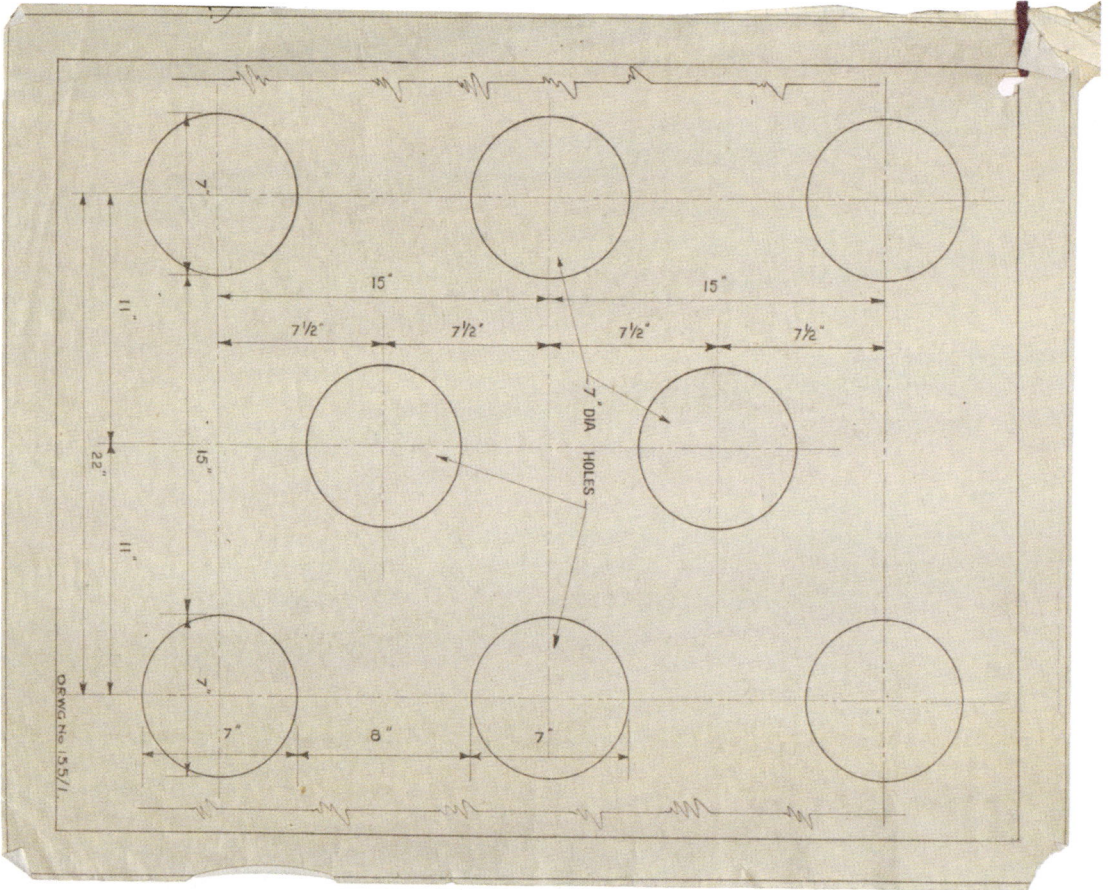

Figure 7. Construction drawing of "Watson's Pots," indicating the suitable spacing of holes fifteen inches on center in staggered rows eleven inches apart, HO217/3/TMM23, The National Archives, Kew, UK.

Watson tested his invention in field studies at an experimental runway constructed at Dodwell Farm, Stratford-upon-Avon. (Figure 6a-6f). Test flights above the area helped to determine construction specifications and ultimately showed that one-fifth to one-third of the total runway surface should be perforated. The most suitable spacing for the holes was fifteen inches on center in staggered rows eleven inches apart. (Figure 7) Besides identifying a suitable soil mixture that included top-spit fibrous loam, farmyard manure, superphosphate, and sand, Watson's studies produced a list of suitable grasses. The idea was to supply turf from nearby pastures that included large proportions of cocksfoot (*Dactylis glomerata*), timothy (*Phleum pratense*), and perennial ryegrass (*Lolium perenne*). Tussock grass (*Poa labillardierei*), yarrow (*Achillea millefolium*), and ribgrass (*Plantago lanceolata*) also fulfilled the purpose well. Taking into consideration the Pacific theater of war and American involvement, Watson extended his list with species recommendations for use in Australia and the United States but also suggested the participation of local botanical experts. To render the runways even less conspicuous, he proposed darkening the concrete with an iron-tannin solution or a tar emulsion. And while the chemical had to be applied before planting, it purportedly had the added benefit of stimulating plant growth.[23]

Watson had initially been employed to research different ground-covering plant species and their reaction to diverse chemicals under various conditions. To camouflage airstrips and deceive the enemy by simulating various land features, chemicals, and paints were applied not only to the concrete runways but to plant cover in rural areas. Wood chips, slag, sand, gravel, aggregate, sawdust, and color pigments were mixed into the concrete to change its color and texture and reduce light reflection. To simulate different field patterns and shadows on extensive vegetated areas, different dyes, including those from tanbark, were applied, and various types of fertilizers were used indiscriminately: sodium arsenate killed grass tops; ammonium thiocyanate turned grass brown and then white; and iron sulfate and tannic acid solutions turned ground cover black.[24] In contrast to killing plants and manipulating their development, Watson's Pots encouraged plant growth. They were part of what was called "living camouflage" or simply "camouflage planting."[25] In general, plants were considered superior to paint and artificial camouflage materials because they changed with the seasons and their masking ability generally improved with age. Furthermore, on infrared film, plants appeared as innocent living matter—they could be used to camouflage camouflage—whereas painted camouflage surfaces were easily detected due to their different reflective properties.[26]

As vertical and horizontal synthetic building materials, the hedge at the Panama-Pacific International Exposition and Watson's Pots were forerunners and inadvertent precedents for today's elaborate green walls and more pedestrian grass pavers.[27] Typical of modernist aesthetics and construction practices, they were both synthetic materials exhibiting a modular character and cultural technologies built upon repetitive planting practices. They combined inert ubiquitous structural materials with live site-specific plants. The plants' primary function was aesthetic, although, in the case of Watson's Pots, their ability to bind dust and retain stormwater was also relevant. As early vegetation technologies, the hedge and Watson's Pots were precursors of more recent iterations of green walls and previous grass pavers, whose ecological functions have since been lauded for enhancing sustainability. Grass pavers flooded commercial landscape markets in the 1960s and, since then, have been used especially on parking lots and driveways.[28] During the past two decades, green walls have increasingly been implemented on interior and outdoor walls of shopping centers, airport terminals, individual stores, hotels, and institutional buildings to improve both microclimates and enhance the marketplace. The USGBC (United States Green Building Council) even promotes green walls as a way of earning points toward reaching LEED (Leadership in Energy and Environmental Design) certification.[29] However, artistic green walls such as those by Patrick Blanc and others have been criticized for their use of non-degradable PVC, and today's mass production of concrete grass pavers not only contributes to CO_2 emissions but also uses one of our most valuable resources, sand.[30] In a market-driven economy, green walls and grass pavers are, at times, literal forms of greenwashing and al too often used to superficially position corporations in an ecofriendly light. Quantifying and modifying plants' ecosystem services can change our behavior by drawing attention to them; however, this does not necessarily resolve the overall disregard and disrespect with which plants are treated more generally. When we render plants in purely economic and technocratic terms, we turn a blind eye to their cultural significance and to the more nuanced relationship that exists between human and non-human nature.

Landscape's Duplicity and Plant Blindness

As the Hedge and Watson's Pots have shown, plants as components of synthetic materials and of landscape designs, more generally, can literally and figuratively be used for cover-ups. They can screen, mask, veil, camouflage, and hide in plain sight not only inert structures or architectures but also the histories and labor that have gone into their creation. In the late 1980s, cultural geographer Stephen Daniels elaborated on the fact that "landscape… does not easily accommodate political notions of power and conflict indeed it tends to dissolve or conceal them."[31] In this way, it could be said that the hedge and Watson's Pots are "ambiguous synthes[e]s whose redemptive and manipulative aspects" reveal what Daniels has called the "duplicity of the landscape."[32]

Plants, in particular, are what render the hedge and Watson's Pots—as well as entire landscapes—duplicitous. In this sense, plants are a force. However, their ability (and that of landscapes more generally) to veil both cultural artifacts and their history and to become cultural artifacts themselves through horticulture, arboriculture, and their synthesis with inert matter, has often rendered them equally invisible. Plants and their agency are frequently overlooked. As we have seen with the hedge and Watson's Pots, they are often considered merely as backdrops and ground cover and not as figures or agents of history.

During the last two decades, plant scientists and philosophers have taken the lead in describing and analyzing what scholars of education Elisabeth E. Schussler and James H. Wandersee have called "plant blindness" in large parts of Western society. This includes "the inability to recognize the importance of plants in the biosphere, and in human affairs; ... the inability to appreciate [plants'] aesthetic and unique biological features...; and ... the misguided, anthropocentric ranking of plants as inferior to animals."[33] Plant scientist Francis Hallé has explained the marginalization and neglect of plants as well as the inferiority attributed to them. For example, we identify with humans and non-human animals that are 'animate,' that are attributed with a soul, and that can move as we do. We identify less easily with plants that are immobile, voiceless, and often thought to be soulless. Throughout human culture, plants have often been associated with the 'weaker' sex. Men hunted animals; women gathered plants.[34] This power relation also becomes clear in the use of language. The word "botany" derives from the ancient Greek word *boton,* describing a herd animal. The animal's fodder, that is, plants, was *botanê.*[35] Plants, therefore, are weak, and if we are "vegetating," we usually find ourselves in a weak state (although the Middle Latin *vegetabilis* means "growing" or "flourishing," and the verb vegetare "to animate" or "to enliven"). If, by contrast, we want to attribute plants with agency and strength, we anthropomorphize them.[36]

As a profession and activity that works with plants because of their beauty and utility, their various aesthetic and spatial effects and their ecosystem functions, landscape architecture suffers the same plant blindness that has beset much of Western thought and society. As philosopher Michael Marder has observed:

> More often than not, we overlook trees, bushes, shrubs, and flowers in our everyday dealings, to the extent that these plants form the inconspicuous backdrop of our lives–especially within the context of 'urban landscaping'–much like the melodies and songs that unobtrusively create the desired ambiance in cafes and restaurants. In this inconspicuousness, we take plants for granted so that our practical lack of attention appropriately matches their marginalization within philosophical discourses.[37]

However, through mutually beneficial collaborative projects with plants that are understood "as active, self-directed, even intelligent beings," landscape architecture is also in the unique position to further the understanding of the relationships between human and non-human nature and work against environmental crises.[38] As Marder has argued, the task is to "maintain and nurture ...[plants'] otherness" because it is the ensuing understanding and respect for them that can avert environmental destruction and exploitation.[39] Plants can, therefore, be regarded as both landscape architecture's weakness and strength. While they help to render landscape architecture and its works duplicitous, it is precisely this duplicity that is worth exploring to further our understanding of the relationship between human and non-human nature.[40]

ENDNOTES

1. Frederick Law Olmsted, *Public Parks and the Enlargement of Towns* (Cambridge, MA: Riverside Press, 1870), 23.

2. See, for example, landscape architect Charles Eliot cit. in Sonja Dümpelmann, " 'Landscape Architect better carries the Professional Idea': On the Politics of Words in the Professionalization of Landscape Architecture in the United States," in *From Garden Art to Landscape Architecture: Traditions, Re-Evaluations, and Future Perspectives*, ed. by Joachim Wolschke-Bulmahn and Ronald Clark (Munich: AVM Edition, 2021): 55-69; Charles Eliot in a letter addressed to Charles Francis Adams, December 12, 1896, in Charles W. Eliot, *Charles Eliot, Landscape Architect* (Houghton Mifflin, Boston and New York, 1902), 630-31.

3. Quotation from *Panama Pacific International Exposition 1915* [exhibition booklet no.1], without page. See, for example, Donald McLaren, "Landscape Gardening at the Exposition," *Architect* 10, no. 1 (1915): 13-14 (14); John D. Barry, *The City of Domes: A Walk with an Architect About the Courts and Palaces of the Panama-Pacific International Exposition* (San Francisco: John J. Newbegin, 1915), 12-13, 36, 107.

4. McLaren, "Landscape Gardening," *Sunset* 34, no. 4 (April 1915): 677; Leonard Carpenter, "Panama-Pacific Exposition: Some of the Horticultural Features," *The American Florist* xlv, no. 1419 (1915): 149-50 (149).

5. *Panama Pacific International Exposition 1915*; Barry, *The City of Domes*, 12-13, 36, 107; *The Panama-Pacific International Exposition. Official View Book* (San Francisco: Robert A. Reid, 1915), [6].

6. John McLaren, "Ornamental Horticulture at the P.P. Exposition," Transactions and *Proceedings of the Third Annual Meeting of the California Association of Nurserymen* (Los Angeles: Kruckeberg, 1913): 104-107; Donald McLaren, "Gardening Features of the Panama-Pacific International Exposition," *Pacific Service Magazine* 6, no. 2 (1914): 39-45 (45); John McLaren, "California's Opportunities in Artistic Landscaping," *California's Magazine* 2 (1916): 139-42.

7. MacRorie-McLaren Co., *Descriptive Catalog and Price List. Season 1915-1916* [1916], 2; MacRorie-McLaren Company, *Descriptive and Price Catalog of MacRorie-McLaren Company. Season of 1924-1925* [1925], without page; Hamilton M. Wright, "The Miracle Workers of the Exposition," *California's Magazine* 1 (1916): 1-11 (5). On John and Donald McLaren also see *Marlea*

Graham, "The other McLaren: Part I," *Eden: Journal of the California Garden & Landscape History Society* 13, no. 2 (2010): 1-7; Marlea Graham, "The other McLaren: Part II," *Eden: Journal of the California Garden & Landscape History Society* 13, no. 3 (2010): 6-13 (8); Tom Girvan Aikman, *Boss Gardener: The Life and Times of John McLaren* (San Francisco: Don't Call It Frisco Press (1988).

8. John McLaren, *Gardening in California: Landscape and Flower* (San Francisco: A.M. Robertson, 1909), 222.

9. McLaren, *Gardening in California*, 217, 220, 294.

10. *Panama Pacific International Exposition 1915*, without page; Barry, *The City of Domes*, 12-13.

11. *The Panama-Pacific International Exposition. Official View Book* (San Francisco: Robert A. Reid, 1915), [6].

12. *Panama Pacific International Exposition 1915*, without page; *Official View Book*, [13]. On the planting of the exposition grounds and the hedge, also see Laura A. Ackley, "John McLaren: Landscape Magician of the 1915 Exposition," *Eden: Journal of the California Garden & Landscape History Society* 18, no. 3 (2015): 10-12.

13. *Panama Pacific International Exposition 1915*, without page. *Official View Book*, [11, 13]; McLaren, "Gardening Features"; Frank Morton Todd, *The Story of the Exposition*, vol. 2 (New York: G.P. Putnam's Sons, 1921), 339; Elizabeth Logan, "The Lotus and the Rose: Californians plant a World in 1915," *BOOM: The Journal of California* 5, no. 1 (2015): 50–61 (55).

14. Wright, "The Miracle Workers," 5.

15. See Sarah J. Moore, *Empire on Display: San Francisco's Panama-Pacific International Exposition of 1915* (2013), 10-11, 13.

16. Logan, "The Lotus and the Rose," 54; *Panama Pacific International Exposition 1915*, without page [2].

17. Frank Morton Todd, *The Story of the Exposition*, vol. 1 (New York: G.P. Putnam, 1921), 307.

18. Barry, *The City of Domes*, 12-13.

19. Todd, *The Story of the Exposition*, vol. 1, 310.

20. Arthur Z. Bradley, "Exposition Gardens" *Sunset Magazine* 34, no. 4 (1915): 665-79 (676).

21. For Watson's Pots, see Sonja Dümpelmann, *Flights of Imagination: Aviation, Landscape, Design* (London and Charlottesville: University of Virginia Press, 2014), 191–92. Guy Hartcup, *Camouflage: A History of Concealment and Deception in War* (New York: Charles Scribner's Sons, 1980), 55; "Watson's Pots," HO217/3/TMM23, The National Archives, Kew, UK.

22. "Watson's Pots," HO217/3/TMM23, The National Archives, Kew, UK; HO217/7/RS7, The National Archives, Kew, UK.

23. Ibid.

24. Dümpelmann, *Flights of Imagination*, 197 98; Merrill E. de Lónge, *Modern Airfield Planning and Concealment* (Toronto: Pitman, 1943), 111–36.

25. Alfred Becker, *Wehrtechnischer Einsatz der Ingenieurbiologie* (Berlin: Volk und Reich, 1944), 13, 21; Robert P. Breckenridge, *Modern Camouflage: The New Science of Protective Concealment* (New York and Toronto: Farrar and Rinehart, 1942); Dümpelmann, *Flights of Imagination*, 184–98.

26. Dümpelmann, *Flights of Imagination*, 184-98.

27. German garden architect Paul Schraudenbach claimed that the grass paver called "bg-Platte System Schraudenbach" which he developed in the 1960s was modeled on cobblestone pavement. See Paul Schraudenbach, "Die 'bg-Platte'," *Neue Landschaft* 1968 13, no. 8 (1968): 371-373. It was tested and considered practicable on Frankfurt airport's runway shoulders, among other locations, where it fixated the ground and created a permeable surface benefitting drainage.

28. American books on home landscaping (see for example William R. Nelson, *Landscaping your Home* [Urbana, Chicago, London: University of Illinois Press, 1975]; Alice L. Dustan, *Landscaping your own Home* (New York: McMillan, 1955) in the 1950s, 1960s, and 1970s do not address this material. "Turf blocks" and "Grid or grass-filled blocks" appear in the first edition of the *Time-Saver Standards for Landscape Architecture: Design and Construction Data* (New York: McGraw-Hill, 1988), 440-6, 840-20. In contrast, in Germany, grass pavers have belonged to a common outdoor building material since the 1960s. See, for example, Schraudenbach, "Die 'bg-Platte'," Josef Metten's patent for a grass paver "Rasterstein zur Befestigung von Grünflächen" filed on August 5, 1969, at the Federal Republic of Germany's patent office, https://www.dpma.de/, accessed 29 January 2021; W. Kolb, "Rasenparkplätze–Ein Vergleich," *Rasen–Turf–Gazon* 4, no. 1 (1973): 14-16.

29. See, for example, https://www.usgbc.org/education/sessions/vertical-living-green-walls-designing-sustainability-11402450.

30. See, for example, Vince Beiser, *The World in a Grain: The Story of Sand and how it transformed Civilization* (New York: Riverhead Books, 2018); Aurora Torres, Jodi Brandt, Kristen Lear, and Jianguo Liu, "A looming Tragedy of the Sand Commons," *Science* 357, no. 6355 (2017): 970-71.

31. Stephen Daniels, "Marxism, culture, and the duplicity of landscape," in *New Models in Geography: The Political-Economy Perspective*, ed. by Richard Peet, Nigel Thrift (London: Unwin Hyman, 1989), 196-220 (196, 206).

32. Daniels, "Marxism," 206.

33. James H. Wandersee and Elisabeth E. Schussler, "Toward a Theory of Plant Blindness," *Plant Science Bulletin* 47, no 1 (2001): 2-9. Also see Monica Gagliano, John C. Ryan, and Patrícia Viera, "Introduction: Plants and Us," in *The Language of Plants: Science, Philosophy, Literature*, ed. by Monica Gagliano, John C. Ryan, and Patrícia Viera (Minneapolis and London: University of Minnesota Press, 2019), vii–xxxiii.

34. For this gender bias, female-gendered nature, and gendered origin myths in Western culture, see, for example, Carolyn Merchant, *The Death of Nature: Women, Ecology, and the Scientific Revolution* (San Francisco: Harper & Row,1980); Carolyn Merchant, *Reinventing Eden: The Fate of Nature in Western Culture* (New York and London: Routledge, 2003); Francis Hallé, *In Praise of Plants* (Portland, London: Timber Press, 2002 [1999]), 23-40.

35. Hallé, *In Praise of Plants*, 25.

36. Hallé, *In Praise of Plants*, 23-40; Also see Michael Marder, *Plant Thinking: A Philosophy of Vegetal Life* (New York and London: Columbia University Press, 2013), 20-28.

37. Marder, *Plant Thinking*, 3-4.

38. Matthew Hall, *Plants as Persons: A Philosophical Botany* (New York: State University of New York Press, 2011), 169.

39. Marder, *Plant Thinking*, 3, 9.

40. On the duplicity worth exploring, also see Daniels, "Marxism," 206, 217-18, 220. Sonja Dümpelmann, "Plants," in *The Landscape Project*, ed. by Richard J. Weller and Tatum L. Hands (San Francisco: Oro, 2022): 52-68.

TRANSCORPOREAL ENCOUNTERS WITH LANDSCAPE:

REINSTATING THE BODY IN THE FOOD CHAIN

AROUSSIAK GABRIELIAN

ABSTRACT

This paper explores the possibilities for trans-species solidarity through the creation of new rituals that force humans out of their exploitative relationship with the more-than-human world and obligate them to collaborate and co-evolve more ethical models for living. Specifically, it focuses on a speculative design prototype for a wearable (and edible) landscape system that functions as a second living skin powered by human waste that doubles as an expanded ecosystem attracting and integrating other forms of animal and insect life. Challenging extractivist environmental paradigms, which have led us to the brink of planetary collapse, the featured project *Posthuman Habitats* shifts away from such exploitative models towards those that value natural systems as sources rather than resources. This work leans on scholarship that challenges the humanist imagination—including Posthuman Feminist Phenomenology, Material Feminism, and New Materialist Environmentalism—and integrates methods adapted from the Futures discipline, Bioart/design, and Worldbuilding. It presents both details of a speculative design prototype (its specific components, how it functions/performs, how it is used, and for whose benefit it is designed) and the ways in which it forces us to shift and adapt our relationality with the multi-species world by dismantling the idea of human exceptionalism and demonstrating the complex webs of life upon which humans depend. The narrative world around which the prototype exists imagines what living more cooperatively and collaboratively with the non-human world might look like while addressing issues of human food insecurity in a not-so-distant future where depleted soils, droughts, and floods threaten agricultural production.

+ trans-species solidarity
+ feminist new materialism
+ speculative futures
+ bioart
+ ethics of care

Figure 1. Diegetic Prototype: Bodily systems and plant ecologies are symbiotic in this speculative prototype. The material of landscape—its moisture, weight, vitality—becomes a second skin. Photography by altrospaziophotography.com.

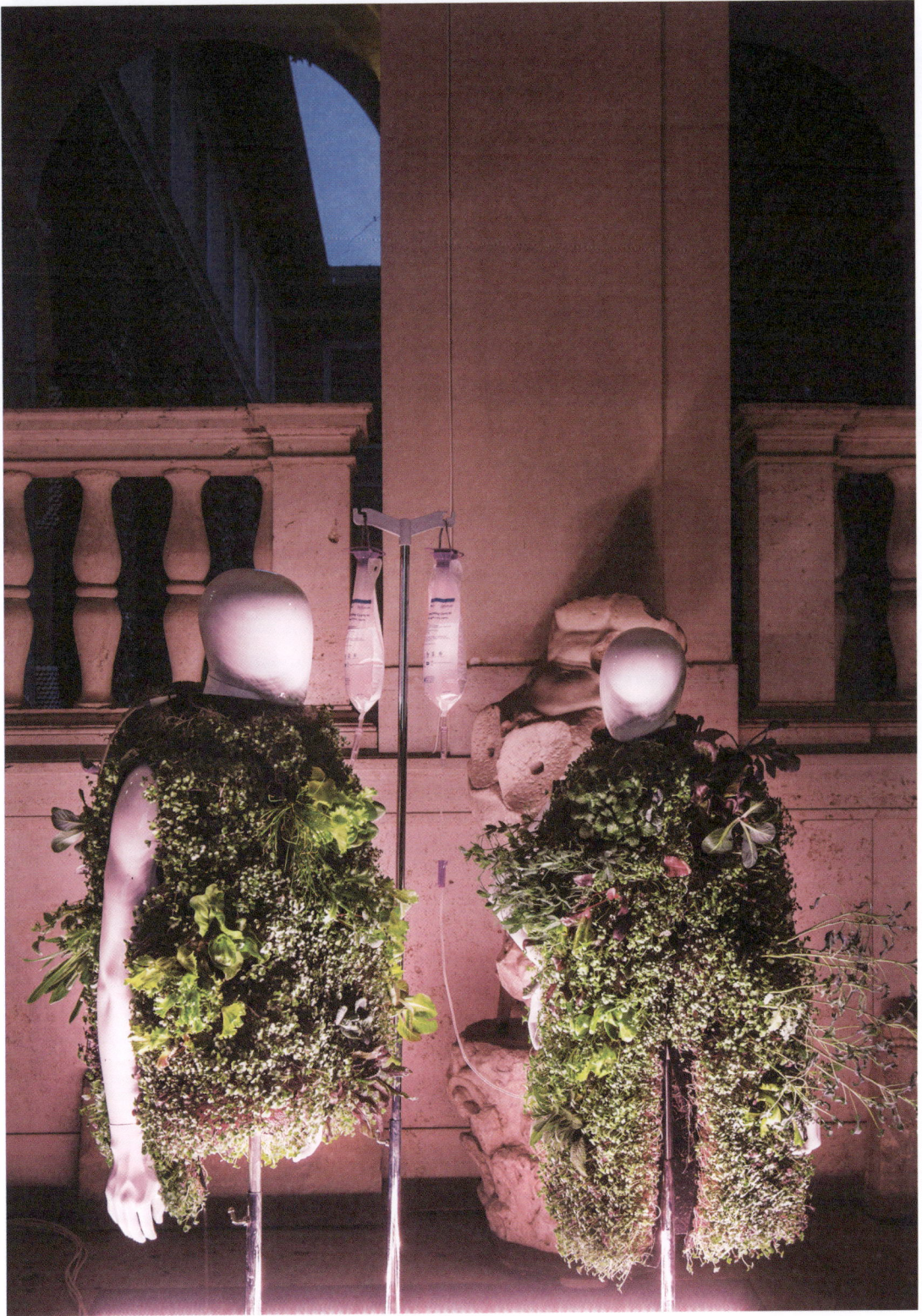

Natural equilibriums will be increasingly reliant upon human intervention... We might just as well rename environmental ecology, machinic ecology... In the future, much more than the simple defense of nature will be required... The creation of new living species—animal and vegetable—looms inevitably on the horizon, and the adoption of an ecosophical ethics adapted to this terrifying and fascinating situation is equally as urgent as the invention of a politics focused on the destiny of humanity.

Felix Guattari, *The Three Ecologies*[1]

In the current geologic epoch of the Anthropocene, in which anthropocenic forces (capitalism, colonialism, and patriarchy) have directly altered atmospheric, geologic, hydrologic, and biospheric systems, the development of design approaches that help steer ourselves, our communities, and our environments toward more ethical futures is critical for planetary survival.[2] Aided by advanced capitalism and bio-genetic technologies, (some) humans have radically disrupted the multispecies world and turned ecosystems into a so-called "planetary apparatus of production."[3] Media scholar McKenzie Wark describes this phenomenon as a series of metabolic rifts, where "one molecule after another is extracted by labor and technique to make things for humans."[4] In her words, when waste is not returned to feed the system's self-perpetuation, "the soils deplete, the seas recede, the climate alters, the gyre widens."[5] Leaning on scholarship that challenges the humanist imagination—including Posthuman Feminist Phenomenology, Material Feminism, and New Materialist Environmentalism—and integrating methods adapted from the Futures discipline, Bioart/design, and Worldbuilding, this paper speculates on our anthropocenic crises in order to imagine future mechanisms for planetary survival that are more ethical, inclusive, and just.[6]

Specifically, this research considers how "the creation of new living species" might be integrated into the posthuman project.[7] Posthumanism does not mean 'after humans' or the end of humanity, but rather the end of human exceptionalism—the assertion of human primacy over the rest of the biotic world. This chapter explores this 'end' through the liberating framework of trans-species solidarity, which includes mutually beneficial interactions across organisms within the multi-species world that we all share. Its theoretical framework builds on the idea of the relationality of the posthuman subject as defined by Rosi Braidotti in her book *The Posthuman*.[8] Braidotti's relational subject operates with a wider sense of interconnection between self and others, where 'others' include the more-than-human. While focusing on humanism and its critique, she suggests broadly that "the posthuman condition introduces a qualitative shift in our thinking about what exactly is the basic unit of common reference for our species, our polity and our relationship to the other inhabitants of this planet."[9] Citing Jürgen Habermas, Braidotti argues "the posthuman provokes elation but also anxiety about the possibility of a serious de-centering of 'Man', the former measure of all things."[10] And she continues to explore the "non-naturalistic" structure of the human in our "globally linked and technologically mediated society."[11]

Donna Haraway's conception of the cyborg provides a foundation for this discourse, focusing on human–animal–machine interrelations. The cyborg is a primary subject of the posthuman project, which consists of assemblages, hybrids, and continuities between animal and human, human and machine, nature and technology.[12] Braidotti's lack of distinction between such categories builds on this concept of the cyborg, while ecocultural theorist Stacy Alaimo's orientation toward bodies and nonhuman natures directly connects these themes back to the environment.[13] Alaimo's concept of trans-corporeality, particularly as framed in *Bodily Natures: Science, Environment and the Material Self*, emphasizes the permeability of bodies as a challenge to the humanist model of discrete individuality. She explores the mobile "interconnections, interchanges, and transits" between human bodies and nonhuman natures – including creatures, ecological systems, chemical agents, and other actors.[14] Arriving at her understanding that "'the environment' is not located somewhere out there, but is always the very substance of ourselves" offers us a profoundly altered conception of the human self and a foundation for the speculation discussed below.[15]

Responding to these positions and particularly to Braidotti's prompt of what a "geo-centered subject" might look like, the speculative design prototype *Posthuman Habitats,* is a complex machinic assemblage in which the human body becomes a habitat for organisms with which it survives in symbiosis.[16] Building on Felix Guattari's introductory statement above, which identifies the "creation of new living species" looming on the horizon, this speculation presents a wearable, edible, and compostable landscape system that allows humans to grow food directly on the body, while using bodily waste as nourishment for a robust habitat of plant, insect, and animal life.[17] (Figure 1)

While it begins with the human as the 'designer' of the system, *Posthuman Habitats* imagines how the newly created ecosystem might evolve and co-evolve into a trans-species continuum. Rather than a 'wearable' (a term used to refer to an architected object, often digitally augmented, that is worn on the human body), this prototype functions more like a systemic prosthetic or a second skin where both the material and processes of landscape unify our bodies with the living world in which we are immersed. In the act of dressing ourselves in living matter, whereby we connect our bodily systems to broader biological, ecological, and geological forces (beyond the human), we lose our sense of discrete self and become one with our habitational field in a kind of transcorporeality of the living world. This 'creation' of bio-technologically mediated bodies directly confronts the foundations of the humanist subject, as previously discussed.

The following presents both the details of the prototype as well as the ways in which it both reconceives human bodies and challenges the resourcist/extractivist attitude toward Nature-as-Other—an attitude which Wark, amongst others, claims to have led us to the brink of environmental collapse.[18] While I situate the project within this context, the work itself is not intended to provide a solution to our anthropocenic crises, but rather to project a mode of environmental existence that dismantles the idea of human exceptionalism and demonstrates the complex webs of life upon which humans depend. It imagines—in one narrative speculation—what living more cooperatively and collaboratively with the non-human world might look like, while addressing issues of human food insecurity in a not-so-distant future where depleted soils, droughts, and floods threaten agricultural production.

While I have formulated my intellectual framework and methodological approach as counter to human-centric thinking, the (human) body is, clearly, fundamental to this speculation. Part of the aims of my larger creative project is to make palpable the psychically distant worlds of the microbiome and arctic ice sheets, and to bring climate stress and adaptation into the physical and corporeal realities of our everyday lives. The visceral, affective dimension of the human body—it's phenomenal presence—is the medium through which I attempt to produce such immediacy. The work asks us to confront the extreme measures that might become necessary for human survival, to jolt us into awareness and action. In this, the focus of the work is not its potential practical application but rather its use in probing the implications of such a design, the kinds of relationships we might need to establish, the kind of society we might need to create, and the kind of new rituals we might need to enact if we were to resituate humans within the food chain. *Posthuman Habitats* comments on and critiques industrial agriculture and the capitalist system that perpetuates our central position within the multi-species world. In exchange, it asks us to recode humans' relationships with the more-than-human world by ritualizing acts of care as part of a multispecies collective.

Posthuman Habitats

The narrative context for *Posthuman Habitats* is a near-future world where food and water scarcity, dwindling urban green space in the face of hyper-densification, stresses on energy and water infrastructure, and soil degradation, threaten the survival of our species and of our more-than-human companions. This context conclusively marks the end to romantic notions of 'nature' and recognizes that our bodies have always already been a part of a deliberately engineered existence. (Figure 2)

Posthuman Habitats are essentially cloaks for edible plant life intended to provide sustenance to the wearer as well as flourish as an ecosystem that attracts and integrates other forms of animal and insect life. The garments of *Posthuman Habitats* promote healthful diets and lifestyles, as the gardens are fed and nourished by

METABOLISM

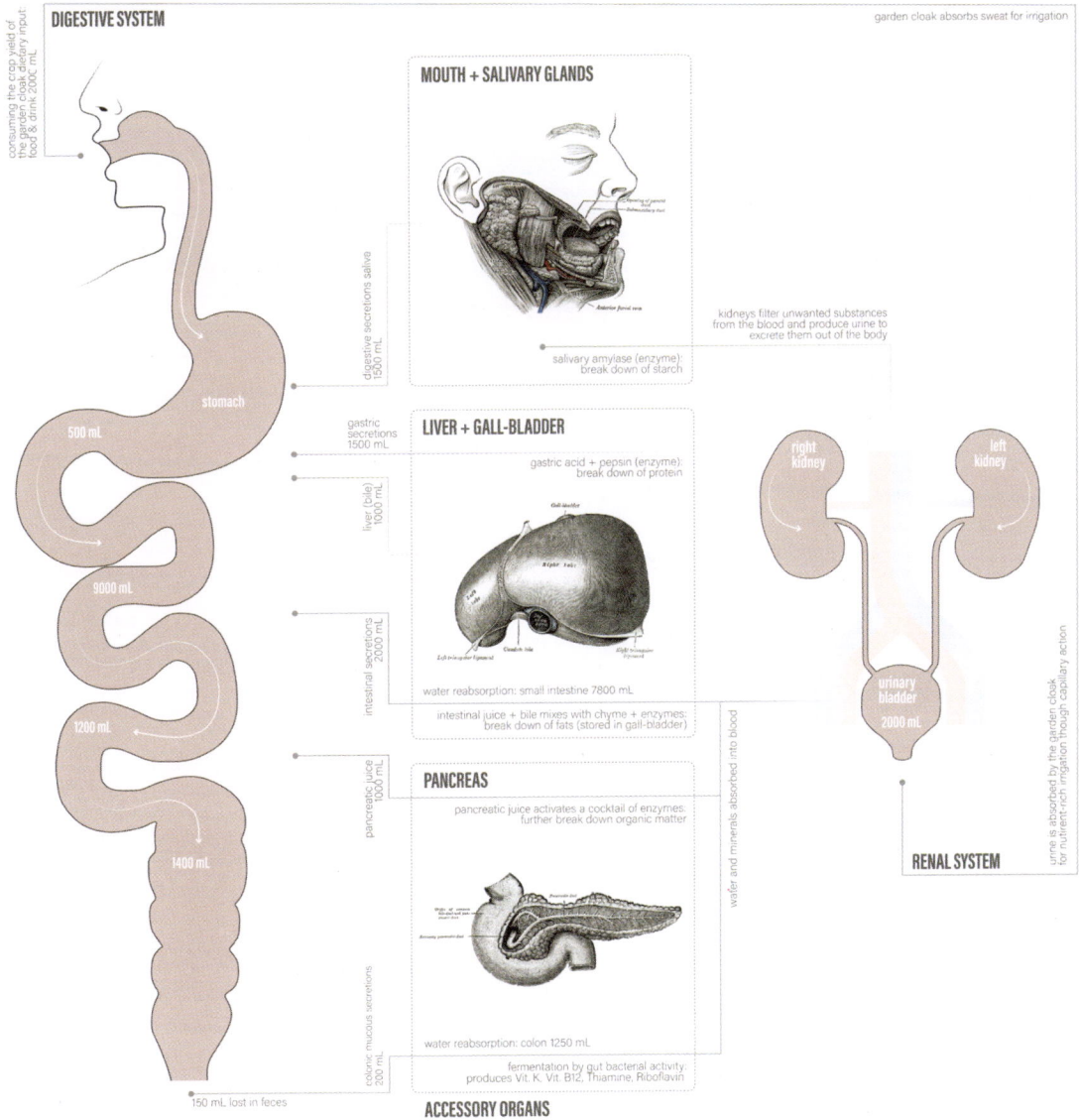

DIGESTIVE SYSTEM

consuming the crop yield of the garden cloak dietary input: food & drink 2000 mL

MOUTH + SALIVARY GLANDS

digestive secretions saliva 1500 mL

salivary amylase (enzyme): break down of starch

kidneys filter unwanted substances from the blood and produce urine to excrete them out of the body

stomach

500 mL

gastric secretions 1500 mL

LIVER + GALL-BLADDER

gastric acid + pepsin (enzyme): break down of protein

liver (bile) 1000 mL

9000 mL

intestinal secretions 2000 mL

water reabsorption: small intestine 7800 mL

1200 mL

intestinal juice + bile mixes with chyme + enzymes: break down of fats (stored in gall-bladder)

pancreatic juice 1000 mL

PANCREAS

pancreatic juice activates a cocktail of enzymes: further break down organic matter

1400 mL

water and minerals absorbed into blood

right kidney

left kidney

urinary bladder 2000 mL

colonic mucous secretions 200 mL

water reabsorption: colon 1250 mL

RENAL SYSTEM

urine is absorbed by the garden cloak for nutrient-rich irrigation though capillary action

150 mL lost in feces

fermentation by gut bacterial activity: produces Vit. K, Vit. B12, Thiamine, Riboflavin

ACCESSORY ORGANS

Figure 2. Metabolic Systems: The habitats activate the digestive and renal systems of the body. The garden cloaks are irrigated by sweat and urine, filtered by the technology of reverse osmosis. Image by Aroussiak Gabrielian.

HABITATS

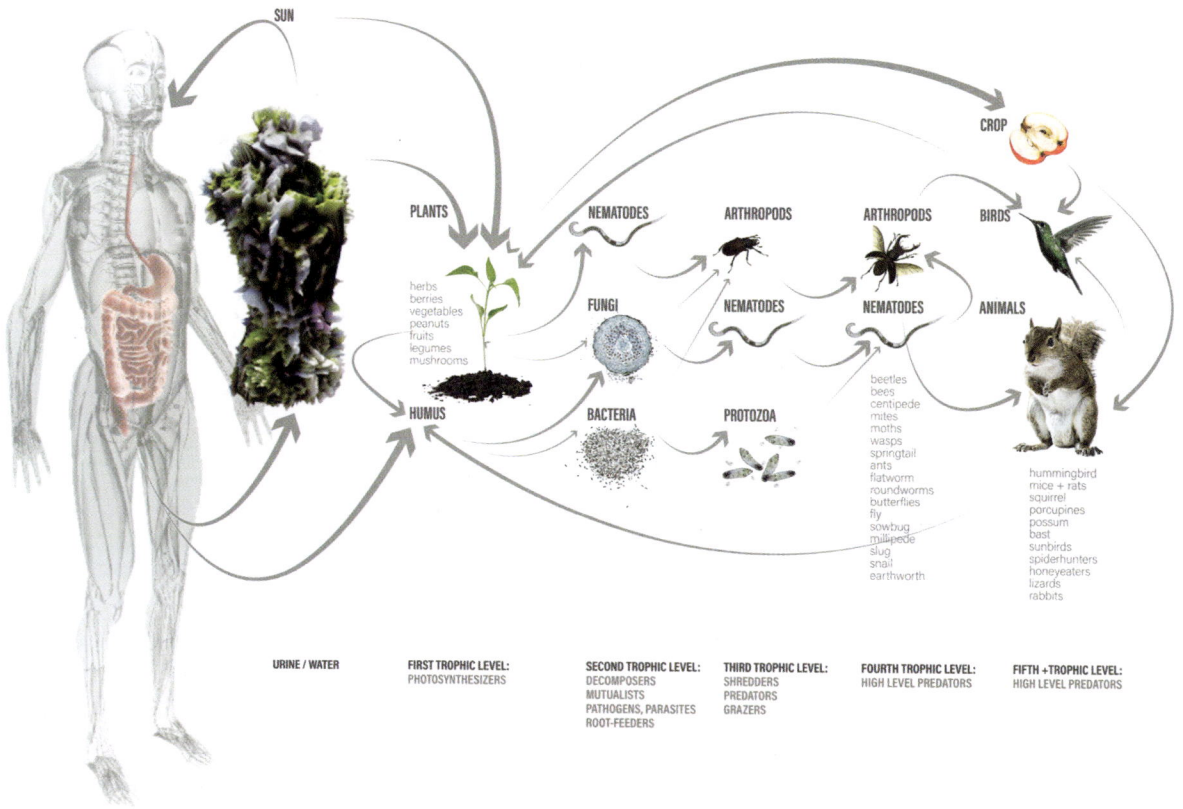

SUN

PLANTS

herbs
berries
vegetables
peanuts
fruits
legumes
mushrooms

HUMUS

NEMATODES

FUNGI

BACTERIA

ARTHROPODS

NEMATODES

PROTOZOA

ARTHROPODS

NEMATODES

beetles
bees
centipede
mites
moths
wasps
springtail
ants
flatworm
roundworms
butterflies
fly
sowbug
millipede
slug
snail
earthworth

CROP

BIRDS

ANIMALS

hummingbird
mice + rats
squirrel
porcupines
possum
bast
sunbirds
spiderhunters
honeyeaters
lizards
rabbits

URINE / WATER

FIRST TROPHIC LEVEL:
PHOTOSYNTHESIZERS

SECOND TROPHIC LEVEL:
DECOMPOSERS
MUTUALISTS
PATHOGENS, PARASITES
ROOT-FEEDERS

THIRD TROPHIC LEVEL:
SHREDDERS
PREDATORS
GRAZERS

FOURTH TROPHIC LEVEL:
HIGH LEVEL PREDATORS

FIFTH +TROPHIC LEVEL:
HIGH LEVEL PREDATORS

Figure 3. Multispecies Habitats: Multifaceted assemblage, the components of which are all required for healthful functioning of the new hybrid ecosystems. Image by Aroussiak Gabrielian.

bodily wastes and inspire outdoor exposure to optimize photosynthesis. Here, bodily systems and non-human ecologies are symbiotic. The material of landscape—its moisture, weight, and vitality—becomes a second skin, one that both insulates and unifies our bodies with the living world in which we are immersed. This mode of environmental existence dismantles human exceptionalism by revealing the complex webs of life upon which humans depend. The new skin biosynthesizes the human body with non-human systems which make up the rest of the habitats. (Figure 3)

The diagram illustrates the multifaceted assemblage of components that are all required for the healthful functioning of the new hybrid ecosystem. The garments integrate all trophic levels or succession of organisms within the food chain. Photosynthesizers, or plants, are the first trophic level, including herbs, greens, fruits, vegetables, and legumes that require sun and water as inputs. Second and third trophic level organisms, including fungi and bacteria, are essential for the breakdown of organic matter to sustain a healthful humus layer that nourishes the entire system. High-level predators of the fourth and fifth trophic levels are largely composed of pollinators (including insects, rodents, and birds), essential to the perpetual regeneration of the skin and optimal crop output. By utilizing processes that help restore, renew, and revitalize energy and material sources within the overall system, the prototype takes its cues from regenerative agricultural practices.

With increased awareness of globalized food industries and their unsustainable carbon footprints, toxic residues, and extractivist and exploitative practices, *Posthuman Habitats* reconnect the food producer with consumers to develop more self-reliant and resilient food networks. Severe drought and diminished soil quality from industrialized farming, as well sea level rise and climate events, will force us all to think harder about the future of food production. The microhabitats proposed here allow the urban dweller to live off-the-grid and outside of capitalist economic frameworks that govern industrial agriculture, providing immediate access to 'landscape' and sources of food.

The garment habitat is stitched from moisture-retention felt typically used in fabric-based green wall technology. Though this felt system is prolifically used for creating vertical gardens throughout the world, its potential for garmenting the body (and feeding the world's human and non-human populations) has yet to be explored. As extensions of our flesh, these microhabitats are completely localized and can be customized to our individual tastes, as well as to the exact nutrient requirements of our bodies. Additionally, sensors embedded within the plant matter regulate the moisture and nutrient content of the system, alerting humans to the vital signs and overall health of our non-human companions who co-exist in the new skin. The garment prompts the wearer to go outdoors if more sunlight is needed, or to add nutrients or water when the system requires. Therefore, while posthuman discourse

Figure 4. Layers: Microhabitat stitched from moisture-retention felt used in fabric-based green wall technology. Photography by altrospaziophotography.com.

is profoundly anti-individualistic, these body enhancements respond to the desire for non-standardization and move toward customization of technologies that exhibit extreme personalization, inclusion, and adaptability. (Figure 4)

Humans may thrive off the crop output, yet are only one small part of the process that sustains the biodiverse habitats. Because habitats that support unique forms of biodiversity are rapidly disappearing from the earth, these garments become a new 'machinic ecology' to which plant and small animal and insect species would adapt.[19] Given that one-third of our food is pollinator-dependent and pollinators are rapidly disappearing, this project is designed to additionally safeguard their survival. The assemblages are not intended to be closed ecosystems but are open to external input and disturbance, particularly through pollinators that introduce new species into the garments and create unexpected hybrids. Cross-pollination sets in motion the hybridization of bodies with the whole environment. (Figure 5)

The recycling of waste is essential for the perpetuation of these garment habitats. Plants convert carbon dioxide into sugar and oxygen through photosynthesis. Urine is collected from the human (via a catheter), stored, filtered, and used to irrigate the plants that provide the base for the system to thrive. Since urine is sterile, low in pathogens, and ninety-five percent water, it is ideal for irrigation. Using a process of forward osmosis developed by NASA, urine is converted into water by forcing it through a semi-permeable membrane that filters out the salt and ammonia. Organic matter becomes compost as it is processed by worms and other insects and nourishes plants. Manure from small animals that occupy the system additionally fertilizes the crops. Furthermore, dead organisms provide food for pollinators and contribute to the humus layer. (Figure 6)

Figure 5. Cross-Pollination: Garments are optimal habitats for pollinators, creatures essential for food production. Image by Aroussiak Gabrielian.

Finally, the project instigates new social rituals around the collective harvesting and sharing of food—the ultimate farm-to-table (or body-to-mouth) experience. Each cloak has the capacity to grow a number of different crops, generating approximately twenty pounds of hyper-local produce. Twenty-two different crops were grown on each garment, including sorrel, cabbage, arugula, purple kohlrabi, broccoli rabe, radish, red leaf lettuce, frisée, green onion, kale, oak leaf lettuce, peanuts, peas, lentil, nasturtium, strawberries, mushrooms, leek, fennel, sage, rosemary, and lemon thyme. (Figure 7) Communal meals require collective harvests and the location of ingredients on the bodies of others. The practice of the harvest becomes re-ritualized as a collective act of labor and a celebration of a closer relationship between acts of production and consumption. Yet rather than a nostalgic desire to return to the 'natural economies' of preindustrial societies, this project speculates on food production in the new planetary landscape of depleted soils and the increasing threat of food insecurity. We collectively ingest this shared harvest, binding us together in a secular act of communion. (Figure 8)

Figure 6. Waste Cycles: The recycling of wastes from small animals and dead organisms provide food for other organisms and contribute to the humus layer. Image by Aroussiak Gabrielian.

POLLINATION

1/3 OF FOOD
POLLINATION
DEPENDENT

INTRODUCED
SPECIES

POLLINATOR:
HUMMINGBIRD

DIRECT
OXYGEN
INTAKE

CROP
PARSLEY

CROP
DILL

POLLINATOR:
MOUSE

WASTE

FILTRATION PROCESS
FORWARD OSMOSIS

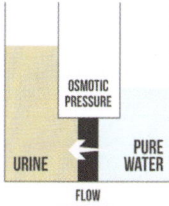

OSMOTIC
PRESSURE

URINE

PURE
WATER

FLOW

SEMI-PERMEABLE
MEMBRANE

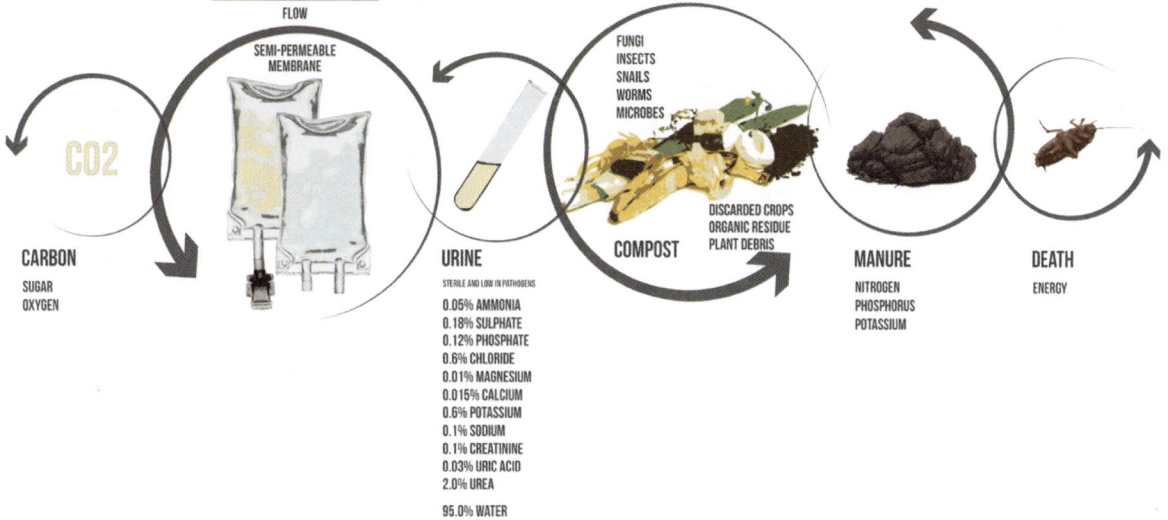

CO_2

FUNGI
INSECTS
SNAILS
WORMS
MICROBES

DISCARDED CROPS
ORGANIC RESIDUE
PLANT DEBRIS

CARBON

SUGAR
OXYGEN

URINE

STERILE AND LOW IN PATHOGENS

0.05% AMMONIA
0.18% SULPHATE
0.12% PHOSPHATE
0.6% CHLORIDE
0.01% MAGNESIUM
0.015% CALCIUM
0.6% POTASSIUM
0.1% SODIUM
0.1% CREATININE
0.03% URIC ACID
2.0% UREA

95.0% WATER

COMPOST

MANURE

NITROGEN
PHOSPHORUS
POTASSIUM

DEATH

ENERGY

TRANSCORPOREAL ENCOUNTERS WITH LANDSCAPE

Conclusion

Posthuman Habitats lives at the intersection of representation and rhetoric, as well as science and speculation. The prototype functions yet, it serves no practical purpose; it provokes our imaginary, asks questions, and initiates conversations around the climate crisis and our position within it. The project additionally asks how we, as geocentric subjects, could use our bodies to feed more than just our kin. Food is a composite web that gathers the technical, ethical, cultural, and affective. According to gastronomic ethnographer Kelly Donati, it is also "the material embodiment of an incredibly complex but largely invisible assemblage of trophic encounters between different living species."[20] Yet humans have intellectually positioned themselves outside the food chain despite being host to millions of microbial organisms that are both nourished by our interiors and exteriors and are essential to our survival. Many feminist scholars, including Val Plumwood and Donna Haraway, have identified that we are less human than we would like to admit. In fact, the microbes that we host on and in our bodies make up a large majority of our cellular and, thus, genetic material.[21] Such microbial life that we depend on for survival remains somehow too psychically and visually distant for us to recognize despite its intimate co-inhabitation. Acknowledging our reliance on these more-than-human forms of life is an essential first step to a co-evolutionary process and a form of collaborative existence that is more ethical and just. The project makes these living species visible through intimate physical contact, altering our behaviors and attitudes toward the multi-species world and necessitating new rituals of care for these companions, who in turn help us move closer to achieving a positive food future.

Figure 7. Crops: Twenty-two different crops were grown on the proof-of-concept prototype. Each cloak produced enough to feed a family of three for three weeks or to make a plant-based meal for 200 people. Photo by Aroussiak Gabrielian.

As humans move blindly towards extinction, we may seek methods of our own survival, but rarely do we conceive of the collective as it poses a threat to our individual liberties. In what Guattari calls "Integrated World Capitalism," we continue to consume depleting natural "resources" (the term itself making clear the hierarchy of man as served by nature), to exploit the "other" (human and non-human), and to make the planet inhospitable and often fatal for our co-species.[22] Recognizing the diverse forms of consciousness and symbiotic expressions of care shared within interdependent webs of life that exist across scales of time and mass, *Posthuman Habitats* provide humans a means to participate in this symbiotic network of exchange, hoping that we might better understand the agency, vitality, and collaborative sociality of the biophysical world (and our place in it), as a model for existence.

Figure 8. The Harvest: Four menu items were prepared by the Rome Sustainable Food Project for this project, shared with visitors during the opening night of the exhibit. Photography by altrospaziophotography.com.

Acknowledgments

I'd like to thank Dr. Holly Willis first for exposing me to the work of Rosi Braidotti and for serving as my dissertation advisor along with Dr. Andreas Kratky, Alex McDowell, Dr. Jeff Watson, and Dr. Steve Anderson, all of whom have shaped my thinking in various different ways both around this work and beyond; the American Academy in Rome for supporting this project (through both monetary and non-monetary means, including all my co-fellows who influenced and sharpened my thinking around this work and beyond; PHILIPS, the lighting company, for donating the grow lights for the project; all of my collaborators below who assisted with the fabrication of the prototype; and finally, Zabel Gabrielian, for being the inspiration for this piece.

Collaborators

Posthuman Habitats was conceived and designed by Aroussiak Gabrielian in 2015 after the birth of her daughter Zabel. The design prototype was fabricated in collaboration with Grant Calderwood (Microgreens researcher, Yale University), Irene Tortora (Fashion Designer, Rome); Chris Behr (Chef, Rome Sustainable Food Project) and his staff in the kitchen, and Alison Hirsch (Co-founder, foreground design agency), while in residence at the American Academy in Rome (2017-2018). Additionally, Emily Scheffler, intern from the Tyler School of Art in Rome, assisted the team, while Leslie Cozzi, Anna Majeski, Jessica Peritz (co-fellows at the Academy) served the delicious finger foods prepared from the crops of the cloak during the opening night of the exhibit in which the prototype was displayed.

Funding

Posthuman Habitats was funded by the American Academy in Rome.

ENDNOTES

1. Felix Guattari, *The Three Ecologies*, trans. Ian Pindar and Paul Sutton (London: Athlone Press, 2000), 66-67.

2. The term the Anthropocene, and even many of its proposed alternatives (Chuthultucene, Capitalocene), have been significantly problematized by Donna Haraway and others, particularly for the use of "anthro" as attempting to speak for all people when clearly, we understand that climate impacts are imbalanced. For a discussion of the term "Anthropocene" and its variants, see Donna Haraway's, "Anthropocene, Capitalocene, Plantationocene, Chthulucene: Making Kin," *Environmental Humanities* 6, 2015, 159-65. A discussion of debates can also be found in Donna Haraway, Noboru Ishikawa, Scott F. Gilbert, Kenneth Olwig, Anna L. Tsing & Nils Bubandt "Anthropologists Are Talking – About the Anthropocene," *Ethnos*, 81 no. 3 (2016): 535-64.

3. Rosi Braidotti, *The Posthuman* (Cambridge: Polity Press, 2013), 7.

4. McKenzie Wark, *Molecular Red: Theory for the Anthropocene* (New York: Verso, 2015), xiv.

5. Ibid.

6. I refer to the work of feminist scholars who challenge the humanist ideal of "Man" as the universal measure of all things. Authors such as Rosi Braidotti who build on the idea of the relationality of the posthuman subject as defined in her book *The Posthuman*. Donna Haraway's conception of the cyborg provides a foundation for this discourse focusing on human-animal-machine interrelations in "A Cyborg Manifesto: Science, Technology, and Socialist-Feminism in the Late Twentieth Century," *Simians, Cyborgs, and Women: The Reinvention of Nature* (New York: Routledge, 1991), 149-81. See also Donna Haraway, *Staying with the Trouble: Making Kin in the Chthulucene* (Durham: Duke University Press, 2016); Stacy Alaimo's eco-cultural orientation toward bodies and non-human natures, in Stacy Alaimo, *Bodily Natures: Science, Environment, and the Material Self* (Bloomington: Indiana University Press, 2010).

7. Guattari, *The Three Ecologies*, 66-67.

8. Braidotti. *The Posthuman*.

9. Ibid., 1-2.

10. Ibid., 2.

11. Ibid., 5.

12. Haraway, *Simians, Cyborgs and Women: The Reinvention of Nature*.

13. Alaimo, *Bodily Natures: Science, Environment, and the Material Self*.

14. Ibid., 2.

15. Ibid., 4.

16. Braidotti. *The Posthuman*, 81.

17. According to Gilles Deleuze and Felix Guattari's *Capitalism and Schizophrenia* project and their marked dismissal of the "root" in favor of the "rhizome" as a mode of questioning and interpretating, both ideas clearly challenge essentialist concepts in favor of hybridization of not just broadly culture-technology in vast systemic webs, but human-machine and man-animal. See Gilles Deleuze and Félix Guattari, *Anti-Oedipus: Capitalism and Schizophrenia* (Minneapolis: University of Minnesota Press, 1983). According to philosopher Alain Beaulieu who focuses on the work of Deleuze and Guattari, nature, in their work, is not classified by genus and species, it "does not define living bodies by their organs and functions," rather it is a series of "machinic assemblages" in which humans participate but have no privilege. See Alain Beaulieu, "The Status of Animality in Deleuze's Thought," Journal for Critical Animal Studies 9, 2 (2011), 79. Using the indistinguishable boundaries between the wasp and the orchid as an example, they introduce assemblages as emergent unities that respect the heterogeneity of their components. For a discussion on this, see Gerald Bruns, "Becoming-Animal (Some Simple Ways)," *New Literary History* 38, no. 4 (Autumn 2007), 703-20.

18. Wark, *Molecular Red: Theory for the Anthropocene*. Many feminist scholars also write on the topic of Nature-as-Other, see Val Plumwood, *Feminism and the Mastery of Nature* (London: Routledge, 1993).

19. Colin Gardner and Patricia MacCormack, "Introduction," in *Ecosophical Aesthetics: Art, Ethics and Ecology with Guattari*, eds. P. MacCormack & C. Gardner (London: Bloomsbury Academic, 2018), 1–28.

20. Kelly Donati, "The Convivial Table: Imagining Ethical Relations Through Multispecies Gastronomy," *The Aristologist: An Antipodean Journal of Food History* 4 (2014), 128.

21. Donati, "The Convivial Table," 130-1; citing Val Plumwood, "Tasteless: Towards a Food- Based Approach to Death," *Environmental Values* 17 (2008), 323-30; Donna Haraway, *When Species Meet* (Minneapolis: University of Minnesota Press, 2008); James Gallagher, "More than half your body is not human" *The Second Genome*, BBC Radio 4, (https://www.bbc.com/news/health-43674270, Accessed March 29, 2021).

22. See Guattari, *Soft Subversions: Texts and Interviews 1977-1985*, ed. Sylvere Lotringer, trans., Chet Wiener and Emily Wittman (Cambridge: MIT Press Semiotext(e), 2008), 229-307.

BIOGENIC TECTONICS:

THATCHED BUILDING FACADES FOR THE GREEN TRANSITION

ANNE BEIM

ABSTRACT

This research project contributes to our understanding of the potential benefits of applying biogenic materials in building construction by examining how organic materials can be more generally applied in the design of architectural facades. Increased use of CO_2-neutral, bio-based materials in building facades precipitates new combinations of building materials, surface treatments, and construction solutions. *Biogenic Tectonics* develops and tests thatched building facades that are environmentally friendly, fireproof, and scalable to an industrialized level. Clay is tested as a sustainable fire retardant, both applied onto the surface of the thatch and as built into the façade's construction. The focus is on fire safety, as it is one of the key challenges when using biogenic materials in construction. The project included a historical review and critical study of radical construction techniques, up-front CO_2 emissions, innovative craftsmanship, and testing of its fire protection capacity using full-scale prototypes and construction mock-ups The project addressed three main research questions: How can architecture be designed to contribute to an 'absolutely' green transition of the construction industry? How can architectural solutions increase the use of biogenic materials in construction to help reduce CO_2 emissions? How can a biogenic architecture that is radically sustainable integrate knowledge from traditional building cultures and craftsmanship alongside today's 'efficient' building processes?

+ radical tectonics
+ biogenic construction
+ CO_2-reduction
+ thatched facades
+ fire protection

Figure 1. Detail of reed for thatching, fire tested. Photo by Anne Beim.

120

An Architecture Made of Reed

The design and construction of the Danish Wadden Sea Center made headlines in 2017, both with professional architects and the public. Located in the southwestern part of Denmark and at the edge of the sensitive coastal landscape of the Wadden Sea, the extension and refurbishment of this local museum were completed by Dorte Mandrup Architects. The rather large building has thatched surfaces from head to toe! Roofs and walls appear as one solid material body with window and door openings cut out of a large body of compressed reed. (Figure 2) This building has, quite simply, changed our perception of how to use reed and thatch by offering us a new way of cladding facades for large-scale buildings. At the Wadden Sea, the thatched roof transcends its 'farmhouse romanticism' by taking its place amongst signature works of architecture in adopting an ambitious, contemporary design language.

Figure 2. The Wadden Sea Center – Dorte Mandrup Architects, National Park Ribe, 2017. Photo by Anne Beim.

And yet, buildings like the Wadden Sea Center that use thatch in their construction are no longer easily permitted. In January 2020, Denmark introduced new certification regulations for fire safety that increased the level of restrictions, professional testing, and documentation required for "alternate solutions," including structures made of biogenic and other renewable materials. As a result, proposing new architectural ideas that include biogenic materials in Danish construction is highly constrained. Small architectural offices, contractors, and manufacturers who cannot afford the cost of expensive testing by authorized agents cannot compete against Danish building regulators. The new rule discourages innovative ideas that promote the use of renewable construction materials and the much-needed green transition of the construction industry.

Biogenic Tectonics explores the idea of achieving 'absolute sustainability' in architecture through the use of renewable materials.[1] Organized

Figure 3. Prototype In-Situ – Reed mounted on site, 2021. Photo by Anne Beim.

as a practice-based collaborative research project, the team includes craftsmen, fire engineers, straw masters, clay masons, and architectural researchers affiliated with the Danish Institute of Fire and Security Technology (DBI), the Information Office of the Thatchers Guild (Straatagets Office), and the Center for Industrialized Architecture – CINARK at the Royal Danish Academy. Fire safety is central when designing constructions that use biogenic building materials. This project investigates how material choices and radical tectonic solutions for building skins and structures can drive sustainable change in the construction industry, building legislation, and architectural practice. The aim is to identify credible proposals for CO2-neutral construction types that have an environmentally low impact, 'close to standard' fire retardant properties, and that are scalable to industry. In this context, the reed is investigated as an annually harvested plant that grows rapidly, absorbs critical greenhouse gases, and maintains biodiversity.[2] (Figure 3)

The History of Thatch Building

Facade thatching has commonly been used in vernacular architecture in one of three ways. Most typically, the straw was nailed to the exterior wall with battens. This protected the lime-washed daub and wattle or adobe stone wall behind it from weathering, acting as a barrier protecting the main structure from deteriorating. Another method employed the twisting gable technique in which several gathered masses of reed were tied together and held in place on the gable by vertically hung bundles of twisted straw acting as strings for mounting additional stacks of twisted straw.[3] The technique was most likely used to provide a more open façade construction for added airflow into the structure (as in the perimeter fencing for pig pens). The third way thatch was used in vernacular buildings was in Lolland's thatched gables, where the roof was extended up on the gable by a small overhang structure that formed a small portico. This may have been a refinement of the much older twisting gable technique. (Figure 4)

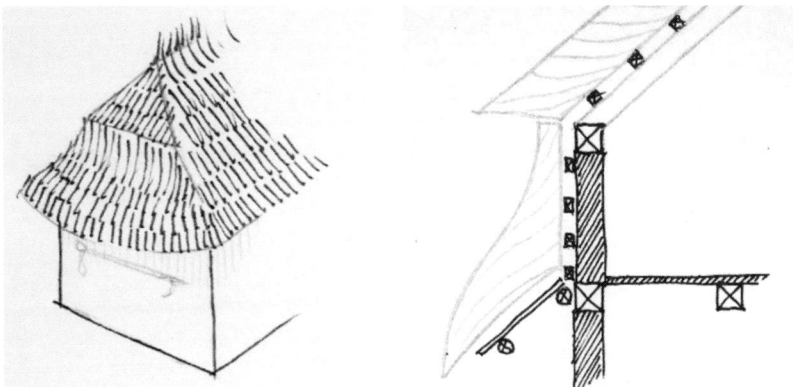

Figure 4. Thatched gable on farmhouse in Lolland. Drawings by Henriette Ejstrup.

In the Netherlands, the use of thatched facades has increased over the past decades. Growing awareness of the construction industry's negative impact on the climate may be a reason; so too might the recognition that thatched facades are part of the Dutch vernacular.[4] Traditional mills are associated with the use of Dutch thatch, and this might 'culturally' be the reason why thatch roofing has been successfully reintroduced into present-day Dutch architecture.[5] Few modern buildings, however, are built with thatch facades in Denmark. While the Wadden Sea Center is one such building that clearly expresses the plasticity of the reed, most other projects with thatched facades are summer cottages, pavilions, and small family homes. These small buildings do not confront the fire safety issues of public buildings such as the Wadden Sea, where fire retardant solutions included the use of a fiberglass membrane and mineral wool within the thatched construction.

In our project, we chose to experiment with the introduction of clay for fire retarding the thatch. Research indicated that clay had been used as a fire retardant in historic buildings. The use of clay set inside wooden floor slabs is described in the earliest Danish building regulations from 1856.[6] Chimneys in vernacular architecture erected with boards of compressed peat sheathed with clay and subsequently treated with lime are the reason why their adjacent wooden structures survived after many years of use.[7] Although we did not find written historical accounts of thatched facades combined with clay for the express purpose of acting as a fire retardant, an interesting source from Southern Denmark indicates that clay was used in combination with straw. A building permit from 1889 required that thatched roofs be sprayed with a substrate of clay over their doors.[8] This indicates that the technique was intended as a fire retardant to secure escape routes. Furthermore, a Danish reference (Byggebogen/ The Building Manual) on applied construction techniques from 1949 describes a technique for dipping reed into clay and ammonia (from animal feces) before the bundles were fixed on a load-bearing structure.[9] Hence, in our construction, we decided to experiment with thatch samples impregnated with clay and with clay boards integrated into the layers of reed.

Fire Testing Assemblies of Thatch and Clay

Anytime one devises a new building assembly in Denmark; fire tests are essential according to our Building Regulations 2018 (BR 2018).[10] Whenever an assembly has materials and construction details that are different from pre-accepted (standard) solutions, these assemblies need to be tested and their performance individually documented. Moreover, the test must focus only on the thatched layer of the façade and not the construction in general.

Initially, it was decided to conduct a broad study of possible mineral-based fire retardants. Choosing the variation and combination of materials and techniques required the experienced observation of craftsmen thatchers, architects, and fire engineers. It also involved the search for documented guidelines. In the end, fourteen different small test objects were fabricated for direct fire testing. The tests were performed as Mini-SBI tests (Single Burning Item).[11] The small rigs for the Mini-SBI tests were symmetrical with inward-angled corners constructed of calcium silicate plates, 200mm (wide) by 600mm (high), with a maximum layer of thatched reed of 50 mm thick. (Figure 5) The reed was fixed with flat metal straps instead of commonly used metal fasteners and screws. This, along with other rig details, introduced challenges in the process that are not typical when testing rigs with industrialized products, as the latter are far easier to cut and measure than assemblies that are crafted on-site.

Figure 5. Mini Single Burning Item (SBI) testing setup, 2021. Photo by Anne Beim.

Figure 6. The lineup of 14 Mini-SBI test items – before fire testing, 2021. Photo by Thorbjørn L. Petersen

Figure 7. Thatched Full scale SBI test – the three test items, 2021. Photo by Anne Beim.

The results of our Mini-SBI tests showed that sprayed moraine clay, dipped adhesive clay, and dipped clay-ammonia mixtures had technical fire-retardant properties equivalent to a Class B rating. We tested our rigs against a baseline FIGRA (Fire Growth Rate) value—which is the measure of how quickly the HRR (Heat Release Rate) develops during a test—of approximately 110 W/s for 'FIRAX', a fire-impregnated Medium Density Fiberboard (MDF) plate. Because FIGRA measures the acceleration of the fire, the lower the number, the better.[12] (Figure 6)

Thereafter, we tested the rig with the best result at full-scale SBI tests (ISO 13785-1:2002 (E)). In this case, the rig was scaled to a 1200mm (wide) by 2400mm (tall) flat construction sandwich panel made of calcium silicate and fireproof medium density fiberboard (MDF) with a layer of thatched reeds, between 220-250 mm thick. The sandwich panels were mounted on a wooden frame 400 mm above the ground, with no façade openings. Three prototypes were developed: a fully impregnated prototype whose moraine clay was injected into the surface layer (50-70 mm) of the structure; a fully impregnated prototype with moraine clay injected into the surface layer (50-70 mm) of the structure and integrated clay boards placed as firestops every one meter, creating relief in the surface of the reed; an unimpregnated ordinary prototype for baseline readings. (Figure 7)

The baseline (third and final) showed a FIGRA value above 1000 W/s, which is significantly more than the baseline, ten times the MDF reference material. The testing of the first prototype (flat surface) was incomplete and inconclusive as it was mounted incorrectly on the wooden rig, causing fire to spread to the reverse side due to the chimney effect. Despite this, unburned pockets of clay and observation of the fire spread indicated that the clay was effective as a fire retardant. The FIGRA value measured was 140 W/s. The testing of the second prototype (relief surface) was successful, indicating a FIGRA value of 25 W/s. It was observed that the clay boards, as well as the relief profile, had a preventative effect on the fire spread.[13]

Prototyping Full-Scale Construction

The materials and methods used in the SBI tests introduced new tectonic strategies and fabrication methods that we developed and tested as full-scale prototypes. We designed and built a one-to-one biogenic prototype of a building corner to study its buildability and architectural character, incorporating the data and knowledge gained during the fire tests. In this case, the supporting structure was made of straw-filled wooden cassettes supplied by EcoCocon.[14] The exterior of the straw elements was sheathed with wood fiberboard (Agepan) and thatched with reed. For this prototype, which was not destined for fire testing, the reed was only sprayed with moraine clay on the surface. Yet, like the test construction, clay boards were also built into the thatched façade as firestops. The interior surfaces were also treated with two layers of clay plaster with a white finish.

With biogenic construction, attempts should be made to tectonically eliminate ventilation gaps between the layers. Contrary to conventional wooden structures, one must ensure permeable construction alongside a high-performance U-value. Avoiding plastic vapor barriers is key. Instead, fiberboard is best introduced both as a wind barrier and as a base plate for fixing the thatched roof. During the construction of the full-scale prototype, discussions arose as to whether the reed should be ventilated, but research in Dutch thatching techniques suggested this was not necessary.[15]

This first prototype was assembled in the exhibition hall of the Royal Danish Academy. It was part of a school exhibition during the fall and winter of 2021, and in March 2022, it was partially disassembled. A smaller piece was placed outside from March to July to observe how it weathers and, more particularly, how the structure and clay impregnation mechanically deteriorate. No significant signs of weaking appeared during this period, but human interaction with the prototype did cause the clay to crumble off. (Figure 8)

Figure 8. The full-scale construction prototype in-situ in the exhibition hall of the Royal Danish Academy, 2021. Photo by Anne Beim.

In its next phase, a new biogenic prototype was constructed for the Terra - Lisbon Architecture Triennale of 2022, this time using pre-fabricated thatched facades conceived according to Design for Disassembly principles. The idea was to design the thatched structure as separate elements to accommodate the low ceiling of the exhibition space and for long-distance transportation. It was also essential that the assembly be installed in the exhibition space by two people using hand tools. Again, EcoCocon straw elements were included, given our aspiration for a fully biogenic stackable construction. Agepan plates were included for similar reasons. The EcoCocon elements and Agepan panels were assembled, and the structure's overall buildability was tested in a workshop where the thatched façade elements (built-in cassettes) were mounted onto the structure via an interlocking system of inclined battens. The fire-retardant clay finished the exterior surface once it was all assembled.

This design differs from the previous prototype in that a ventilated gap was included between the load-bearing structure and the thatched cladding element. And since the prototype was to be displayed indoors, some details were left unresolved, including tectonic solutions now needed for disassembling thatched roofs whose bundles of reed are typically screwed together with metal wires in a continuous joint. There is no tradition of technically dismantling the thatch for further recycling without

Figure 9. The Prototype Prefab exhibited for the Terra – Lisbon Architecture Triennale 2022. Photo by Anne Beim.

downcycling. Hence, a Design for Disassembly principle that offers a radically new tectonic understanding of thatched roof is required. (Figure 9) This would include an industrialized assembly of thatched façade elements, where the thatcher acts as a specialist and inventor of mechanical details rather than a traditional craftsman. Another detail requiring attention is where the thatched façade meets the thatched roof; this is critical due to the weathering of the surface and for shielding the load-bearing structure from water. All these aspects need to be studied more closely, being key details that need to be solved in order for all stakeholders to fully participate in implementing this in the construction industry.[16]

Future Perspectives

The challenge ahead is to define technical standards suitable for thatch construction within the framework of European Union (EU) building regulations and fire requirements. The prototypes developed as part of this project have highlighted radical tectonic concepts with many aspects that still need to be addressed. The project has begun the process of transforming the traditional hand-based craft of thatched reed roofs into an industrialized process, yet this will require new tectonic strategies in the craft of thatching. During the construction of the 'technical' prototypes used in the fire tests, professional partners and aligned disciplines were involved, and it was their participation that enabled an iterative explorative process. Ongoing discussions on buildability, scalability, and technical predictions ensured the project's success. In addition, craftsmen and architects worked closely together to promote designs and strategies that were plausible for full-scale construction.

Our fire tests indicated that existing test standards may not be sufficient when studying biogenic constructions that include craft-based building techniques. New testing standards and methods must be developed to cover the entire construction industry. Issues linked to scalability and actual construction site workflows were also discussed. In the final stages of the last full-scale prototype, the introduction of a horizontal clay slab was suggested as part of the construction to act as a type of firestop. While not part of our original idea, it improved the prototype's fire safety and contributed to protecting the structure. (Figure 10)

In the end, combining clay with a reed in contemporary industrialized construction is sustainable, as both materials are "produced" locally with no energy-intensive processing needed. Moreover, biogenic materials capture carbon while growing and thus store CO_2 from the atmosphere.[17] They result in building constructions with a very small CO_2 footprint, which is a prerequisite for leading the Danish construction industry towards a 70% reduction of CO_2 by 2030. Furthermore, clay has been proven to have positive properties in construction, including a high fire-retardant effect, a reason for which clay has been used in

Figure 10. Close up of the thatched façade impregnated with clay, 2021. Photo by Anne Beim.

Figure 11. Installation of *Thatched Brick Blocks* exhibited in Copenhagen at the UIA Architects World Congress 2023. Designed by architects Rønnow, Leth & Gori (Uffe Leth, Carsten Gori, Philip Lütken, Tobias H. Dausgaard and Gry Jespersen) and CINARK (Anne Beim and Lykke Arnfred). Built by thatchers Thomas Gerner and Stråtagets Kontor, carpenter Ulrik Ahlefeldt. Photo by Franca Trubiano.

combination with biogenic materials throughout history in buildings, both in Denmark and many other regions in the world.

No fully thatched construction solutions with biodegradable fireproofing have been developed, realized, built, and tested as prototypes according to required international standards, neither in Denmark nor internationally. The project is, therefore, considered to be the first of its kind. (Figure 11)

Acknowledgments

Portions of this text were developed in collaboration with Henriette Ejstrup, Architect and PhD, CINARK (historical and vernacular building); Jørgen Kaarup – journalist, former CEO of the Thatchers Information Office; Anders Dragsted – Fire engineer, Danish Fire and Security Institute.

Collaborators

CINARK – Center for Industrialized Architecture is an architectural research center in the Department of Architecture & Technology, at the Royal Danish Academy. The center develops, gathers, and coordinates research and education activities focused on industrialized architecture in a sustainable context. Important to CINARK is identifying and critically addressing the challenges and potentials of special concepts, properties, methods, processes, and products that define a sustainable industrialized architecture. The center collaborates with architectural firms, key players in the construction industry, and relevant organizations. This project involved the participation of Anne Beim (lead), Lykke Arnfred, Henriette Ejstrup, Kenneth Hviid Larsen, Pelle Munch-Petersen and Thorbjørn Lønberg Petersen from CINARK, the Danish Fire and Security Institute (DBI): Anders Dragsted, Robert Firkic, Mads K. Hohlmann; craftsmen: Thomas Gerner (Tækkefirmaet Horneby A/S), Ruud Conijn (Hemmed Tækkefirma A/S) and Lasse Koefoed Nielsen (Egen Vinding & Datter); The Thatchers' Information Office: Jørgen Kaarup and Sven Jon Jonsen.

Funding

The project was funded by the Danish Environmental Technology Development and Demonstration Program (MUDP), between April 2021 and February 2023.

ENDNOTES

1. Paul Lewis, Marc Tsurumaki and David J. Lewis, *Manual of Biogenic House Sections,* (Novato: Oro Editions, 2022), 138-64; Dominique Gauzin-Muller, *Architecture en fibres végétales d'aujourd'hui,* (Plaissan: Edition MUSEO, 2019), 8-9; Adam Weismann and Katy Bryce, *Building with Cob - A Step-By-Step Guide,* (Devon: Green Books: 2006), 80-122.

2. Line Holm Andersen, Petri Nummi, Jeppe Rafn, Cecilie Majgaard Skak Frederiksen, Mads Prengel Kristjansen, Torben Linding Lauridsen, Kristian Trøjelsgaard, et al., "Can reed harvest be used as a management strategy for improving invertebrate biomass and diversity?", *Journal of Environmental Management,* 300, (September 2001): 113637, https://vbn.aau.dk/ws/portalfiles/portal/465648427/1_s2.0_S0301479721016996_main.pdf

3. For additional information on thatching techniques, see https://thatchinginfo.com/glossary-of-thatching-names-and-terms/

4. Dutch Federation of Thatchers, "*Vakfederatie Rietdekkers*", https://www.riet.com (August, 1st, 2023) and Andrew Raffle (Great Britain), International Thatching Society (ITS), https://thatchers.eu/content/holland/ (August, 1st, 2023).

5. M. Van Hemert, M.W. J. Van Rooden and H. Th. D Dijkstra, *Het Weke Dak. Riet en strobedekkingen,* (Den Haag: Rijksdienst voor de Monumentenzorg, 1990).

6. Jesper Engelmark, *Københavnsk etageboligbyggeri 1850-1900: en byggeteknisk undersøgelse,* (Copenhagen: Statens Byggeforskningsinstitut, 1983), 142.

7. Evald Tang Kristensen, *Gamle folks fortællinger om det jyske almueliv, som det er blevet ført i mands minde, samt enkelte oplysende sidestykker fra øerne,* 2. ed. Reprint of the original 1891-1905, (København; Busck, 1987).

8. Building permit from the building archive in Haderslev, Denmark, (Journal no. 2015/89), Ørby 20, 6100, Haderslev June 4th, 1889).

9. Poul Kjærgaard, "Stråtag", *Byggebogen,* (Copenhagen: Nyt Nordisk Forlag, 1949), 348.91.

10. For the Danish Building Regulations 2018, see https://bygningsreglementet.dk

11. According to the European Fire Norms: EN13823. It refers to the dominant fire test method for reaction-to-fire classification of construction products in Europe.

12. FIGRA is used as a classification parameter in the SBI test (EN13823).

13. Anne Beim, "Ler som brandhæmmer - det dur", *Tæk/2,* August, 2021, 20-21. https://straatagetskontor.dk/wp-content/uploads/2021/08/Taek02-2021.pdf.

14. For additional information on this product see, https://ecococon.eu/us/

15. R.A.P van Herpen and M.S. Drost-Hofman, *Brandveiligheid Rieten Gevels.* (Utrecht: Nieman Raadgevende Ingenieurs, 2012), https://www.riet.com/media/vfr/pdf/Brandveiligheid_rieten_gevels.pdf

16. Adam Greenfield, *Radical technologies: the design of everyday life,* (New York: Verso, 2018), 273-303

17. Jørgen Kaarup, "Det ultimative økotag", *Tæk/3, October, 2019,* 3-6. https://straatagetskontor.dk/wp-content/uploads/2019/10/Taek-03-19.pdf

BIO
MATTER
TECHNO
SYNTHETICS

INTERDISCIPLINARY DESIGN METHODOLOGIES AND MATERIAL AUTHORSHIP

MARÍA JOSÉ FUENTES

Matter has evolved throughout history via technology, biology, and human curiosity, allowing novel material systems to emerge. Designers working across a broad range of disciplines discuss herein their research practices and respective interpretations of materials and their properties. In MATTER, we encounter essays on material agency and the role of designers in exploring materiality at varying scales of manipulation. Notwithstanding its ubiquity, however, we learn of the mysteries of matter, its power to grow, and its function as waste.

Laia Mogas-Soldevila speculates on the intersection of design, material fabrication, and engineering. Her design research mines new aesthetic possibilities of bio-based matter developed in a lab environment that seeks functional, environmental, and health benefits. Soldevila rethinks the microscale of cellulose, chitin, silk fibroin, and sericin by reverse engineering their properties and by setting performance criteria that critically expand the designer's authorship and autonomy. Soldevila questions the historical role of the designer by promoting innovation in the design of materials, which is capable of being programmed with new biological behaviors.

Similarly, **Andrea Ling** describes biological decay in which the renewal of waste matter occurs at the microscopic scale. Ling argues for valuing organized decay, especially as it occurs in three of the most abundant biopolymers on the planet. Ling explores how to 'organize' material deterioration using enzymes, bacteria, fungi, and other biological agents that both decompose and compose new matter. Ling recognizes the difficulties of using biologically derived materials, as compared to industrialized materials (environmentally unstable over time, overly responsive, etc.), yet articulates how the exploration of their forms is well worth the effort. Materials should be valued not for their ease of consumption or accessibility but for their responsiveness, renewability, and recyclability, as these are the properties that foster sustainability and circularity.

Mae-Ling Lokko addresses the economic and social conditions associated with the use of materials. Her provocations speak to issues of generative justice, mycelium upcycling, and community integration. The importance of studying how value is extracted, lost, stored, and circulated within a material economy is crucial to understanding how to maximize, translate, and circulate unalienated value. The combinatory framework that Lokko creates in her case studies celebrates an economic model where the use of biomass helps in the creation of jobs and opportunities for those facing social and professional barriers.

Rebecca Popowsky discusses interdisciplinary design practice as collaborative work within material research. She investigates the creation of a more renewable source for sand within manufactured soils, using recycled glass as a substitute. Working alongside professional landscape architects, city officials, and university researchers, Popowsky aims to reduce the carbon footprint of contemporary landscapes using recycled glass waste as a soil-less soil. This alternate material is put through industry tests and analyses, resulting in verifiable data of use when convincing the public of its potential.

Julia Lohman's artistic work operates as a recuperative and restorative form of design where the lifecycle, agency, and origin of seaweed are critical components. As Lohman argues, the ecological and environmental benefits of using kelp-based materials are both positive and joyful. Lohman's ethically conscious writing considers seaweed as a co-designer and organism, working in collaboration with human will. As a biomaterial with the agency, seaweed's temporal frame lies in the future, where speculative collections such as the Department of Seaweed in Oslo, Norway, are necessary and important to understand, control, and create within its biological parameters.

Martina Decker is focused on the ethics and the scale of design in her paper on nanotechnology, showcasing the importance of research at the intersection of nano-matter and advanced technologies. The range of possibilities resulting from the re-arrangement of atoms and the creation of 'nano-factories' speaks to the absolute smallest scale of matter possible in design. When considering concepts such as organic cell reproduction, Decker outlines how working at the scale of atoms has evolved, creating ever more sophisticated models of nanotechnology, including nanorobots.

The authors spotlighted in MATTER integrate their research in material properties with timely questions of sustainability, renewal, economics, and scales of design. The world must critically consider the social and environmental impacts addressed by these designers, as we deliberate outcomes for rehabilitating and reintegrating matter in design today.

REVIVING MATTER:

MODIFICATION, SUBSTITUTION, AUGMENTATION, AND INTERACTION IN EXEMPLAR BIOMATERIAL ARCHITECTURES

LAIA MOGAS-SOLDEVILA

ABSTRACT

The architect's role in choosing, critiquing, and designing with materials is all the more crucial as the field embraces a renewed interest in materiality and materialization. Current research connects rapidly advancing materials engineering with the urgency to build more sustainably and efficiently. This paper participates in this effort by adopting a biomaterial approach that investigates the implications of 'reviving matter'. It theoretically repositions material-based decision-making within the early stages of the design process, and practically synthesizes methods that originate in biomedical engineering with the re-programming of architectural properties. The biomaterial architectures that result in display 'modified' function, industrial 'substitution,' 'augmented' performance, and environmental 'interaction.' Digital design and manufacturing methods used to construct this work embrace both cutting-edge technology and traditional methods of art and craft. They combine engineering and aesthetics whose forms foster both ground-breaking innovation and cultural metrics of quality. This research confers objects with new behaviors and contributes to the fundamental manipulation of the properties of matter by and for design.

+ living materials
+ biomaterial architectures
+ programmable materials
+ reviving matter
+ material-driven design
+ digital fabrication

Figure 1. Detail of *Lachesis 2019*. Copyright Laia Mogas-Soldevila.

Today, there is the potential for far greater control over virtual exploration and physical production of material architectures when new 'designer-biomaterial' interfaces are devised and engaged. This is possible because the distance between digital design, advanced fabrication, and materials engineering is narrowing.[1] This allows architects to directly intervene in how matter is made and how it responds to building more sustainably and efficiently. Consequently, not only is material use being re-evaluated with a focus on reducing environmental impacts, but also the processing of traditional building materials, such as concrete, glass, steel, and ceramics. In this sense, there is growing interest in the many advantages that adopting natural biomaterial alternatives provides for energy and resource preservation.

Natural biomaterials that originate in plants and animals are used to augment, replace, or repair body tissues and organs. Their engineering is one of the main focuses of biomedical science. Biomaterial blends of substances derived from algae, crustacean shells, fish scales, fungi, silk cocoons, plant cell walls, or vegetable gums are processed in water, using low amounts of energy and non-toxic chemicals.[2]All of these are benign to the body when used to sense biochemistries, deliver drugs, or guide the regrowth of, for instance, muscle or bone. Biomaterials help organisms heal and get absorbed into tissue after doing so because their building blocks are the same as those of the body itself.

Globally, strategies are being devised to process biomaterials for uses other than medical ones, opening avenues for the advent of 'biomaterial architectures'. If the production of such biomaterial architectures can be scaled to that of buildings, these substances could be excellent candidates for composing or even replacing architectural building materials and rendering them benign to the Earth. Moreover, biomaterial blends are not only inherently sustainable during their extraction, processing, and end-of-life, but they also offer unprecedented opportunities to design and engineer new functions for said materials by altering their versatile molecular makeup. This vision is described in the case studies that follow.

If architects can re-program glass, steel, and concrete by taking advantage of the incredible properties found in the molecular structures of wood, skin, shell, or bone, living architectures of the future will be able to self-heal, self-clean, produce energy, sense, adapt, and respond to environmental pressures. Biomaterial constructs can be designed to exhibit both strength and flexibility, such as in the structure of trees, differential porosity, smart membranes of skin, and tunable toughness, which is observable in shells. They can also adapt to loads over time, as notable in bone. These fascinating features are the result of growth and self-assembly that occur in living materials and that confer onto simple molecular building blocks geometric and material hierarchies.[3] This is the case with the growth of helixes into fibrils, into osteons, and into the compact or spongy density of which bone is composed; of microfibrils into matrixes, into layers, and into the graded porosity of bamboo; of platelets into fibrils, into networks, and into tiles, as in shell and nacre.

Recent experimental projects that use biomaterials like cellulose, silk, mycelium, or chitin are providing new aesthetic possibilities while re-interpreting fundamental questions across all aspects of architecture. Design is informed by investigations that harness fascinating phenomena that occur in places far removed from traditional building construction. This is notable, for instance, in the study of wood deformation by the Institute for Computational Design and Construction (ICD) in Stuttgart in their hydromorphic and active-bending research pavilions,[4] in that of fungal digestion of waste in the Hy-Fi mushroom brick towers by Ecovative and *The Living*,[5] in the development of organisms as builders at the MIT Mediated Matter's Silk Pavilion,[6] and in the programmed decay in the project *Aguahoja1* of a six-meter tall structure made of chitosan-cellulose printed surfaces.[7]

An Introduction to Reviving Matter

This paper advances research in the use of biomaterials in design by adopting the concept of 'reviving matter'. Material practices, when introduced in the early conceptualization of architecture, can harness the participation of diverse disciplines to 'revive matter' and invent the sustainable materials of the future. Materials can be literally 'revived' by reverse-engineering and providing them with new functions, and they can be theoretically reanimated by devising design processes that originate from materiality.[8] I assert that in order to radically challenge form and function in design, biomedical engineering, and materials science must be embraced using methods borrowed from pharmaceutical and medical industries.[9] The work presented below focuses on biomaterial blends made of cellulose and chitin, the most abundant biopolymers on Earth, and silk fibroin, which makes the fibers of silkworm cocoons. These materials fully biodegrade in the environment and have outstanding properties of not only combined stiffness and elasticity but also functional gradation of flexibility, opacity, porosity, softness, or strength.

Silk cocoons, in particular, are composed of proteins called fibroin and sericin that form long fibers and a sticky matrix, respectively, similar to the organization of carbon fiber and epoxy resin in man-made composites. A silk cocoon is not only strong and protective in shielding itself from predators and the environment, but it is also soft on the inside for comfort, waterproof yet breathable, and extremely elegant in its hierarchical structure and form. Many of the capabilities of natural silk are the result of its self-assembled hierarchy: including its molecules that are cross-linked, its micro-fibrils that are packed into filaments using sericin, and its ability to bundle into a silk thread, of use for millennia since the Silk Road.[10] In biomedical engineering, silk fibroin is obtained by reverse engineering silk fibers into a silk protein solution. The solution can be augmented with new properties and used as stock material to make medical devices using a wide range of fabrication methods such as spin coating, stenciling, inject printing, nanoimprinting, lithography, three-dimensional printing among others. This results in fully biocompatible devices, as demonstrated in the pioneering work of the Silklab at Tufts University.[11] In particular, this includes vein replacements, implantable screws, dissolvable therapeutics, resorbable electronics, and chemical sensors. Importantly, augmentation of material functions, features, or properties is achieved by direct functionalization of certain molecular groups using diverse approaches that can be classified as mechanical, physical, chemical, and biological.

More specifically, the work in this paper transfers medical and pharmaceutical material blends and synthesis methods within the fields of product design and fabrication to develop material forms with augmented functionalities. Results include apparel that heats up with light, alternative leathers that interact with one's body temperature, thin lattices that are tough and flexible, surfaces that ruffle with humidity, and tapestries that change color to monitor surrounding health. Many of these new, functionalized forms can be produced at scale, as later detailed in this paper.

Case Studies in Biomaterial Architectures

The work described below joins a growing effort in the practice of architecture to transition from material selection that typically occurs late in the design process to early-stage design of material properties. The project examples discussed here include fabrication-induced anisotropy 'modification' in lattices, sustainability-driven 'substitution' of animal leather goods, activation-driven 'augmentation' of solid parts, and sensing-based 'interaction' in large soft surfaces. (Figure 2)

'Functional Matter,' 'Computed Craft,' and 'Active Aesthetics' are the author's research sub-fields that guide the design and fabrication of biomaterial structures and investigate the process and implications of imbuing the material definition of conventional objects with new environmentally active functions. A research cycle is applied to (1) program unforeseen material 'functions' within forms, (2) devise format 'viability' constraints, (3) establish 'workflows' for craft and digital fabrication, (4) respond to both material and 'product' parameters, (5) identify desired interface 'effects', and (6) repeat iterative 'characterization' steps to restart the cycle. The four case studies discussed in this paper used this cycle, with each depicting exemplar design strategies for material property augmentation in 'Functional Matter,' combined digital-analog fabrication in 'Computed Craft,' and new forms and effects made possible by programmed behavior in 'Active Aesthetics.' Objects created according to the principles of revived matter display diverse levels of modified function, acquired or augmented properties, and the capacity for environmental interaction.

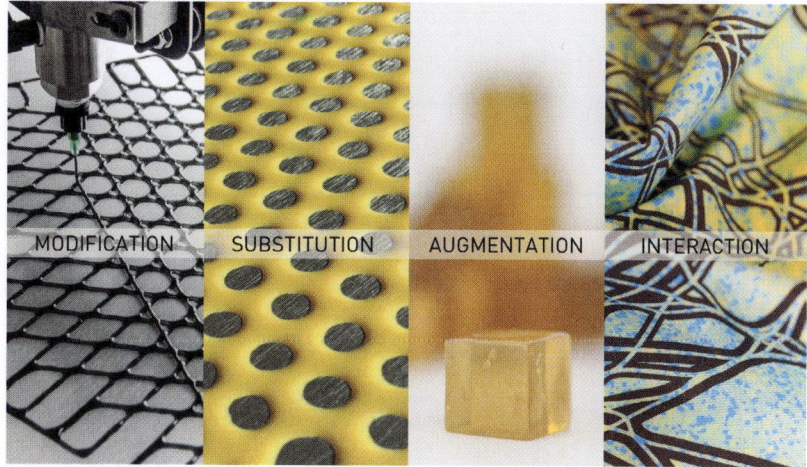

Figure 2. Functional biomaterial architectures include fabrication-induced MODIFICATION of lattices with the design of anisotropy at the microscale to induce structural transformations at the macroscale, sustainability-driven SUBSTITUTION of animal skin goods by leather-like biomaterial alternatives, activation-based AUGMENTATION by programing reactive properties in everyday solid objects, and sensing-based INTERACTION within large-scale surfaces printed with biomaterial inks. Copyright Laia Mogas-Soldevila.

Modification: Micro Design for Functional Macro Transformations

The molecules that makeup wood, silk, skin, shell, or bone—like many life forms—are simple chemicals housed in water and assembled using low energy and ambient conditions. It is within these aqueous environments and through the process of evaporation that their building blocks arrange and that a myriad of interesting properties of flexibility, toughness, or opacity emerge. The design of new ways of manufacturing with aqueous blends and of modifying material properties during industrial fabrication are key processes in assisting the self-assembly of long-chain biomaterials—fibroin, chitosan, or cellulose—into molecular-scale hierarchical formations; materials, which as they solidify, give rise to desirable macro-scale properties usable in man-made products and parts.[12]

To develop fabrication-induced material design—materials whose properties are willingly altered by the process of fabrication—several steps are taken as described above in the research cycle. It starts with defining a target 'function' that programs mechanical behavior in biodegradable materials to induce 2D-to-3D folding, ruffling, or bending. To ensure 'viability' during fabrication, biomaterial blends are made extrudable with additives that guarantee sufficient viscosity and shear thinning. A 'workflow' is devised to distribute matter into hierarchical geometric designs. The molecular composition of these material blends depends on the targeted use of the final 'products', which range from packaging to apparel or interior paneling. In terms of human-perceived 'effects', these material constructs must be light and thin to be easily worn or transported. Further cycles of 'characterization' are needed to ensure product-specific durability.

For instance, induced 2D-to-3D folding in flat extrusions of aqueous chitosan blends is achieved by combining two simple techniques in a design-to-fabrication workflow, as explored at MIT by the author and colleagues during her research at the Mediated Matter Group.[13] These techniques are digitally controlled extrusion-based fabrication and directional tool-pathing. During extrusion, the long chains of molecules in polysaccharides—such as chitosan and cellulose—detangle and orient to the printing direction, which turns the dry material anisotropic to curl in that particular direction.[14] These extrusions follow 2D drawings inspired by naturally occurring geometrical patterns that present 3D formations, such as that of fallen leaves in their natural dehydrated state. In this way, folded constructs at the micro-scale, remind us of three-meter-tall leaves. (Figure 3)

Similarly, when 2D-printing silk fibroin-based lattices, proteins can be modified by cross-linking via shear at the extrusion nozzle.[15] This confers directional stiffness and strength to the sheets of dry material lattices, as in the previous example, but it also introduces flexibility as seen in the folded lattice at the center of the cycle. Importantly, flexibility is present due to the nature of the aligned fiber-like molecular structures of fibroin protein that behave similarly to what we observe at the macro-scale in thin wood branches when woven into baskets. In both examples non-intuitive macro-scale material responses are achieved by the very process of fabrication that arranges matter at the nanoscale.

Substitution: Pliable Biomaterials as Sustainable Leathers

The continued use of traditional production methods is of concern for global warming, as it is for air and water pollution. Transitioning to the manufacturing of leather alternatives with low environmental footprints could offer a key solution to this growing problem.[16] A research cycle is discussed in this section that derives silk-based leather-like materials to support industry goals of reducing negative environmental impacts of apparel and upholstery materials. The process starts with defining a target 'function' that is to substitute animal leathers and confer strength and flexibility to biomaterial-based soft and pliable surfaces. Different blends are researched to ensure 'viability' of fabrication. A 'workflow' is devised to distribute biomaterial blends onto leather-like sheets programmed for differential flexibility with geometric patterning and layering. The composition and responsiveness of material blends is dependent on the final 'product', which range from fashion accessories to upholstery and footwear. In terms of human-perceived effects, these material constructs must be able to drape, be strong, and offer perceptible environmental responses via color changes made possible by thermochromic, photochromic, or pH sensing molecules. Further cycles of 'characterization' are needed to identify post-processing methods that ensure durability as, for instance, the grease coating of animal leather for maintenance. (Figure 4)

Lab-based alternatives to animal leather-goods promise important reductions in the industry's most severe environmental impacts and significant improvements in its ecological footprint.

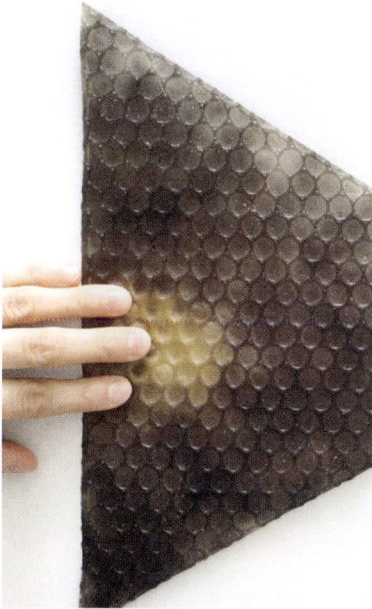

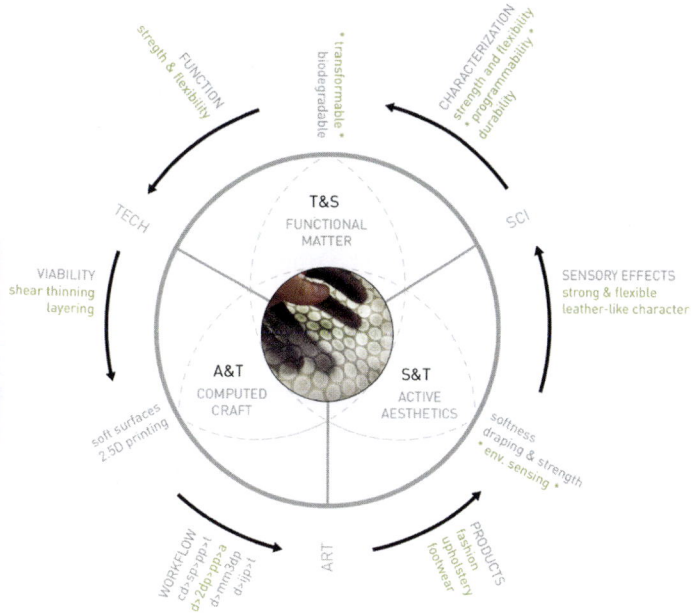

Figure 3. SUBSTITUTION –
Research cycle for deriving
thermochromic leather-like
biomaterial surfaces. Copyright
Laia Mogas-Soldevila.

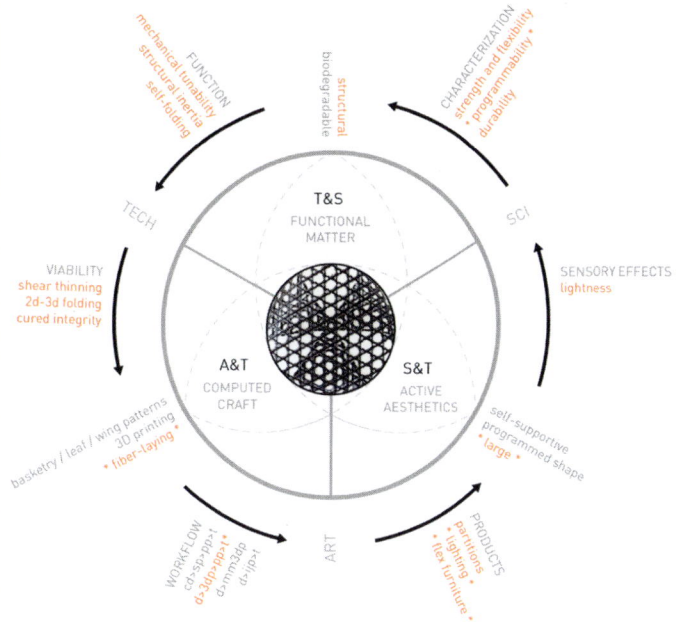

Figure 4. MODIFICATION –
Research cycle for deriving
macro properties via micro-scale
fabrication effects. Copyright
Laia Mogas-Soldevila.

The material blends we fabricated contribute to this, in particular, silk protein-based printed surfaces that are ethically sourced, based in water, and fabricated at room temperature. When processed with mild chemicals they minimize the amount of toxicity and waste. Non-medical formulations of biomaterial blends are scalable at volume, can be digitally manufactured, and are able to withstand the same amount of digital-analogue fabrication methods as traditional leather assemblies. The functional properties of these alternative leathers can be controlled using chemical, physical, and geometrical strategies that directly affect their mechanical properties, resulting in samples that are highly competitive in stiffness and strength compared to existing market products. These silk-based alternative leathers allow for tunability relative to size, flexibility, opacity, and environmental responsiveness. In this way they offer advantages compared to other alternative leathers made of bacterial cellulose, fungal mycelium, or pineapple leaves. Hence, in matching the aesthetics of traditional cow and sheep leather goods, silk leathers are just as luxurious and culturally resonant.

Augmentation: Active Matter for Everyday Products

Some water-based biomaterials used in pharmaceutical and medical research can be dehydrated into solids with resultant properties comparable to those of man-made ceramics. Some of these solids can be functionally augmented to serve multiple purposes.[17] They can, for instance, effectively replace bone screws and slowly dissolve when used for releasing drugs into the human body. In particular, proteins like silk fibroin undergo gelation (gel to solid transition) when the conditions of water removal are controlled. The mechanical properties of the resulting silk solids are conditioned by evaporation: slow evaporation at room temperature results in stiff solids whereas fast evaporation by adding heat can lead to highly fragile structures. The desired performance of many biomaterials can be engineered within the range of these two extremes. During the removal of water, for example, solutions can be doped with molecules that confer new properties such as the ability to respond to stimuli during the solid's life.

As previously explored by Benedetto Marelli in his work on the directed self-assembly of pre-programmed silk solid materials, new instances of compound encapsulation and active response can be designed.[18] To this end, the research steps undertaken in this cycle include defining a target 'function' to signal the environment through solids. Dyes and active molecules are researched and stabilized in fibroin protein. To ensure 'viability' of fabrication, the material is concentrated before being solidified by water removal. A 'workflow' is devised to mold doped protein blocks, machine them, and finish them with desired effects. Shape and post-processing constraints are dependent upon the resultant 'product' scale and possible intended uses such as implantable screws, jewelry, and eyewear. In terms of human-perceived 'effects', these material constructs must be able to control their translucency, and their chromic, photothermic, and magnetic responses to chemicals, impact, and light stimuli. Further 'characterization' is required to better assess their resistance to Ultraviolet (UV) exposure or impact. (Figure 5)

Augmenting materials with the ability to process and respond to information is fueling avant-garde research into active compounds that chemically transform, interact, and adapt to surrounding environmental pressures.[19] Solids programmed at the molecular level with interactive behaviors like the ones presented here contribute to these efforts. In particular, inert silk protein gels are transformed into dynamic materials by combining them with active molecules. This makes them able to report on strain or temperature by changing color, render themselves magnetic, or stabilize pigment and scent. Forms can be explored for daily use objects such as screws, glasses, and earrings with their ability to report activated by impact, heat, piezo-magnetism, or simple exudation.

Interaction: Distributed Environmental Sensing in Soft Surfaces

The material-first design methodology adopted in this final example is targeted at building large-format environmentally interactive soft surfaces that can chemically report on their surroundings. The goal is to easily visualize useful parameters and report interpretable results with color changes. Specifically, bio-based active inks are developed and screen-printed following distributed geometric patterns onto silk scarfs and tapestries, cotton t-shirts, bracelets, and on small-scale cards and large-format art paper.[20] (Figure 6)

Similar to the previous three projects, the research cycle implemented here began by defining a target for the given 'function' to sense and display. (Figure 7) In this example, it is colorimetric dyes reporting on pH levels that are stabilized in fibroin protein.

Figure 5. AUGMENTATION – Research cycle used to derive doped biomaterial solids for conferring new properties to everyday objects. A catalogue of active solids includes control of translucency, magnetism, and chromic signaling to chemical, impact, or light exposure. Copyright Laia Mogas-Soldevila.

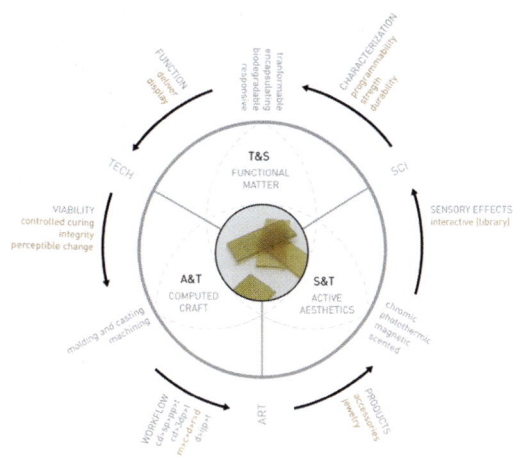

REVIVING MATTER

To ensure 'viability' of its fabrication, the material is transformed into an ink with sufficient integrity to withstand shear during screen printing and abrasion during use and dry cleaning. To design these sensing patterns, a 'workflow' is devised that distributes motifs, post-processes with heat or humidity, and finishes by cutting or sewing to product shape. Resulting 'products' include three-meter tall tapestries, one-meter-long paper prints, cotton t-shirts, fashion scarfs, performance arm bands, and even skin diagnostic patches. In terms of human-perceived 'effects', these material constructs must provide softness and comfort while achieving visible chromic transformation. Further 'characterization' could include more information on abrasion resistance from friction, UV exposure, or the effects of dry cleaning.

The design of these printed geometric patterns is inspired by the complexity of a silk cocoon as seen under a microscope. (Figure 8) Technical inks are transferred onto soft substrates following digitally generated patterns, while using the traditional tools and craftsmanship of screen-printing. The inks take advantage of silk protein's encapsulating and preserving capability to stabilize active compounds. Active molecules are added to silk solutions, mixed with thickeners and made into a printable ink with adequate viscosity for screen-printing and enough resolution for interpretable sensing. (Figure 9) Once distributed, molecules in the inks are bonded throughout diverse scales and substrates such as silk, cotton fabrics, and art papers.

Screen-printed outcomes produced with this technique range from personal performance-monitoring bracelets, t-shirts, large-format tapestries, and canopies capable of reporting on the environment of surrounding architectural spaces. (Figure 10) Once activated, formats change color when detecting pH ranges in, for instance, sweat or precipitation, and results can be respectively mapped to biometrics such as hydration or dietary patterns, or to rain acidity and pollution ranges. They operate at a wide range of sensitivity (from pH.4 to pH.9)

Left: Figure 6. INTERACTION – Distributed environmental sensing in large-scale soft surfaces. Colorimetric chemical sensing within large format surfaces such as tapestries, wallpaper, or canopies that indicate changes in surrounding environmental metrics, such as rainfall acidity or air quality. Copyright Laia Mogas-Soldevila.

Right: Figure 7. INTERACTION – Research cycle for deriving large-format distributed sensing surfaces able to chemically monitor their surroundings. Copyright Laia Mogas-Soldevila.

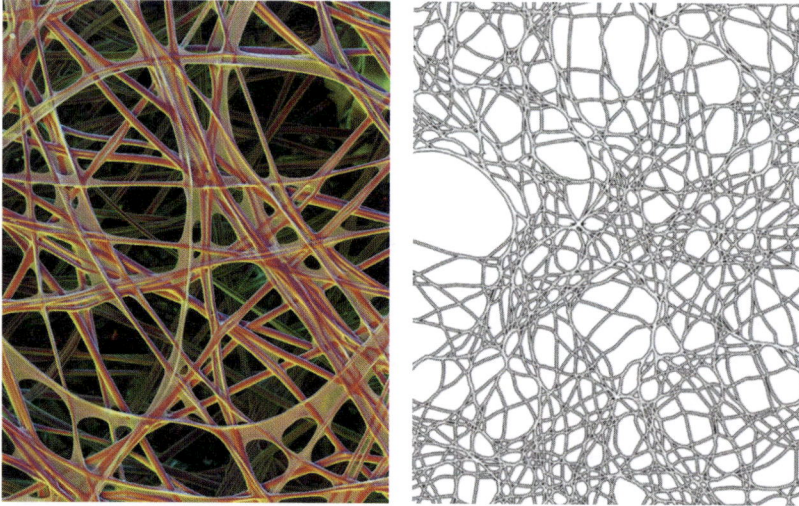

Figure 8. INTERACTION – Cocoon microstructures translated into sensing patterns. Micro-scale silk cocoon motifs inform geometric algorithmic design to distribute sensing inks throughout soft surfaces. Copyright Laia Mogas-Soldevila.

and results can discriminate to 0.2 units. Importantly, these are fully distributed chemical sensors that achieve fine-tuned global monitoring without any hardware or energy input. Characterization of the results reveals stress-stretch curves that do not alter mechanical behavior to distort the drape. Moreover, scratch tests show that the bond between fabric and ink is not superficial but interstitial and capable of with-standing many reversible cycles of sensing and dry cleaning.

Personal health formats can be extended to sheets, pants, or shoes made of both polymers and textiles with great resolution. Large formats can take the form of drapes, canopies, wallpaper, internal partitions, and façade shading systems. Project *Lachesis 2019,* named after the goddess of weaving, features giant tapestries made of silk fabric and silk protein inks that are able to sense and respond to their environment.[21] (Figure 10) Living material inks are processed to encapsulate chemical sensors originally developed for medical diagnostics and infused for the first time into soft architectural surfaces. These silk tapestries have been repeatedly transformed by more than ten thousand visitors at the *Design Does Exhibit* at the Barcelona Design Museum in 2018 and at the Nostos Summer Festival in Athens in 2019. Audiences sprayed fluids emulating tears, sweat, and rain, while surfaces responded with color changes displaying useful information about diet, performance, and pollution. Robust tests like those made possible by project *Lachesis 2019* are promising for architectural-scale distributed monitoring of rainfall acidity, air quality, and other environmental metrics. In this way, these surfaces can become useful signals and material partners in our daily lives.

In sum, material property augmentation is achieved in this work by printing soft everyday surfaces with biosensing inks and turning them into diagnostics that report on crucial environmental metrics in support of the health and performance of our surroundings. These formats combine digital design and algorithmic image post-process-

Figure 9. INTERACTION
– Technical living inks.
Environmental-sensing
biomaterial-based
inks are produced by
mixing components at
room temperature. They
successfully encapsulate
active molecules and are
screen-printed into soft
surfaces matching industrial
ink standards. Copyright Laia
Mogas-Soldevila.

ing with analogue screen-printing fabrication methods. Chemical reporting is embedded into traditional soft forms used on everyday objects that impact and help shape our cultural understanding of aesthetically determined sensing devices.

'New Materialism' in Design

Today's emerging innovations in materials engineering, advanced processing, and digital manufacturing require a new relationship with architecture. They empower designers to take a deciding role in material choice and material critique, and more importantly, a stand on material design. Resulting strategies may be similar to how design was once driven by new materials and technologies in the work of Le Corbusier, Antoni Gaudi, Alvar Aalto, and Frei Otto, but with an unprecedented opportunity to harness recent scientific discovery.

In the 1990s, a post-humanist 'new materialism' theory advocated for a turn to matter as a necessary critical engagement to address material realities for humans and nonhumans alike.[22] This theory worked against inert and non-generative conceptions of matter, calling for a 'return' to matter in all aspects of bodies, spaces, and conditions. For the field of architecture in particular, this article proposes to revive matter, both discursively and practically, and looks beyond traditional building construction and its drive to improve the performance of inefficient materials and systems. Instead, this work investigates the programming of new behaviors by design: by the modification of material properties via fabrication methods; by the substitution of polluting industrial chemicals with benign biomaterial blends; by the augmentation of everyday products with bio-dopants during material synthesis; and, by the distributed monitoring of buildings with living inks instead of energy-inefficient technology. In particular, the design and fabrication of 'revived matter' objects embraces cutting-edge technology and traditional craft, combining engineering innovation and aesthetics, where forms foster both ground-breaking innovation and culturally important metrics of quality.

Acknowledgements

The author would like to acknowledge the help of artist Michael Hecht from the School of the Museum of Fine Arts (SMFA) at Tufts University, photographer Andres Flajszer, craftsman Fernando Suarez at EFS Designs, as well as undergraduate researchers Patricia Blumeris, Joelle Bosia, Trent Turner, and Roger Gu. This research was fundamental to the author's PhD completed in May 2020 within Tufts University's Interdisciplinary Doctorate Program (IDOC) that bridges Engineering, Life Sciences, and Design.

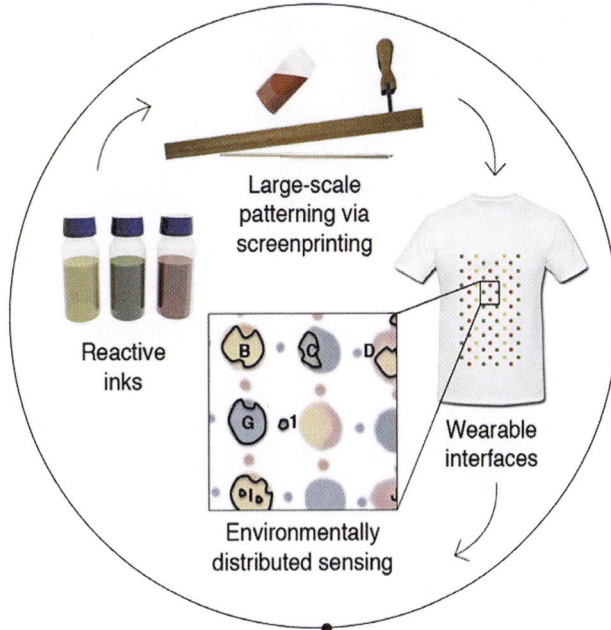

Large-scale patterning via screenprinting

Reactive inks

Wearable interfaces

Environmentally distributed sensing

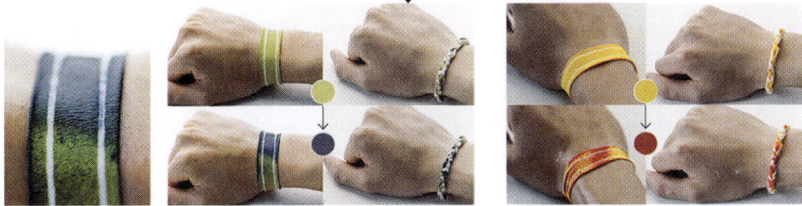

T-Shirt ~1m

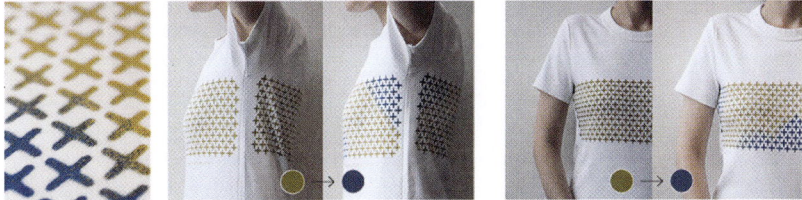

Tapestry ~10m

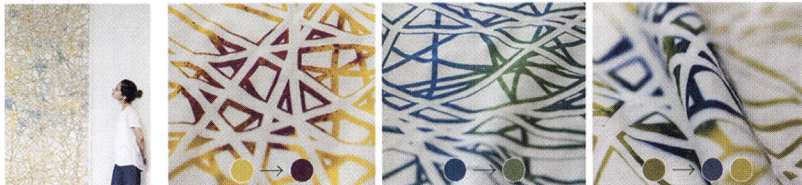

Figure 10. INTERACTION – Distributed sensing in everyday objects. Reactive inks can be patterned via screen-printing onto soft interfaces across scales. Depicted here are athletic bracelets, t-shirts, and tapestries that change color reporting changes in pH. Copyright Laia Mogas-Soldevila.

Figure 11. INTERACTION – *Lachesis 2019* in Athens Opera House during the Summer Nostos Festival. Novel sensing formats enabled by silk-based living materials in drapes, canopies, wallpaper, internal partitions, façade shading systems, or other large architectural surfaces. These distributed monitors communicate metrics such as rainfall acidity, toxic spills, air quality, and pollution. They are useful signals and material partners in our daily lives. Copyright Laia Mogas-Soldevila.

Collaborators

Prof. Neri Oxman, Dr. Jorge Duro-Royo, Dr. James Weaver, Daniel Lizardo from the Mediated Matter Group at MIT. Prof. Fiorenzo Omenetto, Prof. Giusy Matzeu, Dr. Meng Li, Dr. Wenyi Li, Arin Naidu, Dr Marco Lo Presti, and Dr. Giulia Guidetti, from the Silklab at Tufts University.

Funding

Parts of this research were supported by the Thyssen-Bornemisza Art Contemporary (TBA-21), the Stavros Niarchos Foundation (SNF), the 'Design Does Forum and Exhibit' at Museu del Disseny de Barcelona DHUB, MaterFAD Barcelona, the National Science Foundation, the Art and Science seed fund from the Office of the President at Tufts, the U.S. Office of Naval Research, and the Tufts Center for Applied Brain and Cognitive Sciences.

ENDNOTES

1. Martin Bechthold and James C. Weaver, "Materials Science and Architecture," *Nature Reviews Materials* 2, no. 12 (December 5, 2017): 17082, https://doi.org/10.1038/natrevmats.2017.82.

2. Julian Vincent, *Structural Biomaterials* (Princeton University Press, 2012).

3. Peter Fratzl and Richard Weinkamer, "Nature's Hierarchical Materials," *Progress in Materials Science* 52, no. 8 (November 2007): 1263–1334, https://doi.org/10.1016/j.pmatsci.2007.06.001.

4. Sean Ahlquist et al., "Development of a Digital Framework for the Computation of Complex Material and Morphological Behavior of Biological and Technological Systems," *Computer-Aided Design* 60 (2015): 84–104, https://doi.org/10.1016/j.cad.2014.01.013.type and association

5. Eben Bayer and Gavin McIntyre, Method for Making Dehydrated Mycelium Elements and Product Made Thereby (USPTO, issued April 24, 2012).

6. Neri Oxman et al., "Silk Pavilion: A Case Study in Fiber-Based Digital Fabrication," in *FABRICATE Conference Proceedings* (2013): 248–55, https://oxman.com/files/Silk-Pavilion.-A-Case-Study-in-Fiber-based-Digital-Fabrication-(2014).pdf.

7. Jorge Duro, Joshua Van Zak, et al., "Designing a Tree: Fabrication Informed Digital Design and Fabrication of Hierarchical Structures," in *Proceedings of the IASS Annual Symposium: Construction-aware structural design,* (2018): 1-7.

8. Laia Mogas-Soldevila, "Reviving Matter: Design Companions in the Science of Life" PhD Dissertation, Tufts University, 2020. Order No. 27957299.

9. Laia Mogas-Soldevila, "Formalizing Infinite Properties: Beyond Functionalism in Design," *ACTAR UrbanNEXT*, no. Designing Matter (2018), https://urbannext.net/formalizing-infinite-properties/.

10. Fiorenzo G Omenetto and David L Kaplan, "New Opportunities for an Ancient Material," *Science* 329, no. 5991 (2010): 528–31.

11. Danielle N Rockwood et al., "Materials Fabrication from Bombyx Mori Silk Fibroin," *Nature Protocols* 6, no. 10 (2011): 1612–31, https://doi.org/10.1038/nprot.2011.379.

12. Jorge Duro-Royo et al., "Parametric Chemistry: Reverse Engineering Biomaterial Composites for Robotic Manufacturing of Bio-Cement Structures across Scales," in *Architectural Robotics* (Routledge, Taylor and Francis, 2017), 217–23.

13. Jorge Duro, Laia Mogas-Soldevila, and Neri Oxman, "Flow-Based Fabrication: An Integrated Computational Workflow for Digital Design and Additive Manufacturing of Multifunctional Heterogeneously Structured Objects," Computer-Aided Design Journal, 2015, https://dspace.mit.edu/handle/1721.1/112152.

14. Laia Mogas Soldevila et al., "Fabrication-Informed Multi-Functional Anisotropy: Driving Nano-to-Macro Transformations in 3D-Printed Biopolymers.," *Materials Today Chemistry*, 2020.

15. Laia Mogas Soldevila et al., "Additively Manufactured Leather-like Silk Protein Materials," *Materials & Design* in review (2021).

16. Kate Fletcher, *Sustainable Fashion and Textiles* (Routledge, 2014), https://doi.org/10.4324/9781315857930.

17. Benedetto Marelli et al., "Programming Function into Mechanical Forms by Directed Assembly of Silk Bulk Materials," PNAS, 2016. https://doi.org/10.1073/pnas.1612063114.

18. Ibid.

19. Skylar Tibbits, ed., "An Introduction to Active Matter," in *Active Matter* (MIT Press, 2017).

20. Giusy Matzeu et al., Large-Scale Patterning of Reactive Surfaces for Wearable and Environmentally Deployable Sensors," *Advanced Materials* 32, no. 28 (2020). https://doi.org/10.1002/adma.202001258.

21. Laia Mogas, Giusy Matzeu, and Fiorenzo Omenetto, *Lachesis: Drawing the Fabric of Life* (USA: Vimeo, 2018), https://vimeo.com/304908228.

22. Diana Coole and Samantha Frost, *New Materialisms. Ontology, Agency, and Politics*, Duke University Press, 2010, https://doi.org/10.1215/9780822392996.

RETHINKING FIRMITAS:
BIOLOGICAL ENTROPY AS ARCHITECTURAL ORGANIZATION

ANDREA LING

ABSTRACT

This essay rethinks the concept of *firmitas* in architecture through the lens of dynamic renewal. Design has long associated solidity and durability in construction with what is fundamental to good building. Material research in architecture has typically been concerned with the development of longer-lasting, lower maintenance, sturdy materials, while contemporary design practice has kept the consequences of the production and destruction of these invisible to end users. By working with biologically derived materials and agents, I aim to keep these consequences in sight and on-site, designing how objects biodegrade in order to determine how new objects could be made, and how the larger system that encompasses them might persist. *Design by Decay, Decay by Design* is a project that explores how to organize decay and bio-deterioration—in other words, biological entropy—using enzymes, bacteria, fungus, and other biological agents to simultaneously both decompose and compose matter. Bacteria and mold eat away at designed, biologically derived materials leaving metabolites or creating bulk biomass in return. In this, decay is used as a fabrication process that transforms and creates new objects by constantly re-organizing rather than simply destroying them. Biologically derived materials and organisms can be extremely responsive and environmentally variable, making them a challenge to work with inside an industrial system that expects standardization. It is, however, these same characteristics that make biological systems so desirable in an alternate material world. Their ability to respond, repair, and replicate offers access to a form of tunability and self-renewing durability, impossible with industrial materials. In studying and using these types of materials and organisms in structures, I am not only pursuing mutability as a desired quality in the built world but also as the guarantee that the mechanisms of constructive renewal are embedded into that world.

+ decay
+ biodesign
+ biodegradation
+ biofabrication
+ *firmitas*

Figure 1. Cocoon made of laser cut chitosan-cellulose composite; to be used for biotransformation by *Streptomyces*. Photo by Ally Schmaling.

> The building is not a static organization or a structure resembling a machine made of more or less permanent 'construction materials' in which 'energetic materials' provided by nutrition decompose to supply the energy needs of vital processes. It is a continuous process in which both construction materials and energetic substances decompose and regenerate.
>
> Luis Fernandez-Galiano, *Fire and Memory* [1]

The *Gwion Gwion* rock paintings in Western Australia are estimated to be between forty-six thousand and seventy thousand years old.[2] Found in areas of extreme sun and rain, the paintings remain remarkably vibrant. When swabbed, archeologists found that what creates these vivid images are symbiotic colonies of red cyanobacteria and black fungi whose ancestors decomposed and fed on the initial pigments thousands of years ago. These organisms then flourished for generations within the confines of the original drawing, creating a material system that has endured far longer than any man-made system and with far less maintenance. (Figure 2)

This essay rethinks the concept of *firmitas* in architecture through the lens of dynamic renewal rather than through inert robustness.[3] Design has long associated the solidity and durability of construction with what is fundamental to good building. Material research in architecture has typically been concerned with the development of longer-lasting, lower maintenance, sturdy materials, while contemporary design practice has divested itself of the consequences of material waste during the production, construction, and maintenance of architecture, such that both designer and user have a fragmented view of the life cycle of these materials.[4] By working with biologically derived materials and biological agents, I aim to keep these consequences in sight and on-site, designing how objects biodegrade in order to determine how new objects could be made and how their encompassing system might endure. (Figure 3)

Design by Decay: Decay by Design is a series of artifacts expressly designed to decay. Completed as part of the 2019 Ginkgo Bioworks Creative Residency, the project is a response to Ginkgo's prompt to design a world without waste. As an architect and designer, I recognize that most of what I create goes to a landfill. If that is the case, I should design waste I can live with: garbage that retains some desirability as it degrades during our lifespans and within our homes. Can waste be designed as nature designs it, not only as the by-product of material decomposition but also as a form of input for renewal and construction? In biology, one system's entropy can be another system's organization.

In collaboration with scientists from Ginkgo Bioworks, the project here discussed was conceived to organize decay using enzymes, fungus, bacteria, and other biological agents to simultaneously decompose and compose matter. These organisms eat away at designed biological materials, leaving metabolites, or creating bulk biomass in return. Decay can thus be used as a fabrication process,

Figure 2. Bradshaw rock paintings in the Kimberley region of Western Australia, taken at a site off Kalumburu Road near the King Edward River. (Bradshaw Art _ 9 June 2009) Wikimedia Commons, Creative Commons Attribution-Share Alike 2.0 Generic license. https://commons.wikimedia.org/wiki/File:Bradshaw_rock_paintings.jpg accessed February 11, 2021.

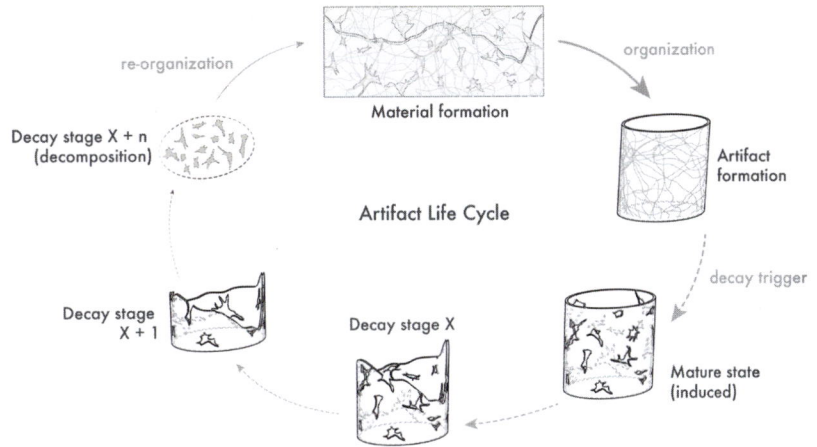

Figure 3. Artifact Life Cycle diagramming formal transformations of material to artifact and back to material with possible transformation mechanisms indicated. Decay states are such that the object can retain some sort of desirability or functionality during the process of decomposition. Drawing by Andrea Ling.

transforming an object, creating another object, or providing future inputs for construction and organization.

The main material system worked with during this residency was a set of biologically derived and biodegradable polymers first introduced to me as a research assistant at the MIT Media Lab.[5] Chitosan (a structural polysaccharide found in insect and crustacean shells and fungal walls), cellulose (a structural polysaccharide found in plant walls), and pectin (a polysaccharide found in fruit skins) are some of the most abundant biopolymers on the planet that can be harvested from waste, as byproducts of shrimp farms, wood pulp mills, and agriculture.[6] All are water soluble with short decay cycles and when made into water-based colloids they can be extruded into forms or cast into sheets of bioplastic, which when dried can be further processed into new forms or used as feedstuff for microbes.[7] The gradation and proportions of these ingredients can be tuned to create a vast design space whose characteristics include opacity, colour, flexibility, and mechanical strength.[8] (Figure 4) Only three structural polysaccharides—minimally processed—ensure these characteristics are achieved, all the while easily accommodating biodegradation.[9] That is, a whole artifact is made of a small library of ingredients which do not require vastly different end-of-life processes to decompose and can be easily taken up again into a new nutrient cycle. This is in direct contrast with many industrial material systems whose complex assemblies require great effort to separate their constituent parts before being taken to different recycling facilities, scrap yards, and landfills for disposal.

The use of biologically derived materials is not new. Bone, wood, grass, and animal skins were used to build our first shelters and artifacts. These simple materials were replaced with metals, ceramics, and more recently, plastics, which exhibit higher levels of strength and conventional durability but are typically more energy intensive, resource expensive, and with higher environmental costs when extracting and disposing of them. Returning to biologically derived materials can potentially reduce our ecological footprint. These materials, however, unlike mineral and petroleum-based ones, tend to be environmentally responsive over time, fluctuating in their dimensions, water content, colour, and other physical properties. They are difficult to standardize and cannot be controlled in the same way we control industrial materials. They do not offer equivalent performance characteristics as conventional material systems and cannot be used as replacements. What they do offer is biocompatibility, resilience, and dynamic capabilities not possible with inert and environmentally agnostic material systems. The variability that makes them so difficult to work with in an industrial context is the same quality that makes them able to respond, repair, and replicate.[10] To partner with these materials effectively requires a different process of design, one that accommodates variation, acknowledges material agency, and pursues mutability as a desired quality in the built world. (Figure 5)

It is in this context that I situate my interest in designing decay as a means of accessing new modes of production. By designing with and for decay, we can return matter and energy back to biology and perhaps bias design outcomes so that we may better live with its consequences. By mediating the decay process through species selection, control of environmental conditions, and templating of nutrients we can actively pursue self-renewable forms and stewardship of the physical world, as well as guarantee that mechanisms of constructive renewal are embedded in this world.

Design by Decay, Decay by Design is comprised of three smaller projects each of which asks how one can harness the responsiveness and temporality of biologically derived materials to shape new things. The first used enzymatic degradation of bioplastics as a means of

Figure 4. Left side, biologically derived material system experiments using Pectin-Chitosan-Cellulose composites. Right side, tripod structures made of Pectin-Chitosan-Cellulose composites using enzymatic degradation (pectinases and chitonases) as subtractive fabrication tool. Photos by Ally Schmaling.

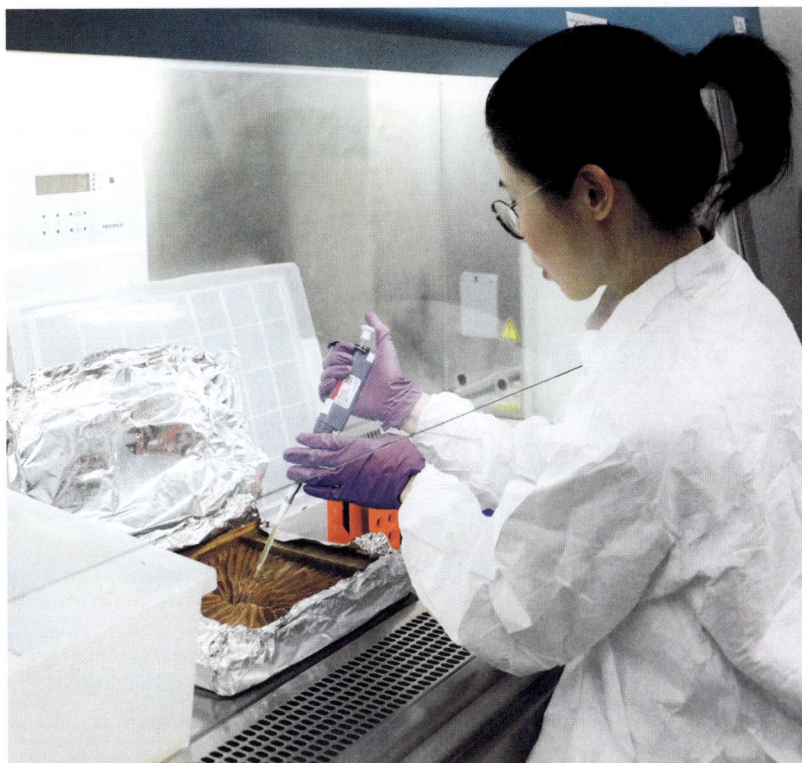

Figure 5. Andrea Ling in Bioworks 1, in the spore room, plating *Aspergillus niger.* Photo by Grace Chuang.

transforming material rather than destroying it. (Figures 1, 4-6, 9-12) This project involved a series of tests mentored by Joshua Dunn, head of design at Ginkgo, and in consultation with Nikos Reppas, foundry director. The second project used different types of fungi, including *Aspergillus niger* (black mold) and *Trichoderma viride* (green mold), in co-cultures to grow on and degrade different materials in a spatially controlled manner. These molds are powerful and resilient decay agents, rapidly colonizing any substrate we provided. (Figures 8, 13, 14) Here tests were conducted in collaboration with of Ming-Yueh Wu, one of Ginkgo's fungi engineers. The final project used different species of *Streptomyces* bacteria to colonize cellulose and bioplastic substrates to change their color. The *Streptomyces* genus encompass a group of common soil bacteria that can facilitate biodegradation but are not considered robust decay agents. They leave evidence of their metabolic activity with the release of vibrant pigments, colouring the materials they colonize. (Figures 15, 16) These tests were guided by Kyle Kenyon and in consultation with Duy Nguyen and Lucy Foulston, all *Streptomyces* researchers.

Cutting with Enzymes

In collaboration with Joshua Dunn, I tested a series of different enzymes—pectinase, cellulase, chitinase, amylase, and a lysing enzyme cocktail—at different dilutions to analyze their effect on samples of materials. These included different concentrations of chitosan, cellulose, pectin-chitosan composites, and chitosan-cellulose composites. Enzymatic degradation was tested both through spot testing (applying small drops of enzymes to wet and dry material) as well as by submerging the composites in buffer solutions that contained

Figure 6. Enzymatic degradation tests showing effects of different dilutions of pectinase, chitanse, cellulase, amylase, and lysing enzymes on material system. Photo by Ally Schmaling.

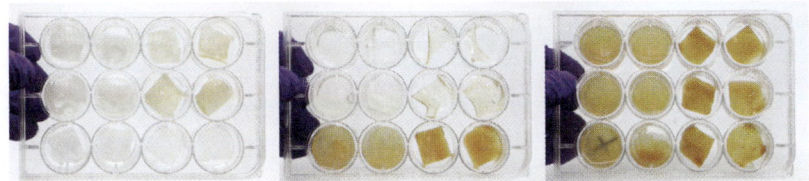

Figure 7. Left side, Large final samples including contaminated wood block originally inoculated with *Streptomyces coelicolor*, pectin-chitosan-cellulose 'painting', and cocoon sample made of chitosan film. Photo by Ally Schmaling.

Figure 8. Right side, Wood blocks inoculated with *Aspergillus niger* and *Trichoderma viride*, as well as two contaminated wood blocks (species unknown). Photo by Ally Schmaling.

Figure 9. Enzymatic degradation tests charting real-time effects of different dilutions of pectinase, chitanse, cellulase, amylase, and lysing enzymes on pectin-chitosan, chitosan, chitosan-cellulose, and cellulose samples. Photos by Andrea Ling.

small amounts of enzymes. (Figure 9) While the submerging method was the most effective at breaking down the material this also meant that the material was completely soaked and without form during the degradation process. The goal was to use enzymes as fabrication tools and to use enzymes in-situ on the material artifacts. While the spot tests on dry materials were not effective unless the materials were then incubated in extremely high humidity, spot tests on wet and semi-wet material were promising. The enzymes were used to degrade pectin, chitosan, and cellulose skins while these materials were being cast, thus shaping the artifacts using selective degradation in certain areas. (Figure 11) Holes and lines were created in areas where the enzyme was applied. As a control, a large pectin-cellulose-chitosan 'painting' was laser cut and dehydrated, and its resolution compared with a similar painting where enzyme was used in lieu of laser cut lines. (Figure 12) The enzyme degraded lines were of poor resolution, a function of the enzyme's diffusion through the semi-wet material and often ate away more material than expected. The remnant material at the lines was a sticky powder and made the intact portion of the bioplastic very difficult to remove from the substrate. (Figure 10) Interestingly, degradation continued even when the colloids were dry to the touch, visible as lines and holes became larger with time. The most compelling

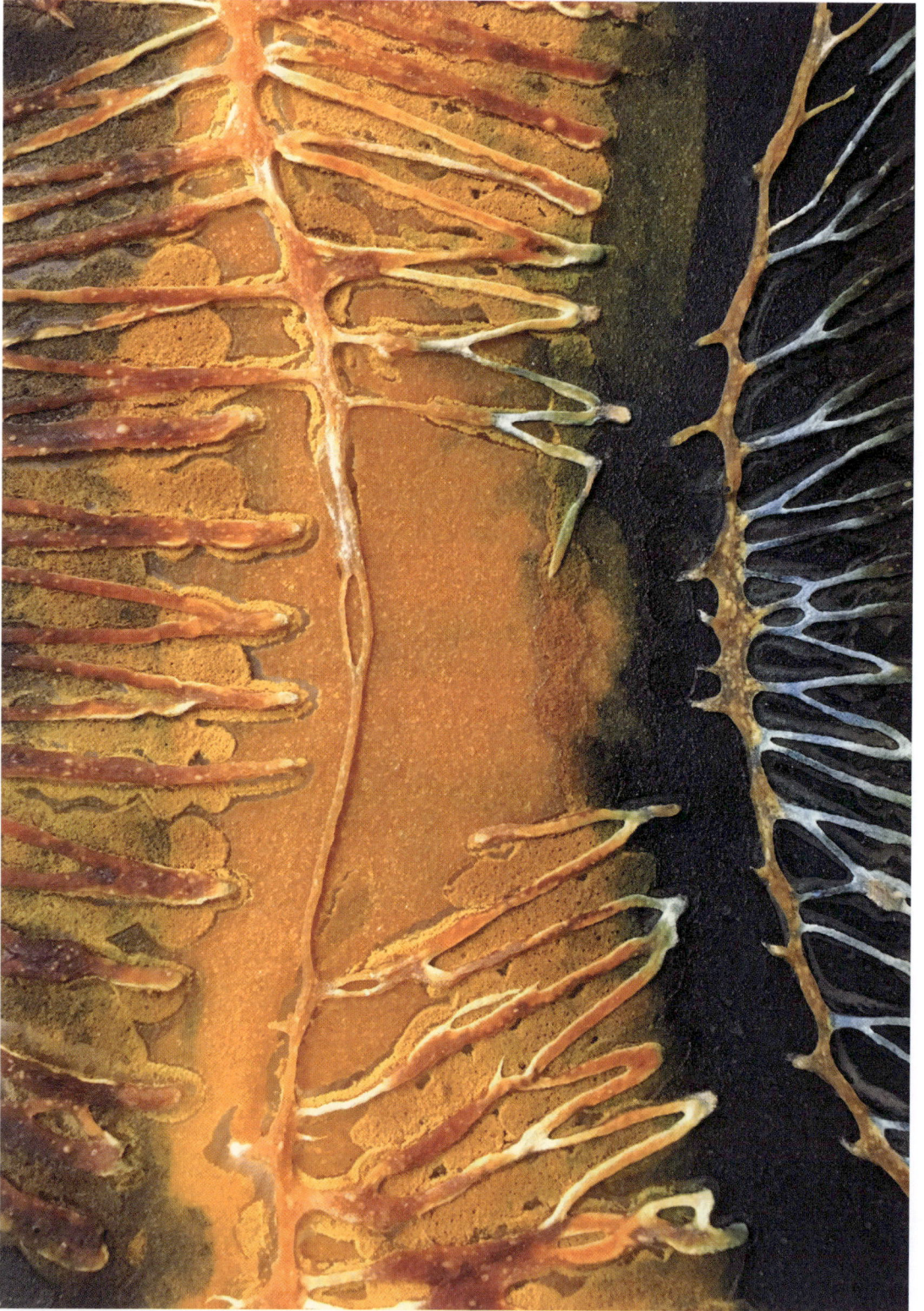

results were obtained when using the enzymes to create holes and perimeter lines for a series of tripod structures. (Figures 4, 6) Here, the material was successfully transformed through degradation into similarly sized objects, that, though not as standardized as industrially produced objects, were similar enough that I could use them as a modular unit for a larger aggregate structure.

Carving with Soil Fungi

Aspergillus niger is a common black mold found in soil and on fruits and vegetables. It produces pectinase and amylase. *Trichoderma viride* is a common green mold found in soil that produces cellulases and chitinases and can be used as a bio-fungicide against other plant pathogenic fungi. Once combined, these organisms smell like humid air. Fungi are extremely effective decomposers, releasing enzymes into decaying material and then absorbing the resulting nutrients. Because they produce airborne spores, the molds must be cultured in a room separate from other labs so as not to contaminate other experiments. Working alongside Ming-Yueh Wu, both mold species were tested on the bio-composites; both were able to grow on these materials, provided there was enough moisture. We templated their growth using acrylic templates into which the nutrient agar was cast to start the fungal cultures and then we pressed a thin sheet of chitosan-cellulose composite onto the template. The fungus then grew according to the pattern on to the bioplastic for a time, eventually eating holes into the material; in one sample it began to over-run the pattern, in another it lost moisture and stayed confined to its original pattern location. (Figure 13) Co-culturing the fungi was also attempted over the span of a month onto sterile large wooden blocks that had been milled to increase the surface area for colonization. (Figure 14) While the *Aspergillus* colonized the block within a week, the *Trichoderma* needed close to three weeks before showing visible green areas. The growth pattern was interesting in that although the *Aspergillus* and the *Trichoderma* were plated into different valleys of the wood block, the growth showed them mixing throughout the block, with the *Trichoderma* also taking over the flat surfaces. While one of the blocks showed evidence of a third species of mold on it (white mold), both the *Aspergillus* and *Trichoderma* were robust enough to withstand the contamination and to continue to flourish. (Figures 8 and 14)

Coloring with *Streptomyces*

Streptomyces are a genus of filamentous bacteria that includes over five hundred species characterized by a fungal like mycelium body and the production of spores for reproduction. (Figures 15, 16) Found predominantly in soil and decaying vegetation, they are a major source of antibiotics and produce geosmin, a metabolite that gives soil its characteristic earthy rain smell. Many species in the genus also produce vivid pigments as metabolites. Natsai Audrey Chieza, Ginkgo's

Figure 10. Close up of pectin-chitosan film enzyme induced degradation. Photo by Andrea Ling.

Figure 11. Pectin-Chitosan-Cellulose 'paintings' during fabrication process, using two different degradation methods. Left image shows the first painting before using a laser cutter and dehydration to subtract material from the object. Right image shows the second painting using enzymatic degradation to subtract material from wet colloid. Photos by Andrea Ling.

Figure 12. Pectin–Chitosan-Cellulose 'paintings' after 20 days, using two different degradation methods. Top right is first painting after using a laser cutter and dehydration to subtract material from the object. Bottom left is the second painting after using enzymatic degradation to subtract material. Photo by Ally Schmaling.

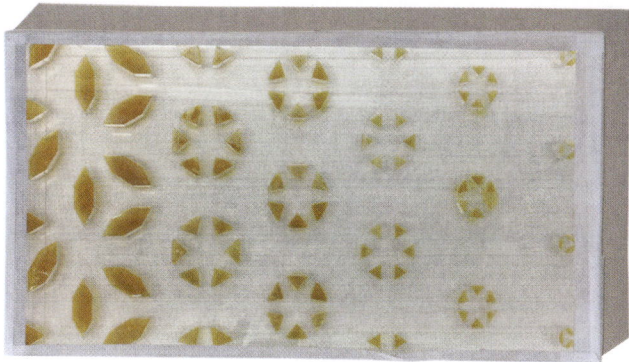

Figure 13. Acrylic templates inoculated with *Aspergillus niger* and then a chitosan-cellulose film is sandwiched into the template as a subtractive fabrication mechanism. Photos by Andrea Ling and Ally Schmaling.

Figure 14. Close ups of wood blocks after five weeks incubation: two are heavily contaminated with unknown species, two have been successfully colonized with *Aspergillus niger* and *Trichoderma viride*. Photos by Ally Schmaling.

first creative resident, used *Streptomyces coelicolor* to naturally dye silk by culturing the bacteria directly onto the textile in different ways to produce different patterns.[11] Ginkgo provided me with different *Streptomyces* cultures to test on my bioplastics to investigate whether they could be cultured to dye these materials, transforming them while decomposing them. Working alongside Kyle Kenyon, we successfully cultured *S. coelicolor* in sterile conditions onto plain cellulose, thin wood substrates and small chitosan-cellulose samples. (Figure 15) However, it was exceptionally challenging to get any *streptomyces* growth onto larger samples of the composite bioplastics. While we knew bacteria could produce cellulases and chitinases and feeding on these materials, sterilizing the composites without damaging them or producing toxic residue was an issue. Specifically, autoclaving dried materials resulted in burning most of them, making them inhospitable to the bacteria. Filter sterilizing the viscous wet colloids was not feasible as the solutions were too thick. Ultraviolet sterilization was only partially effective given the opacity of the material and gas sterilization was not available at the lab. A cocoon structure was made of the chitosan and cellulose colloids, which we planned to culture with bacteria, recognizing that the perimeter edges of the cocoon strips were burnt from the laser cutter and therefore unlikely to be colonized, but that the interior material was still viable. The initial strip tests incubated over thirty days, resulting in the growth of a white mold; however, there was no evidence of the *streptomyces*. (Figures 17 and 18) Attempting to grow the bacteria on thicker wood samples, sterilized intact in the autoclave, also resulted in heavy contamination. In one instance, the resulting wood block smelled so vile it had to be discarded. In another, a white mycelium-like texture resulted and in yet another, evidence was found of a fuzzy brown grey contaminate on top. Herein lies some of

Figures 15. *Streptomyces coelicolor* cultures on cellulose, inoculated with words "Hello Ginkgo" and "Hello World". *Streptomyces coelicolor* cultures on three different species of wood (1.6 mm thick) and chitosan-cellulose film. Photos by Andrea Ling.

the irony of doing this work in the controlled setting of a biology lab rather than simply burying samples in exterior soil. These were wild strain bacteria directed to do what they naturally would do outside, on a dead tree stump, decaying leaves, or rotting fruit, with other micro-organisms present but without outcompeting them. However, inside a lab's highly artificial conditions, which demand sterilization for growing predominantly monocultures which are then sensitive to contamination, this control seemed to hinder rather than enable the growth of this bacteria. Interestingly, other species were able to flourish instead.

Decay – By Design or By Chance?

As someone who wants to design the decay of artifacts to occur in a specific way, I have to reflect on whether these accidents are just as effective as planned forms of decay, and what my priorities are for these artifacts as they change—are they aesthetic, sensorial, or programmatic? These factors must all be weighed, highlighting how bio-designers and scientists are called upon to decide when it is necessary to guide biology with a firm hand and when it is better to let go. As a classically trained architect I am used to prescribing not only the aesthetic quality of a work but also the performance and sequence of how things are assembled, sometimes to submillimeter tolerances. It is very difficult to relinquish control and outcomes to these natural partners. But, in the end, all fabrication tools including industrial ones produce results that are always an approximation of the original design. The most interesting results are produced from the interplay of intent and outcome.

Enzymatic degradation was the process that offered the most industrial and potentially precise method of control over where and when degradation occurred; it also felt the most mechanical and least symbiotic of the three methods used. This is because the enzymes, while derived from some biological specimens, were used in artificial isolation, separate from the living organism and thus without vitality or agency. Working with bacteria provided a recognizable design space for mediating colour and growth patterns while introducing smell as a design element. However, it was much more challenging to work with bacteria than enzymes and the evidence of macro-scale decay was almost non-existent in the samples. The co-cultures of fungi provided the most effective means of decay as the fungi were robust enough to withstand threats of contamination and could grow on different media more easily. However, their appearance remains partly contested, traditionally viewed as materials to be avoided because their presence indicates death and rot.

Future work for this project could include the same tests conducted outside in a variety of different in-situ conditions and a comparison of how each situation guides how and when decay might take place. Biology is often a black box, yet for the purposes of these tests it is not always necessary to know exactly what is going on at

Figure 16. Different *Steptomyces griseoviridis* cultures in various patterns, as formed by spatially patterning antibiotics, on different species of wood. Photo by Ally Schmaling.

Figure 17. Cocoon for biotransformation through bacterial colonization made of laser cut chitosan–cellulose composite, with drawing diagramming potential transformation stages. Photos by Ally Schmaling. Drawing by Andrea Ling.

every level so much as to know that under 'this' condition, 'that' will happen. It is very easy to forget this when immersed in the reductive lens of lab work wherein when trying to control all variables, one often loses sight of the overall goal.

Biology as Value Creator

In her book *The Value of Everything: Making and Taking in the Global Economy*, economist Mariana Mazzucato argues that in our current global economy value has been misplaced onto systems that extract commodity out of resources and that the economy needs to be reformed to reward systems that, instead, create value.[12] Biology is a value creator, the ultimate up-cycler, using death and garbage as fuel for new life. It provides sustained renewal even as its material processes are often seen as inconvenient, challenging, and possessing formal qualities that are difficult to standardize, unpredictable, and uncontrollable. Biological materials need maintenance. And where in an extractive capitalist system, unceasing consumption is required for ever increasing economic growth, a biological system has limits. It

Figure 18. Test of laser cut chitosan–cellulose strips inoculated with *Streptomyces coelicolor* which failed to grow and heavily contaminated with unknown mold species. Photo by Ally Schmaling.

cannot be endlessly mined without any provision of return, for without nurture and restoration biological systems will collapse. Biology's currency is energy and matter, of which in closed loops there is a finite amount. Its efficiency is metabolic, not economic and not lean, and based on redundancy, leakiness, and transience. But if tended to, biological systems can provide in return a far more robust system of growth, renewal, and system longevity than extractive systems.

Design by Decay, Decay by Design does not develop a product, but instead outlines a process by which we can create artifacts where decay is built-in, not only as a means of disposal but also as a potential fabrication process transforming objects in ways that are still acceptable to users. The development of this process seeks not to make a series of artifacts out of shrimp shells and jam that are easy to dissolve or that ensure zero-carbon footprints. Rather it seeks to support an alternative mode of design practice where the process of making and breaking things is not only consumptive but provisional.

By rethinking the nature of waste as something that is both unavoidable and unacceptable, *Design by Decay, Decay by Design* seeks to redistribute value from permanent objects that destroy ecosystems to transient ones that restore them, finding epistemological as well as practical value in designing responsivity, degradation, and renewal into human-made objects. Ginkgo Bioworks, by virtue of being a company that designs life, understands how in synthetic biology value comes from working in symbiosis with the underlying logic of natural systems rather than in trying to subjugate them. This project asks not what I can make biology do to help me, but rather how I can bias it such that it benefits the system we exist in, which includes both me and it. The struggle in such a design practice is to learn to accept the tensions embedded in this mode of practice, where material and biological agency sometimes work in contradiction to what is planned or comfortable. Industry struggles with this as well. However, if we accept this inconvenience, using decay to facilitate renewal offers extraordinary advantages. This includes access to circular systems and the ability to grow, adapt, and reproduce literally out of what rots, encouraging a form of resilience not found in industrial systems. By taking this approach, I aim to shift design away from extractive paradigms with very narrow definitions of efficiency towards softer, more holistic ones that first address ecological consequences to take care of social and economic ones.

> Nothing can dwindle to nothing, as Nature restores one thing from the stuff of another, nor does she allow a birth, without a corresponding death.
>
> Lucretius, *On the Nature of Things* [13]

Acknowledgments

With thanks to Ginkgo Bioworks, Faber Futures, Team Aguahoja, The Mediated Matter Group, MIT Media Lab.

Collaborators

Curatorial Team: Ginkgo Bioworks + Faber Futures, Natsai Audrey Chieza, Dr. Christina Agapakis, Grace Chuang, Kit McDonnell, Dr. Joshua Dunn. Scientific advisors: Ginkgo Bioworks, Dr. Joshua Dunn, Dr. Ming-Yueh Wu, Kyle Kenyon, Duy Nguyen, Dr. Lucy Foulston. Photos by Ally Schmaling, Andrea Ling, and Grace Chuang.

Funding

Funding support received from Ginkgo Bioworks.

ENDNOTES

1. Luis Fernandez-Galiano, *Fire and memory: on architecture and energy* (Cambridge, Mass: MIT Press, 2000), 6.

2. Pettigrew C. Callistemon, A. Weiler, A. Gorbushina, W. Krumbin and R. Weiler, J. Living pigments in Australian Bradshaw rock art. *Antiquity Project Gallery, 84* (326) (December, 2010). Retrieved from http://antiquity.ac.uk/projgall/pettigrew326/

3. Morris Hicky Morgan (trans), *Vitruvius: The Ten Books on Architecture* (Harvard University Press, 1914).

4. Kiel Moe, *Thermally Active Surfaces in Architecture* (New York: Princeton Architectural Press 2010), 28–29.

5. Andrea Ling, "Design by Decay, Decay by Design" (master's thesis, MIT, 2018).

6. For chitin see, Ibrahim Makarios-Laham and Tung-Ching Lee, "Biodegradability of chitin- and chitosan-containing films in soil environment," *Journal of Environmental Polymer Degradation* 3, no. 1, (January 1995): 31–36. https://doi.org/10.1007/BF02067791. For biodegradable chitosan and cellulose plastics see, M. Nishiyama et al., "Biodegradable Plastics Derived from Cellulose Fiber and Chitosan" *Advances in Chemistry* 248, *Hydrophilic Polymers* (1996) 113-23, https://doi.org/doi:10.1021/ba-1996-0248.ch007 ; Sebastian Kalka et al., "Biodegradability of all-cellulose composite laminates," *Composites Part A: Applied Science and Manufacturing* 59 (April 2014): 37–44. https://doi.org/https://doi.org/10.1016/j.compositesa.2013.12.012

7. Jorge Duro-Royo, Laia Mogas-Soldevila, Daniel Lizardo, & Neri Oxman, "Designing the Ocean Pavilion: Biomaterial Templating of Structural, Manufacturing, and Environmental Performance," *Proceedings of the International Association for Shell and Spatial Structures Symposium Amsterdam,* (2015); Laia Mogas-Soldevila, "Water-based digital design and fabrication: material, product, and architectural explorations in printing chitosan and its composites" (master's thesis, MIT, 2015).

8. Joshua Van Zak, Jorge Duro-Royo, Yen-Ju Tai, Andrea Ling, Christoph Bader, & Neri Oxman, "Parametric Chemistry: Reverse Engineering Biomaterial Composites for Additive Manufacturing of Bio-cement Structures across Scales," in *Towards a Robotic Architecture,* edited by Andrew Wit & Mahesh Daas (Oro Editions, 2018); Jorge Duro-Royo, Joshua Van Zak, Andrea Ling, Yen-Ju Tai, Barrack Darweesh, Nicolas Hogan, & Neri Oxman, "Designing a Tree: Fabrication Informed Performative Behaviour," *Proceedings of the IASS Symposium 2018,* edited by C. T. Mueller & S. Adriaenssens (Cambridge: MIT, 2018).

9. Yen-Ju Tai, Christoph Bader, Andrea Ling, Jean Disset & Neri Oxman, "Designing Decay: Parametric Material Distribution for Controlled Dissociation of Water-based Biopolymer Composites," *Proceedings of the IASS Symposium 2018* (Cambridge: MIT, 2018).

10. David Pye, *The nature and art of workmanship* (London: Cambridge University Press, 1968).

11. Natsai-Audrey Chiesza, FaberFutures, *Project Coelicolor,* 2018. Retrieved August 20, 2020, from https://faberfutures.com/projects/project-coelicolor/

12. Mariana Mazzucato, *The Value of Everything: Making and Taking in the Global Economy* (New York, NY: PublicAffairs, 2018).

13. Rolf Humphries (trans), *Lucretius, The Way Things Are, De Rerum Natura of Titus Lucretius Carus.* (Indiana University Press, 1968), 262.

MYCELIUM MATTERS:

PROTOTYPING INTEGRATED MATERIAL LIFE CYCLE DESIGN TOWARDS GENERATIVE JUSTICE GOALS

MAE-LING LOKKO

ABSTRACT

Mycelium—the vegetative part of fungi—has gained acceptance as an effective biomass-to-biocomposite material technology. Feeding on tightly bound lignin and cellulose components of organic matter, mycelium generates resilient chitin-bound green building materials. In the deep past of material systems, fungal organisms have participated in hydrocarbon and carbohydrate material economies. Yet, only in the last two decades has contemporary design explored fungal technologies as a form of modern eco-manufacturing in service to the built environment. New forms of mycelium production make possible more ecological life cycles that capture underutilized biomass feedstocks to meet twenty-first century building demands. Only a small subset of select high-quality biomass resources are, however, at present used to this end. In response, this paper presents a new production paradigm for biomaterial life cycle design whose mycelium technologies facilitate the planning of biomass material flows, distributed and local supply chain design, on-site biomass-to-biocomposite manufacturing, and green-collar workforce planning. Three architectural installations are discussed which explore mycelium eco-manufacturing at different scales. The first, *Carbocycene*, investigates circular material design. The second, *Hack the Root*, investigates mycelium eco-manufacturing at the scale of a large-museum space using deployable room-scale grow chambers. The third, *Agrocologies*, operates at domestic scale of the home, leveraging the kitchen as a landscape for the design and exploration of waste transformation and biomaterial integration. In all three projects, designer and material performance evolve to meet identified generative justice criteria that include the bottom-up creation, translation, and circulation of value across integrated material life cycles in mycelium eco-manufacturing.

+ mycelium
+ agro-waste
+ food waste
+ eco-manufacturing
+ generative justice

Figure 1. Detail of *Hack the Root* panel. Photo credit Mae-ling Lokko.

174

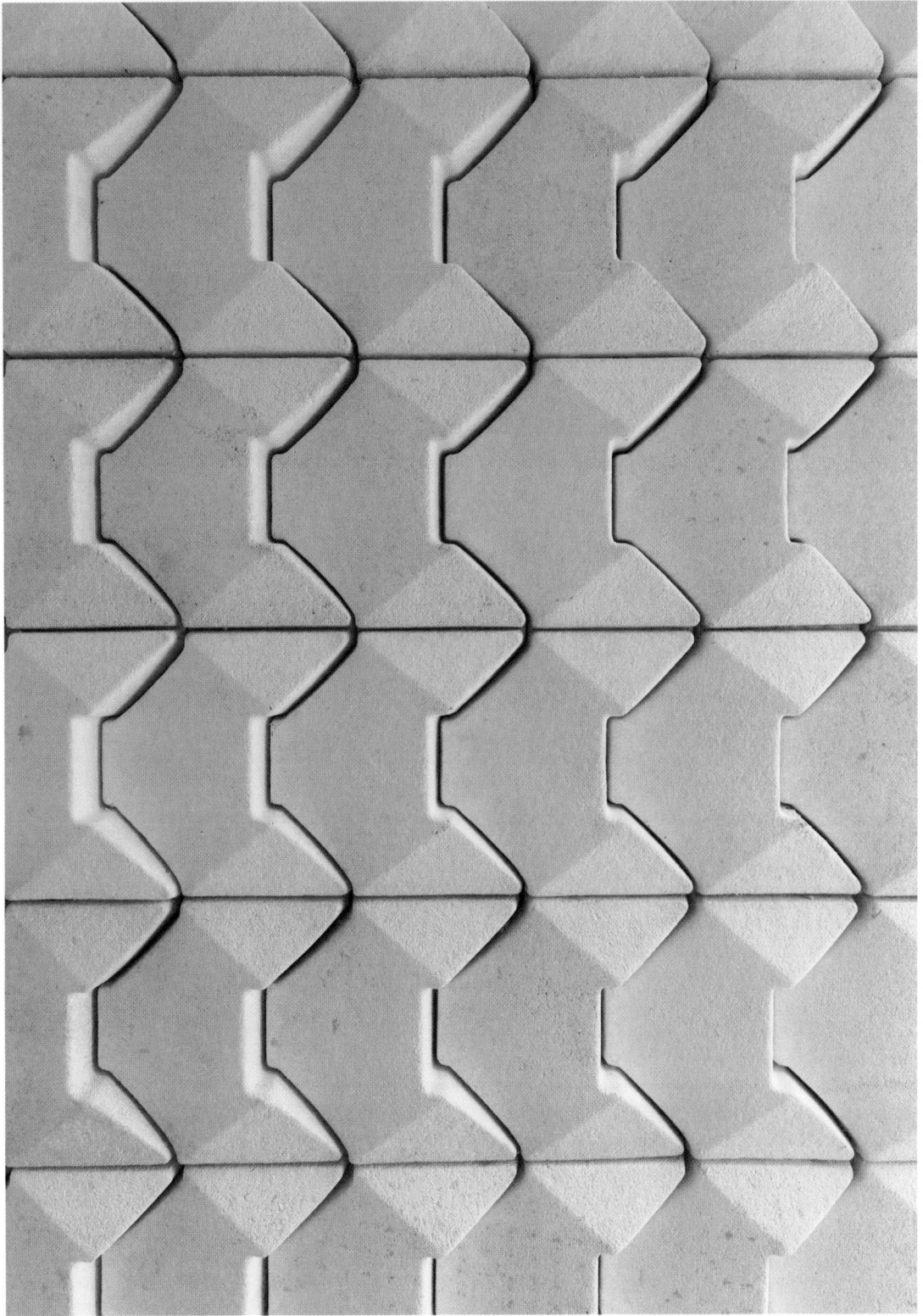

Generative Justice: Biomaterial Pathways to New Economic Models

Over the last three decades, renewable biomaterial streams have gained significant interest and visibility within the building sector. These developments have primarily been driven by quantitative clean-technology goals that champion the reduction of energy and resource inputs alongside lower life cycle emissions. However, innovation is increasingly driven by qualitative goals linked to improved human health, well-being, and worker productivity.[1] Attempts to 'quantify the qualitative' impacts of renewable biomaterials have been useful in translating intangible values into business frameworks, even if their widespread adoption has been limited by lack of affordability, limited access, and deep-seated negative sociocultural and psychological perceptions of renewable biomaterials.[2] While the first two barriers to adoption can be largely addressed through capitalist economies of scale—with better supply chains and distribution networks—the third remains a persistent barrier to reintegrating renewable materials into the built environment. Overcoming negative perceptions associated with renewable biomaterials would involve not only their physical integration within buildings but the entire redesign of the material economy of building, including how value is extracted, stored, circulated, and lost.

More critically, the relationship between affordability, access, and sociocultural barriers in renewable biomaterial chains is currently defined by an economic value system that is designed to alienate the very mechanisms that return value to its generating components. Value, whether in capitalist or socialist systems—the two dominant economic and political models of the twentieth century—is driven by top-down extraction. This alienates value in three ways: first, when the environment undergoes continuous extraction and contamination that results in the degradation of environmental health, over time; second, when alienated biomass resources are created in the form of lignocellulosic by-products from agricultural, forestland, rural and urban food waste systems and downcycled, often prematurely, through combustion or landfills; and lastly, when human labor groups who play inextricable roles in value supply chains do not receive fair returns for their services and are unable to advocate for their participation within value chains.

Differently, generative justice—characterized as bottom-up circulation of unalienated value—affords a fundamentally different approach.[3] Contrary to capitalist biomaterial supply chains that are designed to extract value from the land and to centralize and control production and distribution networks, with minimal transfer of value to labor groups and consumers, the proposed myco-cycling systems discussed in the following three architectural research projects are conceptualized as generative justice economies designed to maximize, translate, and circulate unalienated value.

Distributed Myco-cycling Systems across Community, Urban, and Domestic Scales

The life cycle of all matter in the form of organic systems (including plants and animals), as well as inorganic systems (including mineral, rock formations, and buildings), has essentially been sustained by the presence and activity of fungi.[4] Inasmuch as fungal activity has historically served as the ubiquitous agent for the decomposition of dead matter, it has played a major role in enabling living matter to embark on new conquests.[5] In agricultural and food sectors, the presence of fungal activity has primarily been considered as part of the cycling of carbon, given its role in enhancing photosynthetic plant systems, in the robust maintenance of soil structure, and in its rapid enzymatic decomposition of matter.[6] However, in the last two decades design has capitalized on mycelium conversion of organic lignocellulosic feedstocks to propel a new form of modern eco-manufacturing in the built environment. This now includes biobased packaging, insulation, and biocomposite building material applications.[7]

Because mycelium—the vegetative part of fungal organisms— feeds on tightly-bound lignin and cellulose components in biomass waste to generate resilient chitin molecules in non-toxic bioproducts, its ability to 'grow' materials under low-energy, non-toxic conditions offers new avenues for eco-manufacturing.[8] Today's leading mycelium commercial eco-manufacturing enterprises such as Ecovative Design (USA), MOGU (Netherlands) and Mycoworks (USA) are leveraging mycelium's enormous enzymatic activities and filamentous growth to develop robust, time-efficient grow cycles for target applications by matching mycelium-biomass substrates to agro-waste substrates, nutrient additives, and moisture levels. Further development has involved the optimization of proprietary engineered "molding systems" for building material applications.[9] However, due to the high degree of environmental controls required, such models of eco-manufacturing mycelium-based bioproducts rely on centralized production and the optimization of single-stream agro-waste biomass processing for target applications.

Extending the important technological developments of such current mycelium manufacturing practices, the following three projects explore the future of eco-manufacturing as a highly spatialized, temporal infrastructure in which an on-demand citizen workforce participates in transforming diverse waste streams where they are produced. While all three projects leverage contemporary art, architecture, and design platforms to realize architectural prototypes, all are linked to ongoing interdisciplinary research focused on the formal development of biocomposites from lignocellulosic waste, on emerging bioadhesives, and on the design of industrial, social, and economic infrastructure for the future of distributed eco-manufacturing production.

Carbocycene: Prototyping Community-scale Myco-production for Distributed Biomass Resources

Carbocycene was a design research project at Atelier Luma in Arles, France that focused on a distributed, community-scale model for the transformation of a variety of lignocellulosic waste products using mycelium biobinders.[10] Founded in 2016, Atelier Luma is a think tank, production workshop, and learning network of the Luma Foundation, committed to co-developing new ways of producing and caring for cities and bioregions, using design as a tool.[11] The design research institute is situated within the Camargue region, a unique geological and wetlands landscape covering 220,000 hectares in southeastern France. Bounded by the Alpilles limestone mountains and scrub forest to the North and the salty wetlands of the Camargue in the delta of the River Rhone to the South, the diverse range of biomass feedstocks that originate in the area serve as a broad lignocellulosic inventory for myco-composite development. Today, diverse agricultural activities occur in all twenty-nine communes within the larger Arles region. They employ approximately fifty-eight percent of the population and are coordinated legally by a united 'territorial coherence scheme' which evolves its sustainable development goals in terms of habitat, environment, spatial organization, displacement, and commercial activities.[12]

Given the agricultural and socio-political background of this region, a critical goal of *Carbocycene* is to develop a small architectural-scale pilot project that engages distributed stakeholders in a diverse flow of biomass feedstocks, continuously generated from agricultural, forestland, rural, and urban food waste systems. While current global models for the eco-manufacturing of mycelium-based bioproducts have relied on the optimization of single-stream agro-waste biomass, this project expands its sourcing to include biomass from food waste streams that originate in commercial food and household wastes, in biomass from invasive species that proliferate across wetlands, and in farming environments that generate up to two kilograms of dry lignocellulosic material per square meter.[13]

The myco-transformation of waste materials into low-density building materials and kitchenware offers more favorable pathways to bioconversion, compared to use of local food waste as farm compost or in the upcycling of invasive species biomass into paper. In the case of composting and open-air combustion, the transformation of biomass into myco-composites offers upcycling pathways that sufficiently capture a range of value-generation applications (packaging, acoustic absorption, and foam-based products), otherwise lost in premature downcycling processes. Such applications are critical in displacing high-demand petroleum-based polymers which are widely available, socially accepted, and economically scaled to benefit capitalist business models. Compared to the processing of biomass-to-paper products in decentralized operations, cellulosic pulp extraction requires a water intensive infrastructure that takes typically more than seven days to break the lignin bonds and to retain cellulosic bundles. And if the quality of cellulosic materials is poor, paper production requires the addition of glue additives, followed by labor-intensive pressing. The main advantage of decentralized mycelium degradation over paper processing is that lignin bonds are broken down by mycelium organisms when they generate strong chitin molecules under ambient temperatures (21-24 °C), where high humidity is created by mycelium respiration, and when the whole is maintained in enclosed growing environments over a period of five days. Furthermore, the economic

Figure 2. Local Mushroom foraging at Tour du Valat Research Institute and conservation site with Atelier Luma team, researchers from INRA/Aix-Marseilles University. Photo credit Mae-ling Lokko.

profits per unit of biomass inputs into mycelium composites is greater than for paper products per unit of input mass.

A broad biomass feedstock inventory was identified when visiting farms within the Camargue region and the natural conservation park Les Marais du Vigueirat (where a pilot project for harvesting invasive species was located). (Figure 2) A critical aspect of these biomass waste harvesting operations—also observed in farming and biomass upcycling enterprises in Brazil and Ghana—is their demonstrated capacity to increase the number and range of green-collar jobs for members of society who face social and professional barriers to employment. These programs are valued for their social integration potential, providing technical training and financial opportunities for individuals who seek to enter the job market. In our case, by understanding the relationships between food cooperatives and waste management stakeholders, the biomass inventory was narrowed down to respond to the volume and patterns of biomass waste streams from kitchens (daily), farms (seasonal), and commercial project partners.

The types of mycelium-based components designed, grown, and fabricated at Atelier Luma in Arles included a vertical wall panel made from well-characterized bulk agro-waste harvested from proprietary mycelium strains in custom grow trays, as well as a series of storage, kitchenware, lighting, seed and plant holder objects made using experimental myco-composites that originated in food waste and invasive species biomass (ludwigia, baccharis hamilifolia). Inspired by innovations at the intersection of mycelium and food life cycles, the project's name *Carbocycene* translates into 'carbohydrate-kitchen' and centers on communal scale enterprises. This synergistic project was designed as a multi-component mycelium architectural installation destined for Collaborative Kitchen, a commercial food enterprise in Griffeuille, France. Located a mile from the Luma Foundation, Griffeuille is home to a diverse North African, Arab, and French working-class community. The base mycelium panels were produced in environmental chambers the size of a room and under ambient conditions at Atelier Luma's facility, while the diverse experimental myco-composite forms will be produced in middle school workshops using everyday objects and kitchenware from Collaborative Kitchen. (Figure 3)

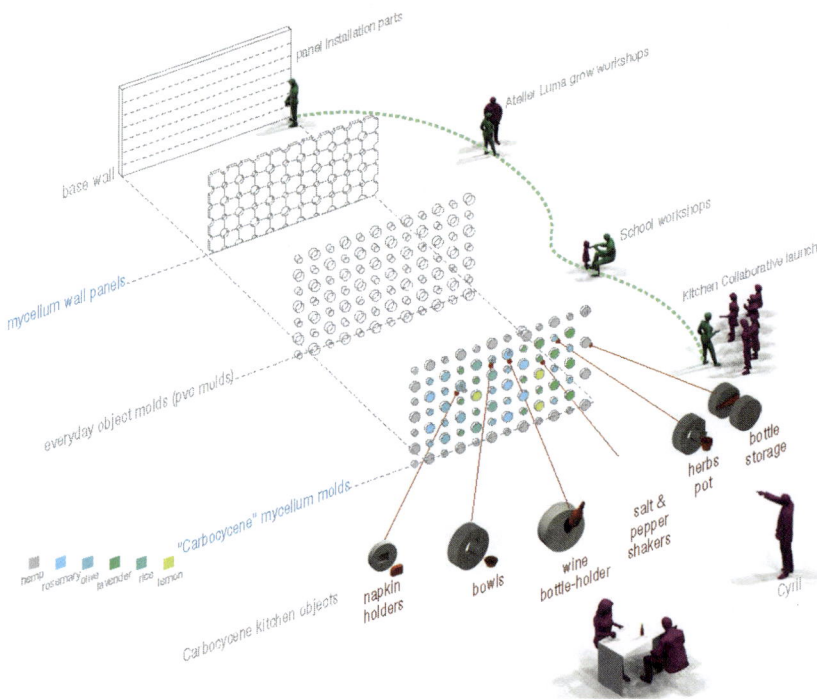

Figure 3. Schematic of Collaborative Kitchen Pilot Project Integrating a Range of Myco-composite Products from Agro, Food and Invasive Species Biomass. Drawing by Mae-ling Lokko.

In collaboration with their chefs, and as part of the education and capacity building aims of the project, a cooking workshop was included in the project's initial scope. The *Carbocycene* menu was conceived to align the design of meal ingredients with the type and volume of ingredient waste needed in the production of the myco-composite ware to be used in the kitchen. Future project goals include identifying a variety of local mycelium strains for degradation of identified biomass streams in collaboration with l'Institut national de la recherche agronomique (INRA) and Aix-Marseilles University. And lastly, in collaboration with Rensselaer Polytechnic Institute (RPI), the project develops the mechanical optimization of myco-composites made from experimental substrates and the scaling up of applications to urban structures meant for shelter.[14]

In developing and characterizing the mechanical and thermal performance of myco-composites generated from local biomass inventories, *Carbocycene* seeds an open-source myco-production framework for lignocellulosic materials. The project links the large inventory of lignocellulosic waste to component ratios and myco-board performance and showcases a critical case study in the mycelium transformation of cellulosic and lignin dominated materials. Biomass feedstocks are a widely available, dynamically spatialized, and diverse resource stream, yet any distributed mycelium enterprise will face seasonal shifts in biomass raw material availability, quality, and quantity.[15] Thus, the long-term goals of this myco-production framework is to provide a knowledge ecosystem to support elastic and flexible value chains for currently unpredictable biomass resourcing.

Hack the Root - Activating Onsite Mycelium Eco-manufacturing and Distributed Urban Production

Commissioned as part of the Liverpool Biennial 2018—the largest contemporary arts festival in the UK—*Hack the Root* was developed at RIBA North, the Royal Institute of British Architect's national architecture center.[16] Liverpool was a globally strategic trading port city during the eighteenth and nineteenth centuries, built on the rise of centralized financial institutions that funded the extraction and exploitation of valuable agricultural and labor commodities as part of the transatlantic trade. In response, the exhibition *Hack the Root* explored the possibility of, and capacity for, decentralized and grassroots production within small material footprints. Intent on 'hacking' contemporary centralized, toxic models of materials manufacturing, *Hack the Root* explored distributed infrastructures and networks for growing biobased materials. Onsite growing, production, and installation of a mycelium 'tunnel' structure was central to the exhibit experience. (Figures 4 and 5)

In addition, located within a two-mile radius of RIBA North, a series of grow-it-yourself (GIY) workshops were staged at Squash Nutrition, a creative food enterprise run by a diverse group of local urban horticulturists and contemporary arts community. Responding to similar interests from Life Sciences UTC (a local high school in Liverpool), another workshop was organized at Farm Urban, a startup company that develops high-tech urban infrastructures for indoor farming. Over the course of four weeks, four workshops enlisted the participation of approximately 200 volunteers from Windsor Primary School, Life Sciences UTC High School, University of Liverpool, as well as from the Liverpool Biennial volunteer community.

During each workshop, participants 'inoculated' hemp-fed mycelium into 'grow trays' in the shape of the designed *Hack the Root* wall panel. Growing protocols and instructions were communicated at each of three rotating stations: the first was focused on inoculation of the

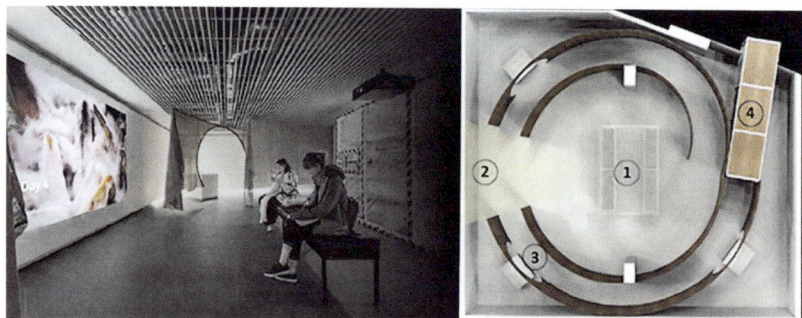

Figure 4. Schematic View of *Hack the Root* at RIBA-North Gallery showing (1) grow chamber (2) educational film (3) mycelium life cycle models (4) entrance tunnel with grown mycelium panels. Photo credit Thierry Bal and Mae-ling Lokko

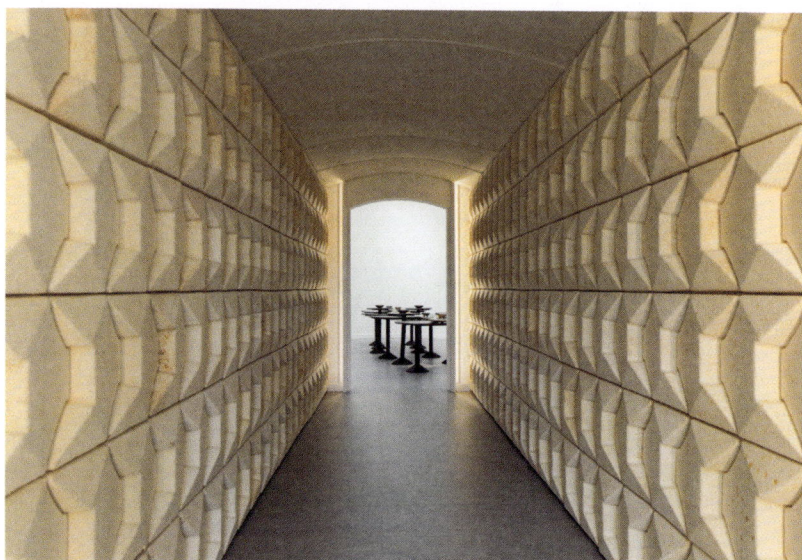

Figure 5. *Hack the Root* entrance tunnel with grown mycelium panels. Photo credit Selma Gurbuz.

mycelium to grow food led by Farm Urban; the second was associated with the design of mycelium panel-inspired geometries led by local artist Nina Edge; the last station was tasked with inoculation of the mycelium to grow *Hack the Root* panels. At this third station, participants broke down pre-inoculated hemp-mycelium wet substrates and added additional 'food' nutrients in the form of plain flour to aid with fungal growth. While this GIY methodology was originally developed by Ecovative Design to grow object-scale mycelium products in workshops and for shipping dried agro-waste and mycelium substrates to consumers interested in growing this at home, *Hack the Root's* goal was to scale up the methodology to an onsite room-sized project within the space of the gallery and that engaged a distributed, diverse workforce to participate in its production. Growing protocols typical of a lab or factory were transferred to an indoor yet public environment. (Figure 6)

Following the workshops, the grow trays were transported back to the RIBA North gallery where they were kept in two (4m x 3m x 2.25m) grow chambers over a four to six day grow cycle. The public could view the harvesting and drying of the mycelium panels in real time. During the incubation period in the gallery, high humidity and temperature (21–24°C) were maintained by covering the grow trays in the chamber and by regulating indoor ambient conditions. After successful colonization of the substrates was observed in the grow trays, the entire mycelium biomass mixture was removed from the trays and placed in a closed humid environment where 'overgrowth'—a process that primarily tar-

Figure 6. *Hack the Root* Workshop, Squash Liverpool. Photo credit Mae-ling Lokko.

gets fungal growth only on the surface of the panel—took place. The overgrown forms were then air-dried in place over a twenty-four to forty-eight hour drying cycle to prevent further fungal growth within the substrate.

Commercial mycelium eco-manufacturing relies on centralized space and energy infrastructure to perform industrial scale substrate mixing, for refrigerating large batches of inoculated agricultural substrate mixtures (kept at 0 to 2°C), and for industrial scale drying at temperatures over 80°C. This project, however, identified under-capitalized infrastructure within the city to provide such services. Unanticipated opportunities were identified amongst food and building sector businesses for sourcing mechanical services for production. Storage of the mycelium-substrate raw materials—typically required prior to a workshop—took place in a large-scale transport and food storage facility, and at inexpensive food storage rates. Project collaborator Dr. Vincent Walsh, facilitated the assembly, design, and integration of airflow control devices within a room-scale mycelium grow chamber, by deploying techniques more typical of companies charged with the removal of building asbestos. However, compared to the stringent air tightness and waste management protocols required during onsite removal of asbestos, the assembly and operation of the grow chambers was economically and efficiently done by a local company, JB-Monitor Asbestos Solutions. Lastly, while a significant percentage of mycelium panels were air-dried onsite, a large batch dried over a weekend using waste heat from an industrial terracotta oven from a facility in Liverpool. (Figure 7)

Following the drying of the first batch, a high degree of variability in mycelium panel strength and surface finishes was observed.

Figure 7. Installation view of *Hack the Root* panels. Photo credit Mae-ling Lokko.

Panels that were made from multiple substrate batch mixes during workshops tended to grow unevenly, whereas those that were made by individually controlling the mixing of flour and wet substrate were generally uniform. As the quality control of panels was dependent on standardization of the growing protocols across a wide range of participant ages and experience, secondary monitoring and quality assurance checks between days two and four were extremely effective measures for ensuring the quality of subsequent batches.

Hack the Root serves as an important prototype for exploring the phases and requirements of on-demand human workforce management within mycelium-biomass life cycle planning, given the scale of urban participation and onsite-growing logistics it deployed. Contrary to current forms of on-demand supply chain storage such as e-warehouses, or on-demand services such as Uber, the study of distributed myco-production enterprises encourages on-demand and 'on the ground' workforce coordination as an inextricably quality of the final product.

Agrocologies - Mycelium Upcycling in Domestic Rituals at the Scale of the Home

Agrocologies was one of five prototypes designed and built as part of the European competition Housing the Human (HtH), organized by the Forecast Platform in Germany and in partnership with five cultural institutions in Germany, Denmark, Italy, Istanbul, and Belgium.[17] Co-funded by the Innovation Program within the Federal Office for Building and Regional Planning (BBR) of Germany and a number of European funding agencies (Creative Europe Programme of the European Union, Future Architecture), HtH focused on examining future scenarios for dwelling through design research. In response, Agrocologies developed a platform for exploring how upcycling can be activated and integrated across the intimate scale of the home. Across four public design-research presentations held in different cities, the idea of 'alienated value' was conceptualized in each HtH host city. The domestic urban dweller was the subject of the alienation, as their role is largely undercapitalized in terms of onsite transformation and management of domestic waste. Addressing widespread punitive systems used to enforce domestic waste management, such as municipal fines and withdrawal of service, the project investigated how the processing of domestic scale waste could be driven by new forms of domestic rituals gathered around food waste.

At its first public presentation during the 4th Istanbul Design Biennial, Agrocologies took the form of an in-situ oyster mushroom installation grown using Turkish coffee waste from the Studio-X gallery. In collaboration with Ghanaian chef Selassie Atadika and chefs from GastronoMetro culinary academy, interactive meals were held with the public which fostered conversations around changing cooking rituals, kitchen infrastructure, and ingredients. Here the mycelium production infrastructure for domestic lignocellulosic waste evolved to incorporate not only custom grow trays that could be integrated into horizontal tabletops and vertical walls, but everyday objects that were largely 'dormant' in household operations.

Relative to agro-waste, food waste management is more challenging due to its inherent chemical and material heterogeneity, variability in supply consistency, and the complex pretreatment that is often required.[18] When compared to physicochemical, chemical, and biological pretreatments used to treat food waste after it leaves the household and enters centralized treatment facilities, the approach in *Agrocologies* was to use domestic equipment to treat food waste at the source and, in so doing, reducing barriers to the practice. A simple protocol involving grinding, boiling, and oven drying the food waste directly in the kitchen eliminated numerous problems that otherwise results from its high water content—problems such as unwanted biological decomposition, odor emissions, high mass, bulk density and a higher frequency of waste collection.[19] Instead, a variety of lignocellulosic food waste that included citrus, banana, onions, and garlic peels were milled, pasteurized, and integrated into agro-waste mycelium substrate mixtures for growing myco-boards that were optimized for improved mechanical performance. Due to the consistent quality and high mechanical strength of myco-boards made with different ratios of lemon peels, this substrate was used to prototype a new catalogue of bio-based, compostable household objects. Pairs of household ware from the IKEA catalogue including buckets, vases, bowls, and cups, were used to form two molds to generate a range of derivative household 'mycelium-ware'. An initial inventory of mycelium-ware was displayed at the Copenhagen Architecture Festival (CAFx) accompanied by a public bartending performance which featured cocktail ingredient waste used in the displayed mycelium ware. (Figure 8)

The final installation of *Agrocologies* at Radialsystem in Berlin staged the real-time performance of kitchen rituals within an urban apartment setting. The ecologies of kitchen and dining room were chosen as interior landscapes for experiencing multiple forms of upcycling, including practices that rethink kitchen infrastructures of waste processing and in-situ growing of myco-composites for household wares. A diurnal cycle included two morning rituals for the preparation of coffee and lemon water, activating food and myco-ritual performances within the kitchen/dining table space. Lemon peels were ground in a coffee blender, steamed, and dried before being added to a tabletop integrated tray used to grow an intricate wall panel. Coffee waste was fed to the oyster mushrooms growing within a cylindrical container, sitting on the kitchen tabletop, positioned as a centerpiece where a vase of flowers may have otherwise been. The scenes performed by actors in the space were loosely based on Chantal Akerman's film from 1975 "Jeanne Dielman, 23, quai du Commerce, 1080 Bruxelles" which explores how rituals repeated within domestic space anchor social activities.[20] Similar to the film, the woman's domestic gestures were given importance in the same way that

Figure 8. Derivative Mycelium Forms from IKEA Catalogue Household ware objects. Photo credit Alexander Brown.

industrial waste treatment participates in domestic waste upcycling rituals. Performances within the architectural space were critical not only in designing new domestic rituals of pre- and post-meal activities scripted around the preparation and harvesting of myco-ware but in future research targeted at the spatialization of the mycelium grow cycle within the kitchen.

Future Propositions for Mycelium Eco-manufacturing

The on-site, distributed mycelium eco-manufacturing processes described in these projects leverage the highly spatialized, diverse nature of biomass as opportunities for engaging its alienated value, and not as properties typically considered liabilities. All three projects, which convert biomass into a range of low to medium density insulation and biobased products, serve as case studies for matching biomass energy conversion pathways to a diverse range of end-use products. Economically, this expansion of the biomass value chain accelerates the substitution of petrochemical-based products with new bio-based products and restructures current business models to align with biomass' spatially distributed properties. For biomaterial enterprises that have championed non-toxic materials and healthy work environments within the walls of the factory, future economic development depends largely on innovation within biomass collection, quality control, and distribution systems outside the walls of the factory. By integrating mycelium-mixed biomass processing with domestic, urban farming, and agricultural waste streams, the proposed model can divert valuable biomass components from the waste away from landfill sites to local eco-manufacturing sites that use a diverse range of biomass for myco-production. The versatility and elasticity of the myco-composite value chain can be characterized as a form of economic resilience, where production can dynamically respond to match market demands from seasonal lignocellulosic inventories and a highly participatory, on-demand workforce. From a generative justice workforce perspective, the eco-manufacturing of mycelium-based systems occurs where waste resources are generated and where people—in the form of new green collar workers who participate in the creation of value through manufacturing activities—reside. This is the value of prototyping integrated material life cycles towards generative justice goals.

Acknowledgments:

The author acknowledges the participation of Atelier Luma and Collaborative Kitchen for *Carbocycene*; Squash Liverpool, Farm Urban, and CAVA for *Hack the Root*; Housing the Human, Radialsystem, and Prinzessinnengarten, Berlin for *Agrocologies*.

Funding:

Carbocycene was funded by Atelier Luma / Luma Foundation and Rensselaer Polytechnic Institute. *Hack the Root* was funded by Arts Council England, the Royal Institute of British Architects (RIBA) North, Liverpool Biennial, and CAVA Liverpool. *Agrocologies* Forecast Platform was funded by Innovation Program of Zukunft Bau, Federal Institute for Research on Building, Urban Affairs and Spatial Development of Germany.

Figure 9. View of *Agrocologies* apartment showcasing mycelium household objects and kitchen upcycling setup, Radialsystem October 2019. Photo credit Camille Blake.

Figure 10. Detail of *Agrocologies* mycelium panel at HtH, Radialsystem October 2019. Photo credit Camille Blake.

Figure 11. Grow-It-Yourself Workshop at Farm Urban at Life Sciences UTC High School, Liverpool. Photo credit Mae-ling Lokko

ENDNOTES

1. For discussions on the quantitative benefits of biomaterials: Luisa Cabeza et al., "Low carbon and low embodied energy materials in buildings: A review," Renewable and Sustainable Energy Reviews 23 (2013): 536-8; Anna Sandak et al., "Biomaterials for Building Skins," In Bio-based Building Skin (Singapore: Springer, 2019), 50-56. For discussions on the qualitative benefits of biomaterials: Vivian Loftness et al., "Elements that contribute to healthy building design," Environmental health perspectives 115, no. 6 (2007): 965-68; Judith Heerwagen, "Green buildings, organizational success and occupant productivity," Building Research & Information 28, no. 5-6 (2000): 3-7.

2. Andrew J. Hoffman and Rebecca Henn, "Overcoming the social and psychological barriers to green building," Organization & Environment 21, no. 4 (2008): 8-18.

3. Ron Eglash, "Of marx and makers: An historical perspective on generative justice," Teknokultura 13, no. 1 (2016), 249.

4. Geoffrey M. Gadd, "Mycotransformation of organic and inorganic substrates," Mycologist 18, no. 2 (2004): 60-70.

5. Stefan Rensing, "Great moments in evolution: the conquest of land by plants," Current opinion in plant biology 42 (2018): 50-51; Xiaoliang Fu et al., "Fungal succession during mammalian cadaver decomposition and potential forensic implications," Nature. Scientific reports 9, no. 1 (2019): 2-5.

6. Gadd, "Mycotransformation of organic and inorganic substrates," 60-66.

7. For biobased packaging, see R. Abhijith, Anagha Ashok, and C.R. Rejeesh, "Sustainable packaging applications from mycelium to substitute polystyrene: a review," Materials Today: Proceedings 5, no.1 (2018): 2140-2143; Greg Holt et al., "Fungal mycelium and cotton plant materials in the manufacture of biodegradable molded packaging material: Evaluation study of select blends of cotton byproducts," Journal of Biobased Materials and Bioenergy, 6 (2012): 433-38. For Insulation, see Zhaohui Yang et al., "Physical and mechanical properties of fungal mycelium-based biofoam," Journal of Materials in Civil Engineering 29, no. 7 (2017): 4-8. For Biocomposite materials, see Wenjing Sun et al., "Fully bio-based hybrid composites made of wood, fungal mycelium and cellulose nanofibrils," Scientific Reports 9, no. 1 (2019): 1-12.

8. Sonia Travaglini, CKH Dharan, and Philip Ross, "Manufacturing of Mycology Composites," In Proceedings of the American Society for Composites: Thirty-First Technical Conference. 2016.

9. Eben Bayer, Gavin McIntyre and Burt L. Swersey. Panel comprising composite of discrete particles and network of interconnected mycelia cells. U.S. Patent No. 8,999,68, filed May 29, 2012, and issued April 7, 2015. See also, Philip Ross, Method for producing fungus structures. U.S. Patent No. 9,951,307, filed August 7, 2016, and issued April 24, 2018.

10. Atelier Luma. "Myco-Structure." https://www.atelier-luma.org/projets/myco-structure

11. Atelier Luma. "Atelier Luma About." https://atelier-luma.org/en/about

12. Arles Crau Camargue Montagnette. "Le Pays d'Arles." http://www. agglo-accm.fr/p%C3%B4le-dequilibre-territorial-et-rural-(petr)-du-pays-darles.html. For development goals see, Pays d'Arles. "PETR du Pays d'Arles." https://www.pays-arles.org/

13. Sophie Dandelot, Regine Verlaque, Alain Dutartre, and Arlette Cazaubon, "Ecological, dynamic and taxonomic problems due to Ludwigia (Onagraceae) in France," Hydrobiologia 551, no. 1 (2005): 131-136.

14. In 2019, an inaugural architecture studio that I taught with Gustavo Crembil at RPI on biomaterials, grew and developed a dome proto shelter, approximately five meters in diameter, made from mycelium cylinders.

15. U.S. Department of Energy, 2016 Billion-ton report: Advancing domestic resources for a thriving bioeconomy, Volume 1: Economic availability of feedstock (Oak Ridge National Laboratory: Oak Ridge, 2016), xvii-xxiv.

16. Liverpool Biennial. "Mae-ling Lokko." https://www.biennial.com/2018/exhibition/artists/maeling-lokko

17. Housing the Human. "Mae-ling Lokko – Agrocologies." https://housingthehuman.com/prototypes/mae-ling-lokko/

18. Anna Bernstad, L. Malmquist, C. Truedsson, and Jes la Cour Jansen, "Need for improvements in physical pretreatment of source-separated household food waste," Waste Management 33, no. 3 (2013): 746-754.

19. Angelos Sotiropoulos, Dimitris Malamis, and Maria Loizidou, "Dehydration of domestic food waste at source as an alternative approach for food waste management," Waste and Biomass Valorization 6, no. 2 (2015): 167-176.

20. Jeanne Dielman, 23 Commerce Quay, 1080 Bruxelles, directed by Chantal Akerman (1975; The Criterion Collection, 1976), 3:21:00, DVD.

SOIL-LESS SOIL:

INTER-DISCIPLINARY AND INTER-AGENCY MATERIAL RESEARCH TO DESIGN, PROCESS, AND MANUFACTURE GLASS-INFUSED SOIL

REBECCA POPOWSKY

ABSTRACT

Landscape architectural discourse is fascinated with process and systems thinking that challenge strict project boundaries in spatial and temporal terms. However, this fascination does not easily translate into professional practice where projects are constrained by property lines, construction schedules, and contracts. Practitioners struggle to act on systems that extend beyond project limits despite the theoretical potential and ethical responsibility of our field to do so. This essay is focused on a case study that argues for material research and specification-writing as effective tools for extending the spatial and temporal boundaries of site-thinking, with a particular focus on hybrid organic–inorganic materials. *Soil-less Soil* is a project-based applied research initiative that aims to test the performance, feasibility, and environmental benefits of using recycled glass fines as a substitute for natural sand in manufactured soils. The project repurposes city-wide waste-stream glass into a soil product suitable for horticultural and green infrastructure projects. It stems from research undertaken by the landscape architecture firm OLIN and carried out by a team of researchers from OLIN Labs, Temple University's Department of Horticulture and Landscape Architecture, the University of Pennsylvania's Ian L McHarg Center for Urbanism and Ecology, the Pennsylvania Recycling Markets Center, Andela Products, and Craul Land Scientists. The project is supported by the William Penn Foundation and the US Environmental Protection Agency. *Soil-less Soil* sits at the nexus of three environmental challenges: global construction-grade sand shortages, limited domestic markets for mixed-color, small-particle-size glass fines/cullet, and soil performance in urban green stormwater infrastructure.

+ soil
+ sand
+ green stormwater infrastructure
+ glass recycling
+ glass cullet

Figure 1. Soil component samples. © Sahar Coston-Hardy/Esto.

In the 2005 edited collection *Site Matters: Design Concepts, Histories, and Strategies*, Carol Burns and Andrea Kahn foreground the landscape architectural understanding of 'site' as 'material terrain', offering an expanded definition of project site that encompasses three physical locations: the area of control, the area of influence, and the area of effect. They argue that all three "exist squarely within the domain of design concerns."[1] *Soil-less Soil*, an applied research initiative that aims to replace natural sand with recycled glass fines as the inorganic component of manufactured soils, offers one instance in which the design practitioner is able to engage with this expanded (three-part) site by instigating shifts in regional resource and waste streams. OLIN Labs, the practice-based research and development group within the Philadelphia and Los Angeles based landscape architecture firm OLIN, is leading an interdisciplinary research team to identify the ecological and economic implications of closing the glass waste loop in cities and making an otherwise negative-value material environmentally productive.

Environmental Context and Material Challenge

Numerous cities and smaller municipalities in the United States (US) are investing heavily in green stormwater infrastructure (GSI) to mitigate combined sewer overflows. This approach has been widely adopted not only because it sustainably improves stormwater management but also because it vegetates urban landscapes and thereby provides a host of ecosystem benefits like transpirational cooling, increased habitat and resources for wildlife, and the trapping of airborne pollutants.[2] While GSI projects contribute immensely to the livability of cities, they also contribute to bettering their ecological footprints. To this end, our research suggests that modifying the soil mixes used in GSI to include post-consumer manufactured glass-sand rather than mined sand could simultaneously reduce the reliance on environmentally damaging materials and provide a sustainable outlet for what is currently a major waste stream.

Large quantities of sand are typically included in soil blends used in GSI systems as sandy loam and loamy sand substrates facilitate infiltration and plant growth, resist compaction by foot traffic, and can serve as a water infiltration medium. Though not widely discussed, sand and gravel mining have a number of negative environmental, social, and human health impacts. Comprising the most heavily extracted material group on the planet, ahead of even fossil fuels and biomass, their mining and transport have enormous detrimental effects.[3] However, because these impacts occur far from cities, they are easily overlooked when considering the urban environment. Moreover, some regions are already facing shortages of sand and gravel, especially of uncontaminated material with predictable physical properties.[4] While it may be possible to replace both sand and gravel in GSI systems with recycled materials, the focus of our current research is strictly sand.

Recycling of household waste continues to pose a major challenge for cities and towns throughout the US. Recycling programs, while effective at collecting waste, often fail to address its processing, storage, and end use which results in large quantities of recyclable materials going to landfills. In the current economic climate with transportation expenses increasing, landfill space decreasing, and access to cheap recycling by former US processors (like China) drastically reduced, the ability to derive real benefit from recycling is a challenge.[5] Even though in theory glass is recyclable, it is often directed to landfills due to technological or market barriers. For one, glass particles smaller than a quarter inch are difficult to clean and separate using current methods. Also, changes in China's recycling policies enacted in 2017 and 2018 have drastically reduced demand for recyclable materials, raising the cost to municipalities nearly tenfold.[6] Because of this market shift, thousands of tons of recyclable material collected in dozens of American cities are now discarded.[7] Only a relatively small fraction of waste colored glass is recycled through re-melting (26.6% nationally, according to the US Environmental Protection Agency (EPA), the majority disposed of in landfills.[8] In 2015, approximately seven million tons of waste glass were landfilled in the US.[9] As a result, vast amounts of colored glass and most sand-sized glass particles (a.k.a. 'glass fines' or 'glass-sand') are never recycled. (Figure 2)

From Project-driven Research to Research-driven Projects

Research into the horticultural use of glass for amending the composition of soil began at OLIN in 2010. We sought to identify an alternate replacement for the sand that was needed in the project specifications for Dilworth Park in the city of Philadelphia. Investigation of sieve gradations was conducted by Tim Craul of Craul Land Scientists, while

Figure 2. Glass sand. © Sahar Coston-Hardy/Esto.

a heavy metals report was obtained from Go Green with Glass, a New Jersey based supplier of glass cullet. The idea was, however, shelved given the lack of a framework in 2010 for further research into the possible use of substitute glass. It was only in 2017 when, during the specification-writing for a fifteen-acre campus housing project at the University of Washington (UW) in Seattle, the OLIN team returned to consider glass. Projects like this one that require the import of large volumes of soil rely heavily on resource extraction. Planting soil for these projects is either mined from greenfields or mixed off-site using a combination of organic and inorganic materials: sand, silt, clay, and compost. OLIN's design team at UW wished to decrease the environmental impact of importing soil by designing a soil specification that reduced or eliminated the need for virgin material extraction. It was to this end that the design team proposed to substitute mined sand with waste glass. To do so, the team needed to ensure that a glass-based soil would support plant growth, that its installation would not negatively impact human health or ground water quality, and that sufficient volumes of material would be available locally at a comparable price to conventional mixes. Thus began a multi-year inquiry into the horticultural and environmental performance of glass-based sand, as well as the cultivation of an inter-disciplinary and inter-agency team of collaborators that could address not only performance, but also economics, policy, and logistical planning.

Over the past four years, OLIN and our collaborators have performed a literature review on the subject, a life-cycle assessment (LCA), an economic feasibility study, lab-based chemical analyses, a mesocosm study, and two greenhouse plant growth studies. These studies have validated the potential environmental and economic impact of our proposed glass processing methods and blended soil. In 2021, a mesocosm study and market analysis was undertaken with funding from the William Penn Foundation to assess the infiltration and leachate characteristics of glass-based soils and the economic implications of material substitution. A Small Business Innovation and Research grant from the US EPA supported the development of a technical plan in coordination with the City of Philadelphia's Department of Parks and Recreation, that shows how public and private initiatives can divert glass waste into soil production within city limits, saving the city money, and reducing landfill volumes.

Literature Review

Despite the current negative economic climate for recycling glass, studies have evaluated the merit of using manufactured glass-sand waste as a landscape construction material. In one case it was found that using recycled glass in sand drains improved permeability and drainage consolidation.[10] Another study, looking at golf course soils in the United Kingdom (UK), showed that recycled glass mixed with peat had greater porosity and hydraulic conductivity than a conventional soil mix.[11] While the horticultural performance findings of this study were promising, the economic outlook for recycling glass in the UK at the time of publication was unfavorable relative to the cost of mining sand. In a 2016 study that compared the performance of natural sand dunes to that of glass-based sand dunes in supporting dune grasses, plants grew equally well in each substrate (as measured by biomass).[12] And finally, in a Life Cycle Assessment (LCA) published in Hong Kong, glass-sand consumed less energy and therefore yielded fewer greenhouse gas emissions than sand mined for the construction industry.[13] Given this range of positive data, and our own efforts at evaluating the commercial viability of glass-based soils, we believe significant environmental and economic benefit can be achieved by replacing mined sand with recycled glass-sand in manufactured soils.

Life Cycle Assessment

An LCA conducted by Anqi Zhang in 2019 as her Masters capstone project at the University of Pennsylvania compared the environmental impacts associated with one ton of quarried sand (excavation-processing-transport) to one ton of recycled glass-sand (collection, processing and transport that avoids landfill).[14] Impacts associated with the 'status quo scenario' were calculated based on a system boundary that began with open-pit mining of sand and ended at delivery of sand to a project site in Philadelphia. The 'proposed scenario' impacts were calculated based on a system boundary that began with collection of waste glass and ended with delivery of fine glass-sand to the same Philadelphia project site. It is worth noting that the assessment of this waste-to-production chain does not include the original manufacture of virgin glass product (of the glass bottle or container), in accordance with comparable LCA publications which begin tracking impacts at the site of waste collection.[15] The LCA indicates that the extraction and associated leachates of sand and gravel mining and the embodied energy of excavation and sieving of aggregates is more energy intensive than crushing and processing glass cullet. The study (performed using Thinkstep Gabi LCA software) utilized Gabi LCA databases where relevant (US) data was available, along with primary-source data collected locally. The LCA model was reviewed by a third-party LCA professional (Christoph Koffler, PhD, Thinkstep) for quality assurance. The results indicate that the extraction and associated leachates of sand and gravel mining and the embodied energy of excavation and sieving of aggregates is more energy intensive than crushing and processing glass cullet. The study found a 67% reduction in greenhouse gas emissions, a 95% reduction in human toxicity, and a 70% reduction in oil-equivalent fossil depletion for the one ton of manufactured glass-sand.

Pilot Growth Trial (2019)

In 2019, OLIN and researchers at Temple University completed a greenhouse planting trial comparing plant growth in soil media containing amended glass versus soil media with mined sand.

Figure 3. Planting trial. Credit OLIN.

The following results were presented by Dr. Sasha Eisenmann and Dr. Joshua Caplan at The Northeastern Plant, Pest, and Soils Conference (NEPPSC) in January 2020.[16] A factorial experiment was run in a climate-controlled greenhouse in Philadelphia, Pennsylvania for eighteen weeks. (Figure 3) Eight individuals of three plant species were grown in six soil mixes (144 plants in total). All mixes were composed of 60% sand, 20% sandy loam, and 20% mushroom compost (by volume). The makeup of the sand portion (glass-sand, mined sand, or a mix of the two) and amendments differed across mixes, as follows:

1. Mined sand (experimental control)
2. 50:50 blend of mined and glass-sand
3. Glass-sand (unamended)
4. Glass-sand + gypsum at 0.5 g/L (≈1 g/pot)
5. Glass-sand + ferrous sulfate at 0.5 g/L (low rate; ≈1 g/pot)
6. Glass-sand + ferrous sulfate at 1.0 g/L (high rate: ≈2 g/pot)

The sand was mined in New Jersey and the mushroom compost supplied by Allied Landscape Supply in Dresher, Pennsylvania (PA). Glass-sand was sourced from Aero Aggregate in Eddystone, PA. Lansdale silt loam was collected from the Temple University Ambler Campus. Mixes were prepared by thoroughly blending all materials in a rotary concrete mixer. Plants were selected to be representative of those typically used in GSI: Iris versicolor L. (northern blue flag); Calamagrostis × acutiflora 'Karl Foerster' (Foerster's feather reed grass); Betula allegheniensis Britt (yellow birch). (Figure 4) Soil pH measurements from unplanted pots as well as measurements of chlorophyll content (via SPAD index) were taken biweekly.[17] Soil pH is a measure of relative acidity or alkalinity and a neutral pH of between 6 and 7.5 is optimal for plant health in this context. Chlorophyll

Figure 4. Iris versicolor. Credit OLIN/Rebecca Popowsky.

content or 'greenness' of leaves is a proxy for the general health of the plant and can be measured without disturbing plants throughout the course of the growth trial. At the end of the study, eighteen weeks after planting, above ground plant material was harvested, oven dried at eighty degrees Celsius, and weighed. (Figure 5) The fresh mass of Iris root systems was also measured, but this was not possible for Calamagrostis or Betula given the fibrous nature of their roots and difficulty in separating roots from the soil.

The results observed included the following findings. To begin with, while glass-based soil mixes can support the growth of young plants, high pH levels (8.0 - 9.0) may inhibit the growth of some plant types. None of the three species evaluated exhibited statistically signif-icant reductions in above ground biomass in the glass-based soil mix. However, sample means were slightly lower for Betula and Iris. Also, above ground biomass of Iris was reduced in mixes with ferrous sulfate and had lower root biomasses than did plants grown in mined sand or a 50:50 blend of mined and glass sand. In addition, our chemical analyses identified two hurdles to using manufactured glass-based soil as a planting medium. First, fine particles and some potentially toxic elements were sufficiently high to warrant a detailed evaluation of leachate (see below for further discussion of leachate analysis). Second, the pH of manufactured glass-sand (9.8) is high for most plants. A pH level that is too high or too low prevents absorption of certain nutrients, such as phosphorous, which can limit photosynthesis, root, and flower growth.[18] The alkalinity of glass comes from sodium carbonate and lime in the glass itself such that all media made from glass-sand will require alkalinity reduction.[19] We examined methods for reducing alkalinity by sieving, washing, and amending the mix with forms of sulfate. Although washing was ineffective, sieving and iron sulfate addition reduced pH

Figure 5. Planting trial harvest. Credit OLIN.

noticeably; while promising, it is unclear if amendments will remain effective through time. Other alternatives such as glass-to-sand ratio adjustments and different compost sources may be more effective in lowering the mix's pH over longer periods. Lastly, while glass-sand may act as an unwanted source of sodium, zinc, and other elements, our analysis shows that sodium levels are sufficiently diluted in mixes and may rapidly leach out of the soil, and that zinc levels are within an acceptable range for plant growth. Further research into soil mix chemistry and the associated effects on plant growth is warranted and is discussed further below. Particle size data suggest that glass-based mixes will have appropriate hydraulic properties, but this is being tested empirically in the mesocosm study described below.[20]

Mescosm Study and Feasibility Brief (2020-21)

OLIN and our academic partners at Temple University are currently working with Dr. William Fleming of the University of Pennsylvania's Ian L McHarg Center for Urbanism and Ecology on a year-long study funded by the William Penn Foundation that seeks to answer the following questions: what are the most effective means for reducing the alkalinity of glass-sand over prolonged periods; how does the performance of media made from glass-sand compare to that made from mined sand with respect to water storage, leachate quality, and chemical stability; how do the current economic costs and benefits of glass versus mined sand compare at the scale of the city, and how does this vary across suppliers; what guidelines and specifications would allow municipalities to produce and utilize recycled glass sand-based soils in typical green stormwater infrastructure projects?

The first two questions (those on reducing alkalinity and media performance) can be answered more cost-effectively and time-efficiently in the laboratory and greenhouse setting than in the field. Although it will be valuable to conduct trials in the field at a later stage, we consider the current project translational. We, therefore, began by performing small-scale experiments that provided direct guidance for developing a trial soil specification to be used in an initial set of installations and subsequently refined further. One set of investigations determined which pre-treatments and amendments reduced the pH of glass-sand mixes over a two-month timeframe. We tested individual and combined treatments of sieving (to remove particles with high surface-area), soaking in weak acid (to remove carbonates), and amending non-alkaline organic materials. This work was conducted in Temple University's greenhouse using complete mixes in irrigated pots (arranged as a completely randomized design), with water matching the pH of rainwater (approximately 5.0). Sieving, acidification, and iron sulfate addition all lowered the pH, though it remained slightly alkaline, indicating that the high pH of glass-sand can be reduced using several methods.

We are in the process of conducting studies in mesocosms (simulating outdoor experimental field conditions) to determine the extent to which the media retains water, leaches sediment/chemical constituents, and remains chemically stable through time. Mesocosms mimic bioretention installations, recreated in one-meter-tall columns and housed in Temple's laboratory. They simulate storms with specified intervals and hydrological profiles. (Figure 6) During fall 2020 and

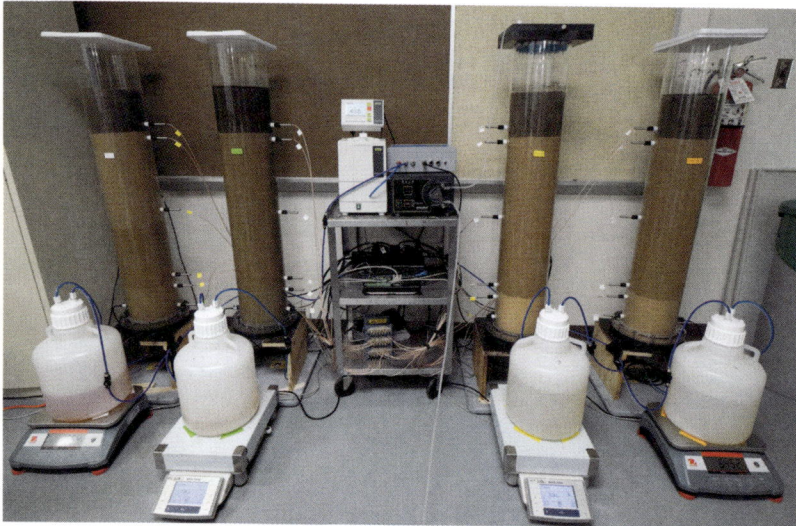

Figure 6. Mesocosm study. Credit OLIN/Rebecca Popowsky.

winter 2021, we constructed a set of mesocosms consisting of vertically oriented polyvinyl chloride (PVC) tubes (one-meter-high x twelve centimeters in diameter) filled with the three layers of media typically used in field installations (top: 6" loamy sand + 3% organic matter; middle: 24" loamy sand alone, base: 6" sand). For the sand component, the type (mined versus glass-sand) and source varied among the columns. Custom-built devices at the bottom of each mesocosm applied a suction that ensured realistic water movement and retention and collected leachate and suspended particles. A watering system irrigated with pH neutral water. (Figure 7) Moisture sensors were installed in each layer and tracked at five-minute intervals.

Initial results indicate that leachate chemistry shifted through time largely as we expected. Specifically, starting concentrations of most metals, nutrients, and other ions were roughly proportional to the amount of glass in the mix. More importantly, concentrations almost always decreased with each subsequent 'storm'. The pH was the most notable exception to this pattern; it remained nearly constant (at pH 8.5 - 9) throughout. At least for the source of glass we used, we concluded that leachate chemistry was not problematic, suggesting the material can safely be used without cleaning. However, our trial specification will recommend testing of all batches of glass-sand to ensure conformity. The most surprising result of the study, so far, is that water moved more slowly through glass-based mixes than mined sand mixes. While this could have positive implications for water retention

Figure 7. Mesocosm detail: irrigation. Credit OLIN/ Rebecca Popowsky.

(and plant availability), it could potentially decrease hydraulic conductivity enough to slow infiltration in a green infrastructure setting. While we hope to evaluate this phenomenon further in a full-scale field trial, our trial soil specification will recommend a mix of mined and glass-sand to optimize infiltration.

Research questions to be addressed moving forward include: what glass-based soil manufacturing process will produce a new material that has properties and costs comparable to currently used analogous materials but with lower lifecycle impacts; what industrial byproducts and/or food waste can be used in combination with glass-sand and compost to meet soil performance specifications; what planning strategies can be developed to facilitate municipal-scale recycled glass processing and soil blending to produce manufactured glass-based soil; what is the cost benefit to cities of producing, utilizing, and selling manufactured glass-based soil as an alternative to conventional manufactured soils?

City of Philadelphia as Testbed

Previous and current studies have focused on horticultural and engineering performance and on cost and environmental benefits of replacing natural sand with manufactured glass-sand in soil blends. Parallel with the material science component of this research, we are also addressing the economic feasibility of adapting public works soil specifications at the municipal scale using existing material procurement sources. The next phases of research, with grant funding by the US EPA, aim to address the design of a scalable manufacturing process (glass pulverizing, amendments, and soil blending) and a technical plan that will allow municipalities to divert waste glass into locally manufactured glass-based soil. Building on our 2020-21 mesocosm study, we will take the most effective soil blends, test them as planting media in a second growth trial, and design a manufacturing process that will produce them at minimal economic and environmental cost. This will demonstrate the potential for cities to repurpose their own waste glass into a commercially viable soil product. To this end, we are building a collaborative network of municipal agencies and industry professionals to develop a planning scenario for the City of Philadelphia to develop a new green industry that turns waste glass into soil. This case study will help us determine the net present value of factors including the savings derived from avoiding landfills and the costs of procuring, processing, and blending components of topsoil made with manufactured glass-sand, versus the standard materials typically used in public green stormwater infrastructure installations.

The procurement analysis which is currently underway entails surveying existing suppliers of glass fines and topsoil, quantifying material costs, reviewing existing processing capabilities, and assessing the presence of additional costs or savings. Results will be summarized in a feasibility brief that will be distributed freely to all interested parties. Initial results indicate that a gap exists between highly processed, costly glass-sand (which is sold as an abrasive for sandblasting), and minimally processed, mixed color glass cullet (which has little to no commercial value and is used primarily for landfill cover or simply discarded). (Figure 8) The manufactured glass-sand product that would be optimal as a soil component would fall between these two existing products in terms of level of processing and cost. The processing cost of glass-sand for sandblasting abrasives is approximately $120 to $150 per ton. To make a viable soil mix using glass-sand, it will be necessary to design a manufacturing process that meets performance specifications while significantly reducing these processing costs. Initial research indicates that the volumes of glass currently being sent to landfill in the city (approximately 90,000 tons annually) could meet the city's entire demand for sand in public works GSI soil installations while finding beneficial reuse for about three-quarters of the city's landfilled glass waste. (Figure 9)

Philadelphia's Parks and Recreation and The Water Department are both heavy consumers of natural sand and bioretention (sand-based) soils. By producing its own manufactured glass-based soil, a city like Philadelphia could reduce or eliminate the need to purchase mined sand. It could also avoid the cost of disposing of mixed color glass cullet through Material Recovery Facilities (MRFs) (about $140 per ton) and/or the cost of sending waste to a landfill ($65 per ton tipping fee plus pick-up and hauling costs). Annually, approximately 100,000

Figure 8. Glass cullet. Credit OLIN/ Rebecca Popowsky.

STREETS DEPT: GLASS

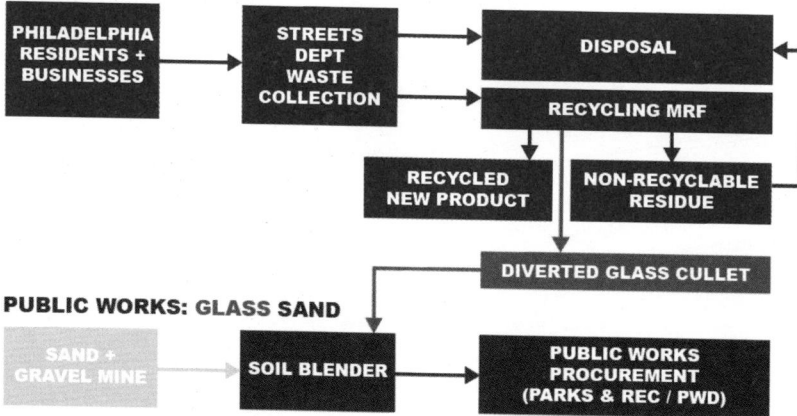

Figure 9. Proposed glass diversion diagram. Credit OLIN.

tons of glass enters Philadelphia's waste stream. Of that, about one third is collected curbside by the City Streets Department and delivered to a privately-owned MRF for recycling.[21] However, since glass currently has a negative market value, all collected glass is simply separated out of the single stream and disposed of as 'residue' in a landfill. According to Charles Raudenbush of Waste Management's Public Sector Services, today no portion of the glass waste collected by the City of Philadelphia curbside is recycled. Little if any glass collected from Philadelphia's commercial sources makes its way to a glass recycling facility that can remelt it into new glass products. Most is disposed of in the landfill or processed into mixed color glass cullet that can be used in low-value applications such as landfill cover.

Engineering a plan that allows a municipality to divert and process waste glass into manufactured glass-sand within existing infrastructural systems requires inter-agency planning. In Philadelphia, for example, collaboration between the Streets Department, Parks and Recreation, Water Department and Procurement Department would be required to install and operate a glass pulverizing and soil blending facility. Policy and waste management guidelines would be required to divert glass waste from single-stream municipal recycling. (Figure 10) Landscape architects are distinctly able to develop such a plan in collaboration with city agencies. We have the capacity to address commercialization potential of 'soil-less soil' by analyzing the costs and benefits of implementing a glass processing program running in parallel with municipal compost and mulch production that is used in public works or sold to private consumers.

By creating a workable, engineered plan for municipalities that effectively repurposes their own glass waste streams by locally producing glass-based soil products, this reduces the environmental impact of sand production and the city's cost for landfilling glass. The market for this product and process includes all urban and suburban regions that are focusing efforts on green space development and water management. The model would be applicable to large urban centers around the country, including island communities, in which waste volumes are high, landfills overtaxed, and natural resources limited.

Conclusion

This paper has situated the *Soil-less Soil* research initiative within a larger discussion of site definition and design agency in proposing material research and specification-writing as mechanisms for expanding the 'areas of influence' of landscape architectural design beyond property lines and site boundaries. The project's research question originated in the specification-writing process of a campus design project and has since grown into a municipal-scale planning and policy initiative. Seen through the lens of Burns and Kahn's expanded domain of design concern, the project aims to bring design thinking and action onto sites that are indirectly impacted by our work in the built environment. Reducing the use of mined natural materials in our project specifications obviously reduces our reliance on landscapes of extraction, the disruption of habitat, and greenhouse gas emissions. When no renewable options exist for the substitution of certain virgin materials, designers can invent and advocate for new options, considering material performance, applicability, and the economic and political constraints that make new markets feasible. The goal of this effort is to instigate a new system for waste management and material procurement that will save public money and reduce the environmental footprint of public-space creation. To realize these goals, a full range of scales and types of research come into play, from the micro-structure of soils to the waste networks of regions, and from horticultural science to commercialization planning. Design practitioners are well-positioned to pull these scales of inquiry together through problem definition, team-building, and real-world, real-time application of research findings in built projects. *Soil-less*

Figure 10. Glass cullet _ OLIN/ Rebecca Popowsky

Soil is one such instance of a design practitioner engaging with this expanded site, not through formal imagination but through *material* research. It demonstrates the efficacy of the material specification as a form of design representation (as opposed to the plan drawing) that speaks to process, systems, and time (as opposed to object and form) by describing not only the 'what' of the design, but the 'how', the 'when', and the 'by whom' of material sourcing, processing, installation, and disposal.

Acknowledgements

This research would not be possible without the guidance of experts in the public, private, and non-profit realms, including Stephanie Chiorean of the Philadelphia Water Department, Daniel Lawson of Philadelphia Parks and Recreation, and Archie Filshill and Theresa Loux of Aero Aggregates.

Collaborators

The team of researchers for *Soil-less Soil* includes OLIN Labs (Richard Roark, OLIN Partners and me—the OLIN Labs Coordinator), Temple University's Department of Horticulture and Landscape Architecture (Dr. Sasha Eisenmann and Dr. Joshua Caplan), the University of Pennsylvania's Ian L McHarg Center for Urbanism and Ecology (Dr. William Fleming), Craul Land Scientists (Tim Craul, CPSS), Pennsylvania Recycling Markets Center (Robert Bylone and Wayne Bowe), Andela Products (Cynthia Andela), Bottle Underground (Rebecca Davies), Engineering & Land Planning Associates (Ed Confair and Clare Moriarty), Circular Philadelphia (Nic Esposito), and Bennett Compost (Jen Mastalerz).

Funding

This project was funded by the William Penn Foundation and the US Environmental Protection Agency (EPA).

ENDNOTES

1. Carol Burns and Andrea Kahn, *Site Matters: Design Concepts, Histories, and Strategies* (New York: Routledge, 2005), xii.

2. Sarah Taylor Lovell and John R. Taylor, "Supplying Urban Ecosystem Services through Multifunctional Green Infrastructure in the United States," *Landscape Ecology* 28, no. 8 (October 2013): 1447–63.

3. Aurora Torres et al., "A Looming Tragedy of the Sand Commons," *Science* 357, no. 6355 (September 8, 2017): 970–71.

4. Ibid., Torres et al.

5. Amy L. Brooks, Shunli Wang, and Jenna R. Jambeck, "The Chinese Import Ban and Its Impact on Global Plastic Waste Trade," *Science Advances* 4, no. 6 (June 2018); Livia Albeck-Ripka, "Your Recycling Gets Recycled, Right? Maybe, or Maybe Not," *The New York Times*, May 29, 2018, sec. Climate. https://www.nytimes.com/2018/05/29/climate/recycling-landfills-plastic-papers.html.

6. Ibid., Brooks, Wang, and Jambeck, "The Chinese Import Ban and Its Impact on Global Plastic Waste Trade."

7. Ibid., Albeck-Ripka, "Your Recycling Gets Recycled, Right?"

8. OLEM US EPA, "Glass: Material-Specific Data," Collections and Lists, US EPA, September 7, 2017. https://www.epa.gov/facts-and-figures-about-materials-waste-and-recycling/glass-material-specific-data.

9. Ibid., US EPA.

10. F. C. Wang et al., "Study of Permeability of Glass-Sand Soil," *Archives of Civil Engineering* 63, no. 3 (September 26, 2017): 175–90. https://doi.org/10.1515/ace-2017-0036.

11. A.G. Owen, L.K.F. Hammond, and S.W. Baker, "Examination of the Physical Properties of Recycled Glass-Derived Sands for Use in Golf Green Rootzones," *International Turfgrass Society Research Journal* 10 (2005): 1131–37.

12. Christopher Makowski, Finkl Charles W., and Kirt Rusenko, "Suitability of Recycled Glass Cullet as Artificial Dune Fill along Coastal Environments," *Journal of Coastal Research* 29. no.4 (July 1, 2013): 772–82. https://doi.org/10.2112/12A-00012.1.

13. Md. Uzzal Hossain et al., "Comparative Environmental Evaluation of Aggregate Production from Recycled Waste Materials and Virgin Sources by LCA," *Resources, Conservation and Recycling* 109 (May 1, 2016): 67–77. https://doi.org/10.1016/j.resconrec.2016.02.009.

14. Anqi Zhang, "Soil-Less Soil Study - A Sustainable Solution for Green Infrastructure Soil Media - Part 1, Life Cycle Assessment" (University of Pennsylvania, May 2019), https://repository.upenn.edu/mes_capstones/78/.

15. Gian Andrea Blengini and Elena Garbarino, "Resources and Waste Management in Turin (Italy): The Role of Recycled Aggregates in the Sustainable Supply Mix," *Journal of Cleaner Production* 18, no. 10 (2010): 1021–30, https://doi.org/10.1016/j.jclepro.2010.01.027; Hossain et al., "Comparative Environmental Evaluation of Aggregate Production from Recycled Waste Materials and Virgin Sources by LCA."

16. Sasha Eisenman et al., "Shifting Sands: Evaluation of Ground Recycled Glass as an Alternative to Mined Sand for Use in Green Infrastructure Soils," vol. 5 (The Northeastern Plant, Pest, and Soils Conference, Philadelphia, PA: American Society for Horticultural Science, 2020), 34.

17. The use of a handheld Soil Plant Analysis Development (SPAD) meter is a rapid and non-destructive approach to measuring chlorophyll in plant leaves.

18. Julie Bawden Davis, "Plant Peril: Soil Too Acid or Alkaline," Los Angeles Times, September 23, 2000. https://www.latimes.com/archives/la-xpm-2000-sep-23-hm-25289-story.html.

19. Yahya Jani and William Hogland, "Waste Glass in the Production of Cement and Concrete – A Review," *Journal of Environmental Chemical Engineering* 2, no. 3 (September 2014): 1767–75. https://doi.org/10.1016/j.jece.2014.03.016.

20. Ibid., Eisenman et al., "Shifting Sands."

21. MSW Consultants and Philadelphia Streets Department, "Municipal Waste Management Plan 2019-2028," https://www.phila.gov/media/20210614135413/Municipal-Waste-Management-Plan-202010.pdf.

SEAWEEDNESS:

THE AGENCY OF MATTER IN THE DEPARTMENT OF SEAWEED

JULIA LOHMANN

ABSTRACT

How can the agency of biomaterials inform and guide practice-led design research? How can designers use seaweed, for example, in developing a method that invites diverse publics to reflect on the relationship that humans have with material and natural worlds? How could seaweed become a model for interacting with nature in a more ethical, sustainable, and regenerative manner? This paper discusses my design practice based in the development of seaweed as a sustainable material for making, and that considers its systemic and ecological impacts. My practice-led research journey began in Japan in 2007 when I changed my view of this biomaterial from that of marine organism to that of muse and method for engaging diverse publics in reflecting on questions of material agency, regenerative design, and ocean health. During my Ph.D. residency at the Victoria & Albert Museum in London in 2013, I founded the 'Department of Seaweed', a transdisciplinary community of practice centred around macroalgae. Through it, I have aimed to demonstrate how co-speculative design with an empathic, more-than-human-centric mindset can connect knowledge across disciplines, engender care for human and non-human species, and inspire action towards regenerative futures.

+ seaweed
+ bio design
+ material agency
+ community of practice
+ co-creation
+ architectural installations

Figure 1. Julia Lohmann harvesting seaweed. Credit Julia Lohmann.

I have been working with seaweed since 2007, when I first discovered it as an artist in residence at the Sapporo Artist in Residency (S-AIR) programme in Sapporo, Japan. Since then, I have researched its ecological and systemic roles in oceans, as well as its material properties and agency. In 2013, as part of my practice-based collaborative Ph.D., undertaken at the Royal College of Art and the Victoria & Albert Museum in London, and funded by the Arts and Humanities Research Council, I founded the 'Department of Seaweed', an interdisciplinary community of practice focused on macroalgae.[1] We have aimed to develop seaweed as a material for making, whilst safeguarding and whenever possible, increasing its positive, regenerative role in marine ecosystems. Seaweed provides a habitat for many marine species, it cleans the ocean of pollutants and excess nutrients, acts as a carbon sink, and contributes significantly to the world's oxygen supply. Kelp, a sub-group of seaweed, is so versatile as a design material that it may even replace many fossil-based materials. At a systems level, seaweed's positive marine impact and material qualities makes it an ideal model for regenerative practices in a world struggling with a climate crisis. Seaweed has become my material, muse, and method. (Figure 1)

Material agency has been widely discussed in academic writing. Bruno Latour's Actor- network theory (ANT), Tim Ingold's anthropology of *Making*, and Jirō Yoshihara's Gutai Art Manifesto all reference how material properties may be brought to bear on the design process.[2] In addition, Elvin Karana's Material Driven Design method (MDD) describes situations where materials are placed at the starting point of the process, asking not what a material is but what it can do.[3] However, as a practitioner who designs and researches, I believe that material agency extends beyond materiality. As with seaweed, material agency radiates outwards, it motivates concepts, processes, and systems of design and production, as well as socio-economic networks and living systems. In my practice-based research, it directs my decision-making process in all stages of design and across many different scales of complexity. This work reflects and builds upon Richard Buchanan's scales of design in outlining a trifold foundation for my practice that incorporates the pillars of knowing, caring, and acting.[4]

I was amazed how in Japanese culture seaweed is primarily understood as a form of food, not as a material. To me, it felt like a kind of marine leather, that I imagined used for book covers, shoji screens, clothing, and a great deal more. I started to explore seaweed as a craft, not through the lens of a lost practice from the past, but as a way of 'crafting' a potential future for seaweed as a natural material for making. As a material, it is highly reactive to water and humidity in its surroundings. It is shrivelled, hard, and brittle when dry, yet flexible and sticky when wet. In the ocean it is shaped by currents, yet it shapes its environment, providing habitats and food for a multitude of species. When harvested, depending on how dry or wet it is, it retains its dynamically changing material properties and qualities, including

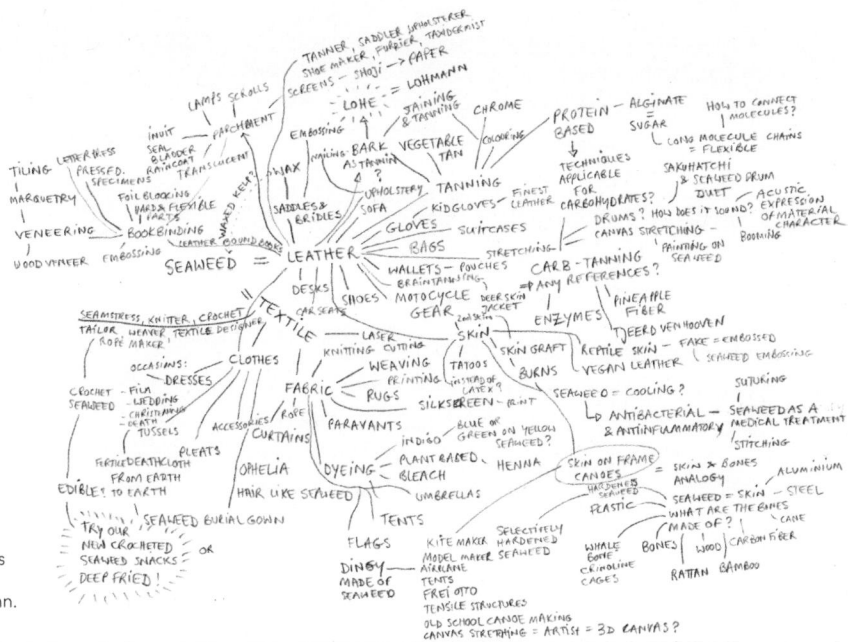

Figure 2. Material analogies between seaweed and leather. Credit Julia Lohmann.

smell and colour. Working with seaweed's evocative aliveness is a rich, multi-sensorial experience. So strong is seaweed's material agency, one feels as if one is connecting with another being.

There is no established canon for working with seaweed. (Figure 2) Because it has little established design history, seaweed helps us rethink the nature of design itself. For me, the key is to explore it as an organism, a non-human co-designer, rather than as a material upon which to impose my will. My aim is to design with seaweed in an open-ended empathic manner based on observation and respect for how the material behaves. In this process, I map analogies with other materials, for instance with paper, plastics, or leather because their qualities overlap with those of seaweed, even if partially. This introduces fields of knowledge, processes, and opportunities for collaboration with practitioners in papermaking, tannery, wood-working, biology, and material science. Examples of such experiments include my work with seaweed veneer and marquetry at the Deutsche Werkstätten Hellerau in Dresden, Germany, seaweed-bamboo struc-tures at the Kyoto Institute of Technology's Design Lab, 'paper-like' marbling of kombu seaweed at the Victoria & Albert Museum, London, and woad-dyeing of seaweed at the BioColour research consortium at Aalto University, Helsinki, Finland.[5] Material analogies also help in establishing qualities, agencies, and aesthetics unique to seaweed which I call 'seaweedness'. It is these I emphasise in this work, for it is not my goal to make seaweed conform to established perceptions and processes.

One of the most striking qualities of seaweed is its interac-tion with light. (Figure 3) It can be translucent or opaque, in colours

Figure 3. Seaweed and light: transluscent seaweed collar, made in collaboration with Moya Hoke. Photography credit Petr Krejci.

ranging from lush dark greens to browns and light, parchment-like tones. This is influenced by the humidity in its surroundings as well as by the changing balance over time of ultraviolet (UV)-degradable chlorophyll-based green and UV-stable fucoxanthin-based brown pigments. As seaweed ages, it fades, changes colour, and continually oscillates between solidity and flexibility, such that its translucency never looks the same. These dynamics make it articulate and legible relative to the human body, for seaweed renders physically and intellectually tangible the conditions in a space that affect humans. After all, we also depend on and react to light and humidity. The Victorians understood these shared qualities when they used fronds of seaweed to forecast the weather. I too wondered how to adopt this way of thinking when using materials for their unique agency.

Figure 4. The Department of Seaweed residency studio at the Victoria & Albert Museum, London, 2013. Credit Julia Lohmann and Petr Krejci.

As part of a practice-based, collaborative Ph.D completed at the Royal College of Art and the Victoria & Albert Museum (V&A) in London, I undertook a six-month residency at the museum. (Figure 4) There, I established the 'Department of Seaweed' (DoS), a transdisciplinary community of practice that gathered people from different professional fields and walks of life to share in a common interest for macroalgae. Our goal was to explore the development of seaweed as a material for making. The title of our collective was inspired by the V&A practice of naming departments after important materials. I proposed that seaweed was just as important as ceramics or silver, even if most things made of seaweed did not yet exist.

Figure 5. Oki Naganode seaweed and rattan sculpture, Victoria & Albert Museum, London, 2013. Credit Julia Lohmann and Petr Krejci.

This positioned the DoS in the future, with its invitation to a diverse public to engage in shared speculations. Operating as an open studio, presenting processes rather than finished outcomes, we invited people to see, smell, and in some cases taste seaweed, and to interact with the half-finished things we made. These objects, without function or purpose—but with a distinctly seaweed materiality—triggered a range of very personal responses. Visitors shared childhood stories of swimming entangled in seaweed, smelly beaches, the umami taste, and of encountering nature's non-human others with whom we share the Earth and upon whom we depend for our survival. These participants became contributors, many suggesting potential other uses for seaweed, providing research links, and volunteering to experiment with seaweed using their own craft skills and ideas. In this context, my role in the residency shifted from author to facilitator, from individual practitioner to collaborative co-speculator.

Wanting to present our findings on seaweed to V&A visitors during the 2013 London Design Festival, we initially planned to install a seaweed-veneered parquet floor in one of the museum galleries. (Figure 5) Upon reflection, we abandoned this idea as it would have meant muting its materiality by gluing it onto wooden panels, taking away its translucency and artificially preserving its deep green colour by lacquering it. Seaweed would have become just another surface. Viewers would most likely have consumed it as a product rather than speculate on the wider potential of seaweed as they had done when visiting the DoS. Hence, we decided to design the opposite of a floor which became *Oki Naganode*, a large modular seaweed and rattan sculpture. This portrait of seaweed materiality and agency was in many ways literally a body of knowledge. The rhythm of its ribbed structure from its large limbs to smaller digits was based on the width of individual blades of Japanese kombu seaweed. These were glued onto a rattan skeleton while still wet. As the kombu dried it shrunk, tightening, and thereby strengthening the skin-on-frame surface of the sculpture in a way that could not be pre-planned. The resulting form, designed empathically, incorporated the material agencies of kombu and rattan, and logically reflected the dialogue between the two materials.

Oki Naganode was intended to make seaweedness experiential, and to demonstrate how seaweed changes over time. (Figure 6) Now part of the collection of the Museum of Modern Art (MoMA) in New York, the chlorophyll content of the sculpture has disappeared, transforming its colour into a translucent tone akin to parchment. The acquisition of *Oki Naganode* was accompanied by debates among conservators on how best to care for an artefact that, compared to other artefacts in the collection, was so transient. Natural materials challenge established ideas about perfection and permanence. They highlight how differently we perceive the world when we are called upon to reflect on our material expectations, our relationships with things, as well as their production chains and origins. Museums

Figure 6. Detail of Oki Naganode. Credit Julia Lohmann and Petr Krejci.

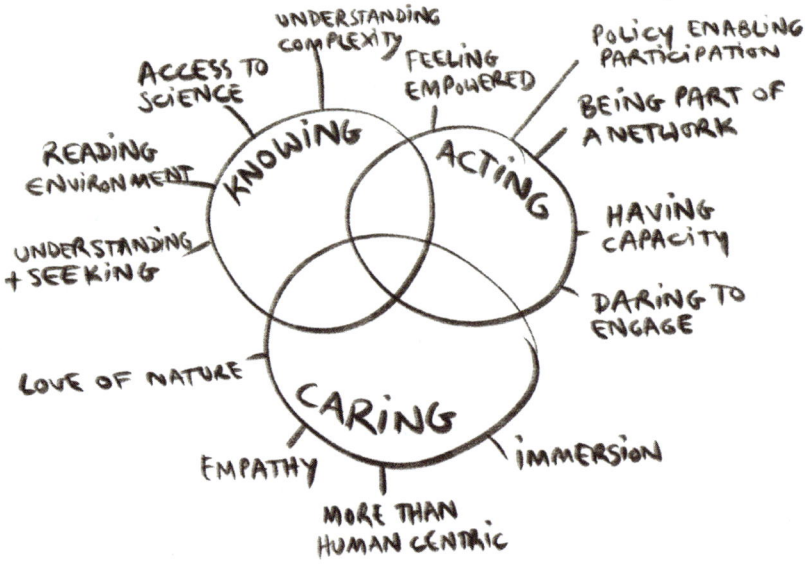

Figure 7. Knowing, acting, and caring framework. Credit Julia Lohmann.

Figure 8. Bundle of seaweed at a fish market in Japan. Credit Julia Lohmann.

are uniquely placed to facilitate these discussions, reflections, and exchanges. As physical repositories in an increasingly digital world, museums can provide much-needed forms of multisensory engagements for a public that is looking to experience new ways of learning in their largely non-commercial spaces. Transdisciplinary communities of practice like the Department of Seaweed can contribute to this, making processes transparent and accessible, and providing a multitude of clues and triggers for inspiration, interaction, and enquiry. Using both physical and virtual infrastructures they can connect partner institutions and spaces.

The "Four Orders of Design", postulated by design theoretician Richard Buchanan in his article "Design Research and the New Learning", offers a very useful framework to reflect on my practice and that of the DoS.[6] Buchanan organizes the ever-expanding field of design according to levels of complexity, beginning with signs and symbols, transitioning to objects and artefacts, interactions and experiences, and finally systems and environment. Design is no longer a delineated discipline designing posters and toasters but actions that bridge disciplines at all levels of complexity. I have extended Buchanan's framework into the micro-scale of materials, molecules, and DNA since we have come to create these too. We now, effectively, design life and living systems, all of which have agency, likely beyond our control. Consequently, I believe that regardless at which level we design we must consider the impact of our decisions on all levels, and on humans as non-humans. I do just this when moving throughout the design process from organism to material, and from material to artefact. Relating to seaweed I ask: What is this seaweed? What agency does it have in the ocean? What happens if it is removed from the ocean? Does it benefit the ocean if removed? Under which circumstances is harvesting beneficial, possibly even regenerative to socio-ecological systems? I actively seek this information as it drives my design processes, from micro to macro scales.

Another design method that I have developed to support this practice is that of 'knowing, acting, and caring', correlated with the 'head, heart and hand' model for transformative learning and eco-literacy first developed by David Orr in 1992.[7] (Figure 7) I use this as a framework for analysing the socio-ecological actors engaged in a design process for change, a process which identifies strengths and weaknesses, opportunities and threats to the status quo. Firstly, we must know of all the complexities at hand, considering as many different perspectives as possible, taking care to consider those that are currently excluded. Secondly, we must care enough to want to act, or be sufficiently empathic to understand why others should care more and might want to act, now and in the future. Again, by 'others' I refer to humans as well as non-human stakeholders, and to act requires a fair, non-hierarchical, and more than human-centric point of view. Thirdly, in situations where to know and to care is overwhelming and depressing, we need to instill an active spirit of "Let's do it and change it!"

Figure 9. Seaweed and light:
transluscent seaweed collar,
made in collaboration with
Moya Hoke. Photography
credit Petr Krejci.

The bridge-building, communicative agency of design is crucial in establishing and balancing out the triad that is 'knowing, acting, and caring'. By observing who cares about what and how, what action is taking place, and whether both align with what we know about the system, we become aware of voices still missing in our collective. We must ask: Whose views are we lacking? Knowing, acting, and caring all need to be present for meaningful change to take place, and creative practices can help communities of practice to build and interweave them in a synchronised way. Caring gives us the impetus to act, action must be rooted in knowing to be guided in the right direction, and knowing turns into change when enacted, instilling a sense of connection and care within a community of practice. Creative practices can offer the tools and mindset to connect scientists and activists, inventors and policymakers, and to engage them in constructive action towards preferable futures.

ENDNOTES

1. Julia Lohmann, *The Department of Seaweed: co-speculative design in a museum residency* (London: Royal College of Art, 2018)

2. Bruno Latour, *Reassembling the Social: An Introduction to Actor-Network-Theory.* Oxford: Oxford University Press, 2005; Tim Ingold, *Making: Anthropology, archaeology, art and architecture,* (London: Routledge, 2013); Jirō Yoshihara, "Gutai bijutsu sengen," *Geijutsu Shinchō* 7, no. 12 (December 1956): 202–04. http://web.guggenheim.org/exhibitions/gutai/data/manifesto.html._

3. Elvin Karana et al., "Material Driven Design (MDD): A Method to Design for Material Experiences," *International Journal of Design* 19, no. 2 (2015): 35-54.

4. Richard Buchanan, "Design Research and the New Learning," *Design Issues* 17, no. 4 (2001): 3–23.

5. BioColour Research Consortium, *BioColour, Future's Pallette,* last modified October 19, 2021, https://biocolour.fi/en/frontpage/

6. Richard Buchanan, "Design Research and the New Learning."

7. Orr, David W. *Ecological literacy: Education and the transition to a postmodern world.* (Albany, New York: SUNY Press, 1992)

NANOTECHNOLOGY IN ARCHITECTURE:
FROM MATERIALS TO NANOROBOTIC ENVIRONMENTS

MARTINA DECKER

ABSTRACT

Nanotechnology has been hailed as the next 'general-purpose technology' with the potential to encourage far-reaching changes that could affect our whole society. In architecture and design, it is starting to find its place in various pathways since initial investigations in the 1950s when Richard Feynman launched the conversation. This chapter introduces the evolution of nanotechnology in material science. It discusses how manipulating material features at the nanoscale improves their capabilities and how nanotechnology imbues materials with novel properties. Besides the introduction of nano particles in traditional building materials such as concrete, several products such as photovoltaics and sensors have also been upgraded through material adjustments smaller than 100 nanometers in size. In reviewing nanomanufacturing and the creation of atomically precise products in adjacent research fields, the reader is introduced to early explorations of nanorobotics for architectural settings, such as the Storrs Holl's Utility Fog. The risks and opportunities of using nanotechnology are also discussed. The ethical development of this disruptive technology must be at the forefront of research not only for the creation of new materials but also for their use and end-of-life strategies.

+ nanotechnology
+ nanorobotics
+ composite
  materials
+ nano-
  manufacturing
+ nano-ethics

Figure 1. Detail of Rendering of Foglets assembling in a Gray Goo Scenario. © Martina Decker.

The Dawn of Nanotechnology

Now that we can truly "arrange atoms the way we want"—as enchantingly foreseen by physicist Richard Feynman (1918-1988) at the 1959 annual meeting of the American Physical Society held at the California Institute of Technology and in his subsequent paper "There's Plenty of Room at the Bottom"—what might one want to make with them? [1] Nanotechnology engages matter at the molecular scale, where measures are less than 100 nanometers. This requires a new approach to making. The products of nanotechnology must be less than 100 nanometers in at least one of their three dimensions to be counted in this category. Long elongated solid nanostructures called nanowires, for example, can be hundreds of micrometers in length but can have a diameter as small as 1 nanometer. Nanoparticles, according to this definition, must be between 1 and 100 nanometers in diameter, as in nanosilver particles that are used for their antimicrobial properties in many commercially available products. [2] So defined, one might say that the entire field of inquiry is bound by considerations of scale, and that it is precisely this circumstance that affords nanotechnology its impressive range. Disciplines such as chemistry, materials science, physics, biology, engineering, medicine, among others have been impacted by innovations in nanotechnology.

Nanomaterials in Architecture

In architecture we have enjoyed the fruits of nanotechnology for over a decade now. During this time, the field has introduced and integrated emergent materials and devices that have been perfected at the nanoscale. [3] To approach material design at the molecular scale is a powerful tool since nanomaterials behave differently from their macro-sized counterparts. We all know the shiny optical quality that is the color gold, but gold nanoparticles, when suspended in a solution, will appear to the eye to be red purple. [4] Size dependent optical properties are by far not the only interesting behaviors that change at the nanoscale. The surface area to weight ratio of nanoparticles, in comparison to larger particles of the same material, is vastly increased at the nano scale. (Figure 2) This increase of area per unit volume is the reason why many materials may be inert in their bulk form (meso scale) but very reactive at their nano scale. Aluminum is such a material. The lightweight metal is inert at the meso scale, but so reactive when in the form of nanoparticles that it has been used in rocket propulsion for decades. [5]

Surface to volume ratio can also play an important role in improving the performance of ordinary Portland Cement. By altering the cement granulate size using a high energy milling process, researchers have tested the role of particle size—at the level of micrometers and even nanometers—for improving the compressive strength of the final concrete product. Ultrafine mixtures of Portland cement result in an enhanced overall compressive strength as well as a modified

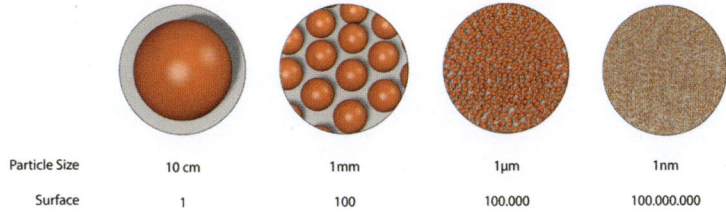

Figure 2. Schematic representation of particle size and surface from the macro to the nanoscale. © Martina Decker.

Particle Size	10 cm	1mm	1μm	1nm
Surface	1	100	100.000	100.000.000

curing time.[6] Concrete is also imbued with many other properties when adding nanomaterials to the mix. Carbon Nanotubes (CNTs), for example, can improve crack resistance and fracture toughness while titanium dioxide (TiO2) nanoparticles can facilitate a photocatalytic pollution remediation.[7]

And yet, concrete is most certainly not the only building material that has undergone a nanotechnological evolution. Examples of improved or reinvented materials include glass that can be rendered self-cleaning,[8] textiles that are dirt repellant and antimicrobial,[9] high performance insulation in the form of aerogels with exceptional thermal, acoustic, and optical properties,[10] and paint additives that prevent lasting, damaging effects of graffiti on surfaces.[11]

Nano-enhanced Devices in Architecture

The operation of many devices, or device components, has also been enhanced through atomic precision. Photovoltaic (PV) technologies, for example, have come a long way since the first commercially viable crystalline silicon (Si)-based cell was developed at Bell Labs in 1954.[12] Since then, the performance of PV technologies has been enhanced by nanotechnology products such as nanocomposites, nanostructured

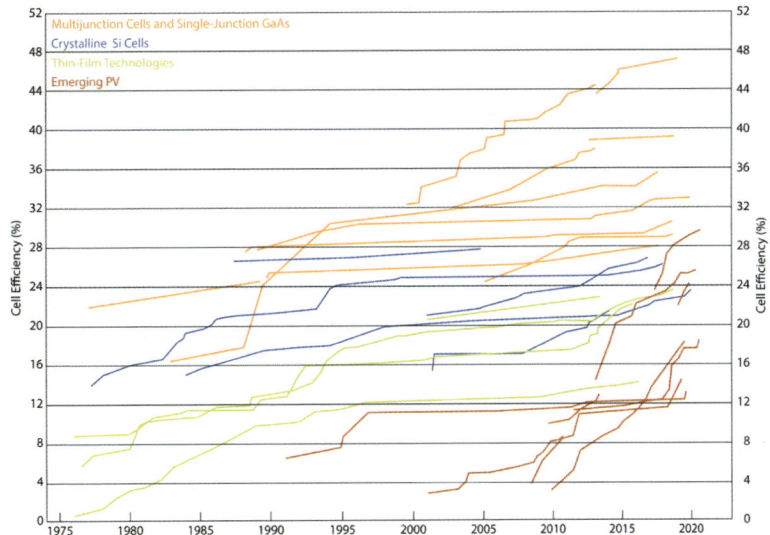

Figure 3. Reported timeline of selected solar cell energy conversion relative to the rise in number of papers relating to research in nanotechnology. Data by reported National Renewable Energy Laboratory (NREL).© Martina Decker.

polycrystalline materials, nanowires, nanotubes, nanoparticles, and quantum dots, as well as with the integration of quantum wells.[13] Today's best solar cells convert the energy of light directly into electricity with an efficiency of 47% in laboratory settings, steadily improving since Norio Taniguchi first coined the word nanotechnology in 1974.[14] (Figure 3) This has impacted architecture and building integrated photovoltaics (BIPV). We will remember that BIPVs serve as the outer layer in façade systems and are prized for their ability to generate on-site energy. This is crucial since in 2020, 65% of the electricity produced in the power sector was lost during the generation, transmission, and distribution process. Roughly 75% of electricity produced in the United States is consumed in residential and commercial buildings and the integration of high efficiency BIPVs could help our society wean itself off the use of nonrenewable resources.[15]

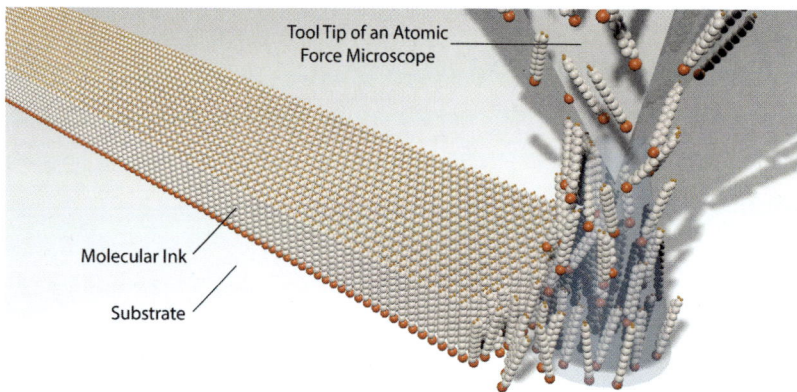

Tool Tip of an Atomic Force Microscope

Molecular Ink

Substrate

Figure 4. Dip Pen Nanolithography uses a scanning probe lithography technique. Molecular ink can be deposited with high precision onto a substrate with the tool tip of an atomic force microscope. © Martina Decker.

Another area of investigation revolutionized by nanotechnology is that of nanosensors for monitoring urban ecosystems and smart city infrastructures.[16] Nanosensors have several advantages over conventional sensing technologies. Given their small size, they can translate information from the nanoscale to the macroscale and they can be embedded in many architectural building components with a very high spatial resolution. The mentioned carbon nanotubes (CNTs), for example, that introduce crack resistance and high compressive strength in concrete, can also double as sensors that monitor the very cracks they are asked to minimize. Embedded in a cement matrix as CNT networks, their excellent conductivity creates a charge that can be sent through the network to detect cracks in the cement matrix.[17] The collected data can give us important information on the condition of our built environment and inform the decision-making process around operation, maintenance, and repairs of buildings. Other examples of nanosensors are those developed to monitor humidity, corrosion, toxins, pathogens, and pollutants in service to increased health, safety, and welfare.[18]

Nanomanufacturing

The production and synthesis of materials and devices at the nanoscale require a new approach to making. Since our eyes cannot observe anything smaller than 0.1 millimeters, which is one thousand times larger than the largest structural feature in nano-products, highly sophisticated microscopes are essential for conducting research at the nanoscale. Scanning Probe Microscopes (SPMs), for example, allow us to 'feel' the surface of a material sample with a probe tip, discerning information in a manner not unlike the way fingers are able to read raised Braille characters. These minute sensing probes can also actively manipulate a specimen by pushing, pulling, or depositing nanoparticles—be they molecules or atoms—to form nanoproducts. (Figure 4)

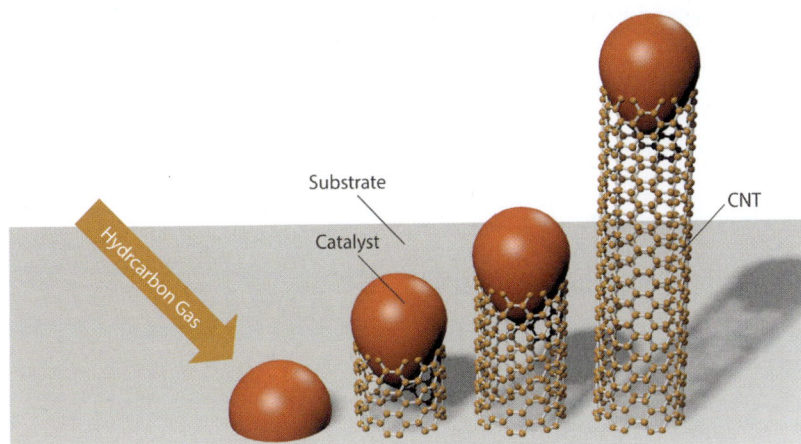

Figure 5. Schematic representation of CVD in the production of CNTs. © Martina Decker.

Even though the methods involved when manipulating matter using SPMs can be very precise, they do not lend themselves to low-cost and high-throughput nano-manufacturing. A different number of manufacturing methods at the nanoscale are being developed to service the growing nano-market. Chemical Vapor Deposition (CVD), for example, has become the dominant method for producing CNTs. The long hollow tubes self-assemble out of metal catalyst particles when carbon-based gasses are introduced in a reaction chamber that is heated to a very high temperature. (Figure 5) The size of the catalyst nanoparticles will determine the diameter of the nanotube while the reaction time will regulate its length. CVD is one of several methods for producing carbon nanotubes that allow for the growth of large quantities at a time.[19]

While these innovations are impressive, organic self-assembly offers an excellent model for manufacturing at the nanoscale. Organic cell reproduction is a marvelous and meticulous example of the precise arrangement that is possible when manipulating atoms.

Research into DNA replication, the essential phase that ensures that the genetic code is effectively duplicated before cell division, has resulted in the development of molecular building blocks for nanotech applications. This research field is called DNA Origami. The DNA molecule can be enticed to take on complex shapes and forms. In this process the complimentary base pairs that usually assemble into a double helix can be interconnected by staple strands (also made of DNA). By way of these connections, the molecules assemble into shapes that deviate from the usual double helix, enabling the synthesis of highly complex geometries, such as pyramids, hemispheres, spheres, ellipsoids, or flasks.[20] The resulting molecular structures have a variety of applications. These DNA nanostructures are inherently biocompatible since they are made of the exact building blocks found in every living cell of our bodies. They are used, therefore, in drug delivery systems as vehicles that transport cancer drugs directly to a tumor.[21]

In addition to medical applications, DNA origami also offers a fabrication path for nanoelectronics; these can be used for integrating nanosensors in smart city infrastructures. Indeed, complex origami structures can be molds, scaffolds, or platforms in the creation of conductive miniaturized circuits. This can be achieved by combining the bottom up-self-assembly of DNA origami with top-down nano fabrication techniques of metallization, that include electrochemical and sputter deposition.[22]

Atomically Precise Products

Atomically Precise Manufacturing (APM) has been a 'future vision' since the 1980s when American engineer K. Eric Drexler introduced the idea of molecular engineering at the National Academy of Sciences.[23] After exploring the fundamental principles of APM in his work at the Massachusetts Institute of Technology (MIT), Drexler described his future vision that included the implementation of "Productive Nanosystems."[24] His nanofactories were desktop-sized molecular manufacturing units in which complex products such as cars and laptop computers could be built one molecule at a time. Certain atoms would be selected and then delivered to tool tips, which in turn would precisely position other atoms, assembling them into molecules and ever-larger building blocks. These blocks would then be bonded together and assembled into larger components, ultimately amassed into an atomically precise product such as a laptop computer with excellent processing power. These future visions inspired a whole community of scientists, including longstanding experts like American nanotechnologist Robert A. Freitas Jr. and computer scientist Ralph C. Merkle, to further develop the technology for a future desktop appliance that could produce atomically precise products.[25]

The concept of molecular assemblers that operate one atom at a time remains highly controversial in the scientific community, even if we have achieved groundbreaking progress in creating near atomically precise product components such as the previously described photovoltaic devices or nanosensors.[26]

Nanorobotics

Nanorobots, also called nanobots, take advantage of the physical phenomena that prevails at the nanoscale, even though their overall size might be in the realm of micrometers.[27] They originated in the realm of science fiction. They were designed to sense their environment, process information they collected, and adaptively act on available data to perform their tasks. While the entertaining scenario popularized by the 1966 science fiction film "Fantastic Voyage" in which a team of medical 'micronauts' and their submarine were miniaturized and injected into a patient is unlikely, analogous procedures in medicine in which nanoscale devices are used, are far more likely.[28] As previously noted, nano-devices deliver medicine with great precision and can perform cell-by-cell surgery by killing cancer cells through an intelligent drug delivery system.[29]

In addition to these applications wherein a few precious nano-robots can critically impact health outcomes, nano-machines hold even greater potential when considered as a collective mass and programmed to self-replicate. In his book *Engines of Creation* (1986), Drexler describes nanorobotic devices with the capacity for sophisticated programming that enables them to both self-replicate as well as to assemble towards the macroscale.[30] Even if such devices were to only replicate one at a time, the collective structure could grow exponentially with a constant supply of raw materials.

In one of the earliest examples of this envisioned use of molecular machines in the built environment, nanotechnologist John Storrs Hall introduced, in a report published by NASA in 1993, a design for a nanorobotic environment called "Utility Fog," whose field of multiple nanorobotic devices of varying capabilities were referred to as "Foglets." These Foglets were conceived as a near-invisible addition to the environment, floating and permeating spaces on Earth and discretely taking on functions in architecture. They could simulate any kind of object – from furniture to entire buildings. As their name suggests, Foglets would fill the air and gradually decrease visibility once one entered the region of Utility Fog. One would rely on holographic goggles to simulate a virtual reality of the user's choice. As Storrs Hall described:

> Fog can simulate air to the touch but not to the eyes. The best indications are that it would look like heavy fog. Thus, the Fog would need to support a pair of holographic goggles in front of the eyes of an embedded user.[31]

By changing the arrangement of the Foglets and by modifying the density of the Utility Fog, any object of any shape or color could be created. This nanorobotic invention could also advance automation in space, entirely creating and maintaining environments for homesteaders on the moon. The original robot that was published in Storrs Hall's paper "Utility Fog: a universal physical substance," featured a spherical body roughly forty-five micrometers in diameter which incorporated twelve telescopic arms arranged like the faces of a dodecahedron. At the end of each arm, mechanical nano-grippers allowed the individual Foglets to hold on to each other and to their environment.[32] (Figure 6)

Figure 6. Sketch of Foglet detail: Gripper © Martina Decker.

Designs for today's state of the art nanorobotic devices are in stark contrast to the mechanical, hard-bodied appearance of Storrs Hall's Utility Fog Foglets. Since they make full use of nano scaled physics, they need not rely on mechanical gripping mechanisms. The mentioned medical nanorobots, for example, that seek out and kill cancer cells use specialized DNA strands to bind themselves to proteins found near tumors. This mechanism triggers the drug's release without relying on mechanical gripping mechanisms.

One could easily dismiss these early attempts to envision a trajectory for architectural nanorobotic environments as science fiction that will never come to pass, yet this is what was said about many inventions in the past that emerged from science fiction and that turned into science facts. Forms of communication popularized by Star Trek in the 1960s, are today's cell phones, albeit far more powerful. Luke Skywalker's prosthetic hand is a reality for many people that have lost limbs, even though they might not look or feel quite like their original body parts. And yes, we have nanorobots that can aid in the cure of cancer today, but they do not look anything like Storrs Hall's Foglets, being far smaller than predicted.

The question remains what would we do architecturally with our swarms of Foglets if we had them? Could we send swarms to create hurricane barriers to protect our cities when a storm approached, or assemble them into shelters after an earthquake? Could we have them linger in the atmosphere to absorb carbon dioxide and use their molecules as raw materials to build new products?

Nano Consequences for the Macroworld

Many new technologies hold potential risks that need to be studied at the same time as we investigate their benefits. Fears relating to nanorobotic environments, such as the possibly catastrophic consequences of "Gray Goo", have been discussed not only in scientific communities but beyond. (Figure 7) Drexler describes a scenario in which self-replicating nanorobots could potentially consume our entire biosphere and turn earth into an undifferentiated mass of grayness.[33] This nightmarish scenario is highly unlikely, yet Gray Goo has become a favorite cautionary tale in nanotechnology.[34] The more pressing question, however, is whether manufactured nanomaterials pose a risk to our environment.[35]

One of the most widely used nanoparticles found in numerous anti-microbial products are the mentioned nanosilver particles. While generally believed to be nontoxic to humans, the safety of using nanosilver particles has yet to be fully assessed as they are known to wreak havoc on microbe communities.[36] Found in medical tools, clothing, sports equipment, and even personal hygiene products and toys, they are known to leach out into the environment through use and at the end of life.[37] Nanosilver contaminations have a negative, toxic effect on aquatic organisms and in municipal landfill bioreactors where we rely on microbes to break down our refuse for biogas production.[38] The careful use of these antimicrobial substances should be on our minds as designers as well as consumers, given the current risk assessment of nanosilver. While their nano-enhanced performance is invaluable in hospital settings, do we really need them in socks, so they won't smell bad?

And yet, many disruptive technologies have been developed with greater safeguards in place. The so called "Precautionary Principle" advocates for halting development processes and erring on the side of caution before leaping into action.[39]

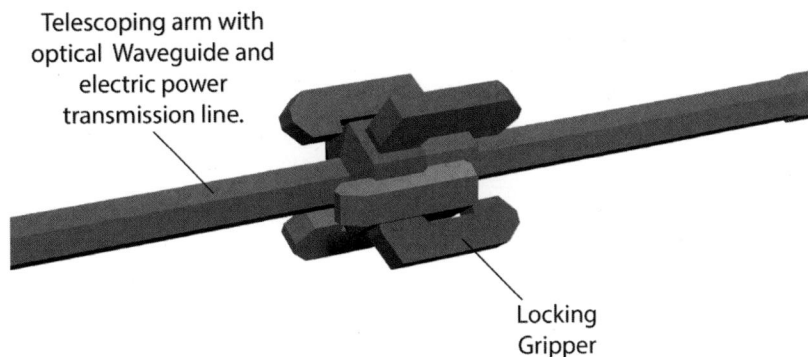

Telescoping arm with optical Waveguide and electric power transmission line.

Locking Gripper

Figure 7. Rendering of Foglets assembling in a Gray Goo Scenario. © Martina Decker.

Its most strict interpretation could completely halt any kind of nano-technological development for fear of negative impacts, while a more lenient form of the Precautionary Principle could simply slow down innovation. The Center for Responsible Nanotechnology promotes a balanced approach, while pointing out that idleness when it comes to nanotechnological innovation holds severe risks. For if we are too cautious, we might not find urgently needed solutions to the great challenges of our times. We might also hand over exploration of nano-technology to less responsible groups and be entirely unprepared to deal with the outcomes.[40]

Nano-Ethics

From the inception of new nano-products we must continue to eval-uate and reevaluate their benefits and dangers. Architecture and building industry mechanisms have not been satisfactorily updated with the introduction of nanotechnology. Material Safety Data Sheets (MSDS) for example, still lack crucial detail when it comes to engi-neered nanomaterials. Studies that analyze the knowledge gap in new construction and demolition work are underway, while they advocate the cautionary use of nanomaterials and reinforce already existing health and safety protocols. They recommend the creation of detailed material records of help in decision-making during the refur-bishment or demolition of nano-enhanced buildings.[41] Furthermore, the Environmental Protection Agency (EPA) has started to control nanoscale materials under the Toxic Substances Control Act that seeks to promote innovation while ensuring the safety of the novel substances.[42]

Inventions in nanotechnology demand a new field in ethics: Nano-Ethics.[43] Albeit in its infancy, it has already been scrutinized and challenged. The very fragmentation of nanotechnology means that this practice encompasses many disciplines, drawing its ethics from the sciences, medicine, technology, and environmental studies.[44] Considerations of equity, for example, are of great importance since the benefits of nanotechnology—such as clean energy and medical innovations—are most needed in developing countries.[45]

Technological Revolutions and Ecosystem Engineering

All examples described in this essay reinforce the idea that nano-technology is here to stay. It fulfills the three main requirements of the league of 'general-purpose technologies' (GPTs), having touched society at large with 'pervasiveness, continuous improvement, and innovation spawning'.[46] Nanotechnology is widespread, influencing a vast number of fields and industries. Improvements afforded by nanotechnology have led to productivity growth in the photovoltaics industry.[47] Nanorobotic innovations in medical applications would have been impossible without nanotechnology. Some researchers have called it a "revolution" with the capacity to reshape our global economy and societies.[48]

Agricultural and industrial revolutions have given us powerful tools and techniques to engage in ecosystem engineering. We have the ability to transform our environment on a global scale, that includes increasing the earth's carrying capacity.[49] Only 23% of our planet's landmass and 13.3% of our oceans can be still defined as "wilderness", untouched by humans.[50] (Figure 8) It is also the very same two technological revolutions that can be directly linked to anthropogenic climate change and the loss of biodiversity. The nanotechnology revolution holds the promise of helping us address some of the negative effects of previous technological advancements. A new subfield called "Environmental Nanotechnology" is developing innovations that will revolutionize how we handle for example agriculture, pollution, water safety, food safety, energy production, and energy storage.[51] Some of the harmful consequences of our actions on this planet could be disrupted, alleviated, or even rectified through nanotechnologies with the capability to produce fundamental transformations at the scale of atomic elements.

Figure 8. Percentages of wilderness and transformed terrestrial and aquatic environments. (Data by Allan 2017 and Jones 2018) © Martina Decker.

70.8% of earth's surface area is covered by water.

23%
Wilderness

77% Modified by Humans

23% of the world's landmass is classified as wilderness.

13.2%
Wilderness

86.6% Modified by Humans

13.2% of the world's ocean is classified as marine wilderness

ENDNOTES

1. Richard P. Feynman, "There's Plenty of Room at the Bottom" *Journal of Microelectromechanical Systems* 1, no. 1 (1992): pp. 60-66, https://doi.org/10.1109/84.128057.

2. Malcolm Xing et al., "Nanosilver Particles in Medical Applications: Synthesis, Performance, and Toxicity," *International Journal of Nanomedicine*, 9, no.1 (2014): 2399. https://doi.org/10.2147/ijn.s55015.

3. Fernando Pacheco-Torgal and Said Jalali, "Nanotechnology: Advantages and Drawbacks in the Field of Construction and Building Materials," *Construction and Building Materials* 25, no. 2 (2011): 582-90. https://doi.org/10.1016/j.conbuildmat.2010.07.009; George Elvin, "Nanotechnology in Architecture," *Nanotechnology for Energy Sustainability*, (November 2017): 967-96. https://doi.org/10.1002/9783527696109.ch39.

4. Olivier Pluchery, "Optical Properties of Gold Nanoparticles," *Gold Nanoparticles for Physics, Chemistry and Biology* (2012): 43-73, https://doi.org/10.1142/9781848168077_0003.

5. Luigi T. DeLuca, "Overview of Al-Based Nanoenergetic Ingredients for Solid Rocket Propulsion," *Defence Technology* 14, no. 5 (2018): 357-65, https://doi.org/10.1016/j.dt.2018.06.005.

6. Roukaya Mejdoub et al., "The Effect of Prolonged Mechanical Activation Duration on the Reactivity of Portland Cement: Effect of Particle Size and Crystallinity Changes," *Construction and Building Materials* 152 (2017): 1041-050, https://doi.org/10.1016/j.conbuildmat.2017.07.008; Armando García-Luna, and R Bernal. Diego. "High strength micro/nano fine cement." In 2nd International Symposium on Nanotechnology in Construction, vol. 13. 2005.

7. Ahmed Hawreen, José Bogas, and Rawaz Kurda, "Mechanical Characterization of Concrete Reinforced with Different Types of Carbon Nanotubes," *Arabian Journal for Science and Engineering* 44, no. 10 (2019): 8361-376, https://doi.org/10.1007/s13369-019-04096-y; Anantha-Iyengar Gopalan et al., "Recent Progress in the Abatement of Hazardous Pollutants Using Photocatalytic Tio2-Based Building Materials," *Nanomaterials* 10, no. 9 (2020): 1854, https://doi.org/10.3390/nano10091854.

8. Walter A. Daoud, Neil Shirtcliffe, and Paul Roach, "Superhydrophobic and Self-Cleaning," in *Self-Cleaning Materials and Surfaces: A Nanotechnology Approach* (Chichester, England: Wiley, 2013), 3-28.

9. Andrea Zille et al., "Application of Nanotechnology in Antimicrobial Finishing of Biomedical Textiles," *Materials Research Express* 1, no. 3 (2014): 032003, https://doi.org/10.1088/2053-1591/1/3/032003.

10. Ruben Baetens, Bjørn Petter Jelle, and Arild Gustavsen, "Aerogel Insulation for Building Applications: A State-of-the-Art Review," *Energy and Buildings* 43, no. 4 (2011): 761-69, https://doi.org/10.1016/j.enbuild.2010.12.012.

11. Martin Heinisch and Dan Miricescu, "Innovative Industrial Technologies for Preventive Anti-Graffiti Coating," *MATEC Web of Conferences* 121 (2017): 03009, https://doi.org/10.1051/matecconf/201712103009.

12. Daryl M. Chapin, Calvin S. Fuller, and Gerald L. Pearson, "A New Silicon p-n Junction Photocell for Converting Solar Radiation into Electrical Power," *Journal of Applied Physics* 25, no. 5 (1954): 676-77, https://doi.org/10.1063/1.1721711.

13. Loucas Tsakalakos, "Nanostructures for Photovoltaics," *Materials Science and Engineering: R: Reports* 62, no. 6 (2008): 175-89.

14. NREL, "Best Research-Cell Efficiency Chart," NREL.gov, December 14, 2021, https://www.nrel.gov/pv/cell-efficiency.

html. See also Norio Taniguchi, "On the Basic Concept of Nanotechnology," in *Proceedings of the International Conference on Production Engineering* (Tokyo, Japan: Japan Society of Precision Engineering, 1974).

15. "U.S. Energy Information Administration - EIA - Independent Statistics and Analysis," U.S. Energy Information Administration (EIA), accessed January 25, 2022, https://www.eia.gov/totalenergy/data/monthly/pdf/flow/total_energy_2020.pdf.

16. Baoguo Han et al., eds., *Nanosensors for Smart Cities* (Amsterdam, Netherlands: Elsevier, 2020).

17. Mohamed Saafi, "Wireless and Embedded Carbon Nanotube Networks for Damage Detection in Concrete Structures," *Nanotechnology* 20, no. 39 (September 2, 2009): 395502-1-395502-7, https://doi.org/10.1088/0957-4484/20/39/395502.

18. Zhenyu Li et al., "Highly Sensitive and Stable Humidity Nanosensors Based on LiCl Doped tio2 Electrospun Nanofibers," *Journal of the American Chemical Society* 130, no. 15 (March 15, 2008): 5036-037, https://doi.org/10.1021/ja800176s. See also Loïc Exbrayat et al., "Nanosensors for Monitoring Early Stages of Metallic Corrosion," *ACS Applied Nano Materials* 2, no. 2 (January 11, 2019): 812-18, https://doi.org/10.1021/acsanm.8b02045.

Peter J. Vikesland, "Nanosensors for Water Quality Monitoring," *Nature Nanotechnology* 13, no. 8 (2018): 651-60, https://doi.org/10.1038/s41565-018-0209-9.

19. J. Robertson et al., "Chemical Vapor Deposition of Carbon Nanotube Forests," *Physica Status Solidi (b)* 249, no. 12 (May 2, 2012): 2315-322, https://doi.org/10.1002/pssb.201200134.

20. Dongran Han et al., "DNA Origami with Complex Curvatures in Three-Dimensional Space," *Science* 332, no. 6027 (2011): 342-46, https://doi.org/10.1126/science.1202998.

21. Qian Zhang et al., "DNA Origami as an in Vivo Drug Delivery Vehicle for Cancer Therapy," *ACS Nano* 8, no. 7 (June 25, 2014): 6633-643, https://doi.org/10.1021/nn502058j.

22. Xinpei Dai et al., "DNA-Based Fabrication for Nanoelectronics," *Nano Letters* 20, no. 8 (July 27, 2020): 5604-615, https://doi.org/10.1021/acs.nanolett.0c02511.

23. K. Eric Drexler, "Molecular Engineering: An Approach to the Development of General Capabilities for Molecular Manipulation," *Proceedings of the National Academy of Sciences* 78, no. 9 (September 1, 1981): 5275-278, https://doi.org/10.1073/pnas.78.9.5275.

24. K. Eric Drexler, "Molecular Machinery and Manufacturing with Applications to Computation", (Thesis (Ph.D.) Massachusetts Institute of Technology, Dept. of Architecture, 1991; K Eric Drexler and John Burch, "Productive Nanosystems: From Molecules to Superproducts," Internet Archive (Lizard Fire Studios, October 27, 2004), https://archive.org/details/NanoFactory.

25. Robert A Freitas and Ralph C Merkle, "What Is a Nanofactory?", Nanofactory collaboration, June 14, 2006, http://www.molecularassembler.com/Nanofactory/.

26. Edward L. Wolf and Manasa Medikonda, *Understanding the Nanotechnology Revolution* (Weinheim, Germany: Wiley-VCH, 2012).

27. Sylvain Martel, "Swimming Microorganisms Acting as Nanorobots versus Artificial Nanorobotic Agents: A Perspective View from an Historical Retrospective on the Future of Medical Nanorobotics in the Largest Known Three-Dimensional Biomicrofluidic Networks," *Biomicrofluidics* 10, no. 2 (2016): 021301, https://doi.org/10.1063/1.4945734.

28. *Fantastic Voyage,* directed by Richard O. Fleischer (Twentieth Century Fox, 1966).

29. Suping Li et al., "A DNA Nanorobot Functions as a Cancer Therapeutic in Response to a Molecular Trigger in Vivo," *Nature Biotechnology* 36, no. 3 (February 12, 2018): 258-64, https://doi.org/10.1038/nbt.4071.

30. K. Eric Drexler, *Engines of Creation the Coming Era of Nanotechnology* (New York, United States: Anchor Press, Doubleday, 1986).

31. John Storrs Hall, "Utility Fog: a Universal Physical Substance," *Interdisciplinary Science and Engineering in the Era of Cyberspace* (1993): 115-36.

32. Hall, "Utility Fog: a Universal Physical Substance," 118.

33. Drexler, *Engines of Creation.* 127

34. Ann Dowling et al., "Nanoscience and Nanotechnologies: Opportunities and Uncertainties," *The Royal Society,* (2004) accessed January 25, 2022, https://royalsociety.org/-/media/Royal_Society_Content/policy/publications/2004/9693.pdf. See also Marie-Hélène Fries, "Nanotechnology and the Gray Goo Scenario: Narratives of Doom?," *Revue De l'Institut Des Langues Et Cultures D'Europe, Amérique, Afrique, Asie Et Australie,* no. 31 (March 1, 2018): 1-17, https://doi.org/10.4000/ilcea.4687.

35. Richard Owen and Richard Handy, "Formulating the Problems for Environmental Risk Assessment of Nanomaterials," *Environmental Science & Technology* 41, no. 16 (August 1, 2007): 5582-588, https://doi.org/10.1021/es072598h.

36. Thomas Faunce and Aparna Watal, "Nanosilver and Global Public Health: International Regulatory Issues," *Nanomedicine* 5, no. 4 (June 8, 2010): 617-32, https://doi.org/10.2217/nnm.10.33.

37. Troy Benn et al., "The Release of Nanosilver from Consumer Products Used in the Home," *Journal of Environmental Quality* 39, no. 6 (2010): 1875-882, https://doi.org/10.2134/jeq2009.0363.

38. Zhuang Wang et al., "Aquatic Toxicity of Nanosilver Colloids to Different Trophic Organisms: Contributions of Particles and Free Silver Ion," *Environmental Toxicology and Chemistry* 31, no. 10 (2012): 2408-413, https://doi.org/10.1002/etc.1964. Yu Yang et al., "Nanosilver Impact on Methanogenesis and Biogas Production from Municipal Solid Waste," *Waste Management* 32, no. 5 (2012): 816-25, https://doi.org/10.1016/j.wasman.2012.01.009.

39. Steve Clarke, "Future Technologies, Dystopic Futures and the Precautionary Principle," *Ethics and Information Technology* 7, no. 3 (2005):121-26, https://doi.org/10.1007/s10676-006-0007-1.

40. Chris Phoenix and Mike Treder, "Applying the Precautionary Principle to Nanotechnology," *Center for Responsible Nanotechnology,* (2004): 1-5.

41. Beatriz María Díaz-Soler, María Dolores Martínez-Aires, and Mónica López-Alonso, "Potential Risks Posed by the Use of Nano-Enabled Construction Products: A Perspective from Coordinators for Safety and Health Matters," *Journal of Cleaner Production* 220 (2019): 33-44, https://doi.org/10.1016/j.jclepro.2019.02.056. See also Alistair Gibb et al., "Nanotechnology in Construction and Demolition: What We Know, What We Don't," *Construction Research and Innovation* 9, no. 2 (June 3, 2018): 55-58, https://doi.org/10.1080/20450249.2018.1470405.

42. John C Monica, "Examples of Recent EPA Regulation of Nanoscale Materials Under the Toxic Substances Control Act," *Nanotechnology Law & Business* 6 (2009): 388-406. See also EPA, "Control of Nanoscale Materials under the Toxic Substances Control Act" (Environmental Protection Agency), accessed January 25, 2022, https://www.epa.gov/reviewing-new-chemicals-under-toxic-substances-control-act-tsca/control-nanoscale-materials-under.

43. Chris Toumey, "Early Voices for Ethics in Nanotechnology," *Nature Nanotechnology* 14, no. 4 (April 3, 2019): 304-05, https://doi.org/10.1038/s41565-019-0422-1.

44. Armin Grunwald, "Nanotechnology — a New Field of Ethical Inquiry?," *Science and Engineering Ethics* 11, no. 2 (June 2005): 187-201, https://doi.org/10.1007/s11948-005-0041-0.

45. Anisa Mnyusiwalla, Abdallah S Daar, and Peter A Singer, "Mind the Gap: Science and Ethics in Nanotechnology," *Nanotechnology* 14, no. 3 (February 13, 2003), R9-R13, https://doi.org/10.1088/0957-4484/14/3/201.

46. Vernon W Ruttan, "General Purpose Technology, Revolutionary Technology, and Technological Maturity," *Staff Paper P08-3,* (2008): 1-25, https://doi.org/10.22004/ag.econ.6206.

47. T. K. Manna and S. M. Mahajan, "Nanotechnology in the Development of Photovoltaic Cells," *2007 International Conference on Clean Electrical Power* (July 16, 2007): 379-386, https://doi.org/10.1109/iccep.2007.384240.

48. Wolf, *Understanding the nanotechnology revolution.* IX. See also K. Eric Drexler, *Radical Abundance: How a Revolution in Nanotechnology Will Change Civilization* (New York: BBS Public Affairs, 2013), 50.

49. Menno Schilthuizen, *Darwin Comes to Town: How the Urban Jungle Drives Evolution* (New York: Picador, 2018). See also, Bruce D. Smith and Melinda A. Zeder, "The Onset of the Anthropocene," *Anthropocene* 4 (2013): 8-13, https://doi.org/10.1016/j.ancene.2013.05.001.

50. James R. Allan, Oscar Venter, and James E.M. Watson, "Temporally Inter-Comparable Maps of Terrestrial Wilderness and the Last of the Wild," *Scientific Data* 4, no. 1 (December 12, 2017): 1-8, https://doi.org/10.1038/sdata.2017.187. See also, Kendall R. Jones et al., "The Location and Protection Status of Earth's Diminishing Marine Wilderness," *Current Biology* 28, no. 15 (August 6, 2018): 2506-2512, https://doi.org/10.1016/j.cub.2018.06.010.

51. Hongqi Sun, "Grand Challenges in Environmental Nanotechnology," *Frontiers in Nanotechnology* 1 (December 20, 2019), 1-3, https://doi.org/10.3389/fnano.2019.00002. See also, Gregory V. Lowry, Astrid Avellan, and Leanne M. Gilbertson, "Opportunities and Challenges for Nanotechnology in the Agri-Tech Revolution," *Nature Nanotechnology* 14, no. 6 (June 5, 2019): 517-22, https://doi.org/10.1038/s41565-019-0461-7.

Joerg Lahann, "Nanomaterials Clean Up," *Nature Nanotechnology* 3, no. 6 (May 30, 2008): 320-21, https://doi.org/10.1038/nnano.2008.143; Ilka Gehrke, Andreas Geiser, and Annette Somborn-Schulz, "Innovations in Nanotechnology for Water Treatment," *Nanotechnology, Science and Applications,* January 6, 2015, https://doi.org/10.2147/nsa.s43773; Bhupinder Sekhon, "Food nanotechnology- an overview," *Nanotechnology, Science and Applications* (May 4, 2010): 1-15, https://doi.org/10.2147/nsa.s8677; Abdul-Sattar Nizami and Mohammad Rehan, "Towards Nanotechnology-Based Biofuel Industry," *Biofuel Research Journal* 5, no. 2 (June 1, 2018): 798-799, https://doi.org/10.18331/brj2018.5.2.2; Mohammad H. Ahmadi et al., "Renewable Energy Harvesting with the Application of Nanotechnology: A Review," *International Journal of Energy Research* 43, no. 4 (November 14, 2018): 1387-410, https://doi.org/10.1002/er.4282; Peiwen Li, "Energy Storage Is the Core of Renewable Technologies," *IEEE Nanotechnology Magazine* 2, no. 4 (December 2008): 13-18, https://doi.org/10.1109/mnano.2009.932032.

BIO
MATTER
TECHNO
SYNTHETICS

CONVERGENCE OF ART, COMPUTING AND REPRESENTATION

MARTA LLOR

Stemming from the Greek word *téchnē*—whose definition is typically allied with art, craft, and skill—my choice of the term, 'Techno' intersects advancements in computing technologies with their uses in design, speculative representations, and questions of machine agency. The convergence of art practice, computation, and representation is not new, having been explored for more than a century. From Sigfried Gideon's (1888-1968) conceptualization of mechanization to contemporary robotically-aided construction and their graphic simulations, representation is a means of communication which binds the seemingly disparate disciplines of art and computing. However, this collection of authors speaks to the agency of machines in design, to epistemologies of communication between machines and humans, and to representational approaches that support computationally generated design processes.

Behnaz Farahi explores the sensing capacity of material systems in projects that speak to biological processes integrated with computational technologies at the interface of human emotions. The expressivity of matter is explored as encoded information that distinguishes, at a molecular level, sensations and behavioral responses. Synthetically re-creating human involuntary reactions, Farahi re-frames the relationship between affective computing and humans through the articulation of matter. Her research communicates in ways that transcend two-dimensional graphic modes of representation, by materializing three-dimensional objects capable of adapting their behaviors to varying sets of stimuli.

Collaborative design takes on a new meaning in **Stefana Parascho**'s piece wherein robots are entities that—more than mere tools—innovate alongside designers. Her design-build projects demonstrate the need for a shared representational language to emerge in which intuition-based design and digital technologies coalesce. Design is here a logical sequence of decisions built upon previous ones, which depend on machine intelligence that is able to 'see' future outcomes as part of the design process. Parascho calls for a shift to human-robot collaborative design practices and to the development of a shared visual language that asks questions beyond optimization.

Dorit Aviv's essay argues that architecture must embrace the thermal sensations experienced by humans. Her research highlights how architectural elements that mediate exterior climate and interior comfort have the capacity to influence thermodynamic flows with the goal of helping humans reach optimal thermal sensation and delight: alliesthesia. This shift towards thermally effusive materiality requires computational technologies and material research whose data collection and thermal imagining informs architectural design. Aviv's gradient-based representations of thermal flows encourage a fluid approach to designing with materially integrated mechanical systems that help designers think through climate adaptation.

Viola Ago's research highlights the importance of visual representations for working through ideas rather than being mere means to an end. Expressed through the drawings of artists and the words of philosophers, Ago discusses the recent history of form-based representations concerned with movement and motion in human and non-human actors. Her own drawings, which reveal the processes used in their production, transcend two- and three- dimensional space. While her research echoes that of other authors who have reflected on the potential of visual representations to synergistically choreograph the collaborative actions of designer and machine.

Jacqueline Wu challenges reality and notions of artificial nature by culling and curating photogrammetrees. Her work critiques and re-thinks the culture of data visualization in which digital images of trees are produced and consumed via technological, spatial, and social frameworks. Wu challenges the digital aesthetics of commercial technologies by discussing the utilitarianism of the representations they produce. Her research exists at the threshold of the artificial and the physical, as objects that are originally physical become digitally abstracted, culled and re-digitized in the physical world. Wu highlights the disconnect between our desire for high-fidelity outputs and the high degree of abstraction that subtends computing technologies.

Techno's cohort of authors stem from a variety of backgrounds and research fields. They all find common ground, however, at the intersection of computing, art, and design. Their work identifies synergies best manifest in their shared call for the development of a representational language that transcends human communication in its bridge to robotics, data, and the digitial. Techno highlights the need to integrate machines as entities that work alongside us in the design process, and that collaborate in the emergence of a new representational language.

EMOTIVE MATTER:

APPLICATION OF AFFECTIVE COMPUTING AND NEW MATERIALISM IN DESIGN

BEHNAZ FARAHI

ABSTRACT

Emerging material systems are inherently different from traditional materials in that the substrate of matter is now a blend of physical and digital information. In this context, matter can behave as a machine and the machine as matter. This paper argues that the integration of emerging digital technologies into matter should not be solely for pragmatic reasons such as efficiency, but as a means of addressing larger psychosocial questions, such as human emotion. As such, through the lens of 'affective computing' and 'new materialism' the research presented in this paper, develops a theoretical framework for a new empathic approach to the design of materials, where the roles of emotions and matter are re-assessed. Emotive Matter is a research initiative which develops bio-inspired material systems equipped with computational technologies that can interface with human emotions. *Opale*, a project conceived and developed by the author, illustrates the implementation of Emotive Matter for the design of a soft robotic wearable.

+ expressive matter
+ bio-inspired
+ affect-sensing
+ affective computing
+ soft robotics

Figure 1. *Opale*: An emotive soft robotic wearable move based on emotions of people around. Photographers Nicolas Cambier and Kyle Smithers.

Introduction

What is Emotive Matter? Are material systems capable of recognizing human emotions and provoking emotional responses in users? In other words, can materials be imbued with sensing and actuating capabilities that foster affective loops between users and their environment? [1] In most material research, matter is commonly investigated solely in terms of 'performance' and 'functionality'; seldom is it associated with human emotions. Moreover, most advances in artificial intelligence (AI) and affective computing have been limited to the digital environment and have not been integrated into material research. By bringing together insights from 'affective computing' and 'new materialism', the research discussed in this paper reappraises the relationship between emotions and matter, while exploring the potentially dynamic effects which results when the two are combined. Firstly, it provides the theoretical background on expressive matter, interpreted through the lens of new materialism by specifically examining the role of 'material expressivity' in both living and non-living systems and by addressing the implementation of active materials in design practices. Secondly, through the lens of affective computing, it illustrates how computational material systems can sense and decode the emotional state of users by tracking their physiological responses. Illustrated through *Opale*, a project developed by the author, this paper demonstrates how one can track and respond to a user's emotional states using soft robotic and computer vision technologies such as facial expression tracking.

Expressive Matter

> Physical information pervades the world and it is through its continuous production that matter may be said to express itself.
>
> Manual DeLanda, *Matter Matters* [2]

The relationship between form (*morphe*) and matter (hyle) is central to the world of design, even if typically, it is 'form' that is privileged over 'matter'. According to Tim Ingold, Aristotle united form (morphe) and matter (hyle) in his hylomorphic model, yet in Western philosophy matter is commonly seen as subservient to form. [3] An alternative *morphogenetic* model, however, theorized and articulated by philosophers Gilles Deleuze and Felix Guattari, asserts that process (matter) can be privileged over representation (form). Emerging as we are from a period of postmodernism which favored representation over process, we are under the obligation to counterbalance this condition by privileging process over representation. In so doing, matter is allowed to express itself, and form is reappraised as the result of material forces. With their emphasis on process, Deleuze and Guattari consider not only the materiality of forms, but also biological, geological, and social processes. They draw upon the notion of double articulation which refers to the processes by which an object is produced out of other objects.[4] As Manuel DeLanda states in his article "Deleuze, Materialism and Politics" when discussing the idea of double articulation:

> The first articulation concerns the materiality of a stratum: the selection of the raw materials out of which it will be synthesized (such as carbon, hydrogen, nitrogen, oxygen, and sulfur for biological strata) as well as the process of giving populations of these selected materials some statistical ordering. The second articulation concerns the expressivity of a stratum.... This second articulation is therefore the one that consolidates the ephemeral form created by the first articulation and that produces the final material entity defined by a set of qualities expressing its identity.[5]

DeLanda draws upon Deleuze and Guattari's notion of "expressivity," which is not related to linguistic concerns, but rather to the color, sound, texture, movement, and geometrical forms of matter. DeLanda refers to these qualities as "fingerprints" which can be used to identify and determine the properties of a given material through the process of spectroscopy.[6] Spectroscopy was first developed in the nineteenth century and refers to the study of the interaction between matter and electromagnetic radiation which produces an expressive result for a given material through emission, absorption, and similar other processes. These expressive patterns are what scientists call information; even if, this information has nothing to do with semantics, but with linguistically meaningless physical patterns in which matter can express itself.

Besides the material expressivity of single atoms that can produce distinctive patterns, these fingerprints can express themselves in more complex forms, such as a genetic code or DNA. As DeLanda elaborates:

> Groups of three nucleotides, the chemical components of genes, came to correspond in a more or less unique way to a single amino acid, the component parts of proteins. Using this, correspondence genes can express themselves through the proteins for which they code.[7]

The underlying patterns of information that define living organisms lead to the emergence of certain expressive behaviors that are actually functional, such as when a bird puffs up its feathers in order to protect its body from cold temperature or to impress a potential mate for sexual selection. We can also see expressivity in the muscle functions of various animals. Muscles are made of elastic fiber-like stretchable material, which can contract and expand in size.

These behaviors can lead to various expressive forms of movement, from walking to trotting to galloping. This can also be seen in various human facial expressions such as those that result from micro muscle contractions that allow us to shift from a smile to a frown. In this case, material expressivity and muscle functions at a micro scale can change the form and behavior of an organism at a macro scale.

Figure 2. *Opale*: Anger. Photographers Nicolas Cambier and Kyle Smithers.

One example of material expressivity in non-biological materials can be seen in the 'formal' transformations of highly complaint materials in soft robotic systems. Heavily influenced by the way living organisms move and adapt to their environment (as in, for example with biological muscles), soft silicon rubber materials can go through extreme shape and texture changes when exposed to air pressure. This, and similar phenomena which belong to the field of soft robotics, define a sub-branch of robotics wherein robots are fabricated from compliant, soft materials including soft actuators, flexible sensor/circuits, and soft bodies.[8] Soft robots are composed of deformable materials such as fluids, gels, and elastomers that adapt their shape and movement, behaving similarly to the way in which an octopus can squeeze itself into a narrow opening.[9] In the last few years, research into new soft robotics has gained a lot of attention. For instance, *The Soft Robotic Toolkit* is an initiative which shares open-source knowledge about the design, fabrication, modeling, characterization, and control of soft robotics.[10] (Figures 1 and 2)

My own research explores 'pneumatically actuated materials' as part of the field of soft robots. Pneumatic systems use air or gas to transform the volume of an elastomer or a pliable material like a balloon. One of the advantages of pneumatic actuators over conventional rigid-body robots is safety in terms of human-robot interaction; another is their flexibility and efficiency, as they can be stored in a small space and yet be deployed in a comparatively large installation. Being safe and flexible, pneumatic actuators can also be worn, either as a medical device or an expressive fashion item. The researcher, inventor, and entrepreneur Jifei Ou notes, "The compliance of soft actuators and the actuation speed and strength of pneumatic actuation make it well-suited for wearable applications."[11] (Figure 3)

Figure 3. Graded material using multi-material 3D printing in combination with soft robotics. In this prototype, the group of cells in the center were assigned a soft material, and as we move further away from the center, materials were assigned harder materials. This resulted in a bulging behavior, in which the movement is greatest in the center and dissipates as we get closer to the edges. Image courtesy of Behnaz Farahi.

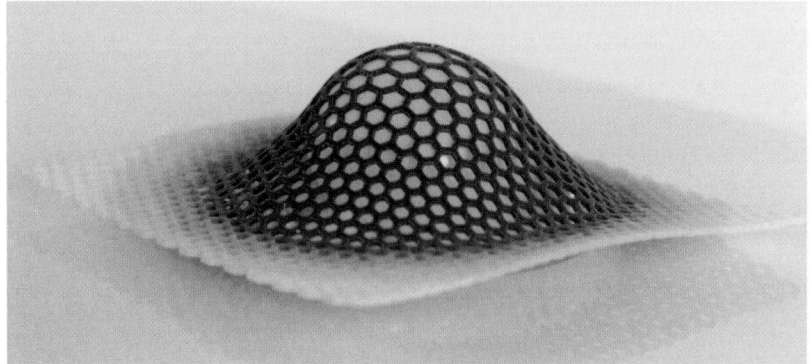

Emotion: Affective Material

Can material systems recognize human emotions? There are many different theories and perspectives on the question of our emotions, depending on whether they originate in psychology, neurology, or the cognitive sciences. Charles Darwin (1809-1882) claimed that the body reveals emotions as an outer expression of an inner emotional state. As psychologists Ursula Hess and Pascale Thibault state, "Darwin had no doubt that the expressive behavior that he described was part of an underlying emotional state, that is, that emotion expressions derived their communicative value from the fact that they were outward manifestations of an inner state."[12] Meanwhile American philosopher and psychologist William James (1842-1910) viewed bodily motor actions as emotions.[13] Around the same time, in 1887, Danish physician Carl George Lange (1834-1900) proposed a similar theory to James's which emphasized bodily responses. Their theories were later combined into what is now known as the James-Lange Theory of Emotions, which argues that emotion is a form of physiological, visceral arousal prompted by an external event. The Cannon-Bard Theory of Emotion (1927), by contrast, argues that emotions can arise independently of physiological responses and take a variety of manifestations. For instance, seeing a snake might trigger the feeling of fear and a racing heartbeat at the same time.

Then, in 1962, Schachter and Singer devised a theory that combined both the James-Lange and Cannon-Bard theories, suggesting that emotional experiences require both physical reactions and cognitive processes. Although they believed that bodily arousal might occur prior to experiencing emotions, they suggested that physical reactions alone could not be responsible for emotional responses. More recently, Portuguese American neuroscientist António Damásio has added a certain clarity to these theories of emotions by differentiating between the conscious experience of feeling versus emotion. He observed that emotions are complex reactions of the body to certain stimuli, such as when our heart rate increases when we are afraid. According to him, "This emotional reaction occurs automatically and unconsciously."[14] Even if for Damasio this is quite different to how feelings operate, he further elaborated that "feelings occur after we become aware in our brain of such physical changes; only then do we experience the feeling of fear."[15] He differentiated between emotions and feeling in his book *Descartes' Error* as follows:

> To feel an emotion it is necessary but not sufficient that neural signals from viscera, from muscles and joints, and from neurotransmitter nuclei—all of which are activated during the process of emotion—reach certain subcortical nuclei and the cerebral cortex. Endocrine and other chemical signals also reach the central nervous system via the bloodstream among other routes.[16]

The first attempts to measure physiological responses were found in the work of physiologist Etienne-Jules Marey (1830–1904) who had long been fascinated with the study of movement in humans and animals. In 1865, he invented the first portable sphygmograph, a mechanical device placed above the radial artery that was able to detect a pulse wave and record it on paper.[17] However, it was Manfred Clynes who first studied the relationship between physiological responses and emotional states. In 1972, Clynes invented a machine for measuring emotions called a Stenograph. In his experiments, he asked subjects to use touch and finger pressure to express a sequence of emotions— anger, hate, grief, neutral, love, sex, joy, and reverence—while listening to twenty-five-minute cycles of music and tracked their responses on his stenograph. The aim of this research was to show that it was possible to use music to change a negative emotional state into a positive one.

In his book *Sentics: The Touch of the Emotions*, Clynes outlines his findings on emotional perception by elaborating on the notion of 'sentic' forms. As he puts it:

> How remarkable it would be if one could experience and express the spectrum of emotions embodied in music originating from oneself – without the crutch of a composer's intercession, without being driven by the composer; and to do so moreover whenever we wish, not when circumstance may call them forth. This, indeed, has become possible through the development of sentic cycles.[18]

In 1995, Rosalind Picard, director of the Affective Computing Group at the MIT Media Lab, coined the term 'affective computing' to refer to systems that sense, recognize, respond to, and influence human emotions.[19] The interdisciplinary field of 'Affective Computing' operates between psychology, cognitive science, physiology, and computer science. It is based on the thesis that many otherwise hidden affective states manifest themselves through a variety of corporeal and physiological responses, which can be measured and analyzed. Picard notes, "A variety of physiological measurements are available which would yield clues to one's hidden affective state."[20]

Over the past few years, many start-up companies have invested in the development of affective computing systems. For instance, the company *Affectiva*, co-founded by Picard, has been working on software for recognizing emotions based on various physiological responses. The company has also developed a wireless bracelet called *Affectiva-Q Sensor* for affect detection, which can track skin conductance, motion, and environmental temperature. Other companies, such as *Beyond Verbal*, have been working on the sentiment analysis of speech. Microsoft has released a series of versions of *Kinect*, a device equipped with an infrared sensor which is able to track not only the motion of the body but also the heartbeat.[21]

Furthermore, applications of affective computing have prospered with the development of social robots, such as *Kismet* and *Leonardo*, developed by Cynthia Breazeal. Uses of social robots have included customer services, marketing, healthcare, and education. Picard describes this vision of a future life with affective computers as follows:

> For example, a computer piano tutor might change its pace and presentation based upon naturally expressed signals that the user is interested, bored, confused, or frustrated. A video retrieval system might help identify not just scenes having a particular actor or setting, but scenes having a particular emotional content: fast-forward to the "most exciting" scenes. Your wearable computer could pay attention to things that increase your stress, so that it might help you learn better strategies to boost your immune system when needed, practicing preventive medicine.[22]

The human body is the place where all emotions are embodied, where they undergo a series of physiological responses. Bodily arousal, which typically happens with changes in muscle tension, heart rate, skin temperature, and moisture, is a good indication of one's inner emotional state. For computational systems to be affective they must first 'sense' a user's emotional state by tracking its physiological condition. These systems can interpret affective cues, similar to how we can understand someone's emotions by reading their facial expressions. But how to identify and code such cues? American psychologist Paul Ekman believes that certain types of emotions are not culturally specific but shared by all. Ekman defines a certain class of emotions as 'basic emotions' and argues that facial expressions associated with basic emotions are innate and universal.[23] According to Ekman, joy,

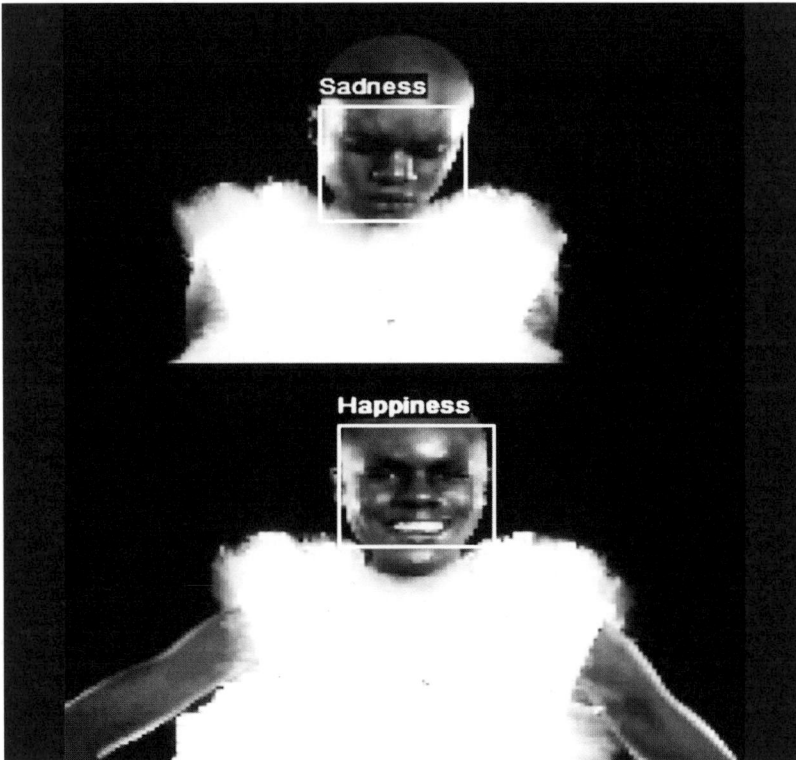

Figure 4. *Opale* (2017): Using facial tracking technology, we can detect a range of facial expressions such as happiness, sadness, anger, surprise and neutral. Image courtesy of Behnaz Farahi.

distress, anger, fear, surprise, and disgust reveal facial expressions that are clearly visible and communicable.[24] His team has produced an atlas of emotions including thousands of distinctive facial expressions and they've released a facial action coding system (FACS) to help identify human expressions with their underlying muscle movements. And while computer vision technologies (such as RGB cameras and Infrared Receiver (IR) sensors) can be used remotely to capture bodily gestures and even facial expressions, other in-contact sensors (such as Electroencephalography (EEG), Galvanic Skin Response (GSR) and Heart Rate Variability (HRV) sensors) can measure physiological data such as brain activity, skin galvanic resistance, and heart rate variability to help us to better understand our affective states.

Case Study: *Opale*

How might clothing sense aggression, and, accordingly, assume a defensive posture? This section describes the design process used for *Opale*—an emotive garment which can recognize and respond to the emotional expressions of people around it. Opale is a soft robotic wearable fitted with an electro-mechanical system that controls dynamic behavior so as to respond to the facial expressions of onlookers. Human hair and animal fur are some of the most inspiring natural phenomena in terms of both their morphology and their

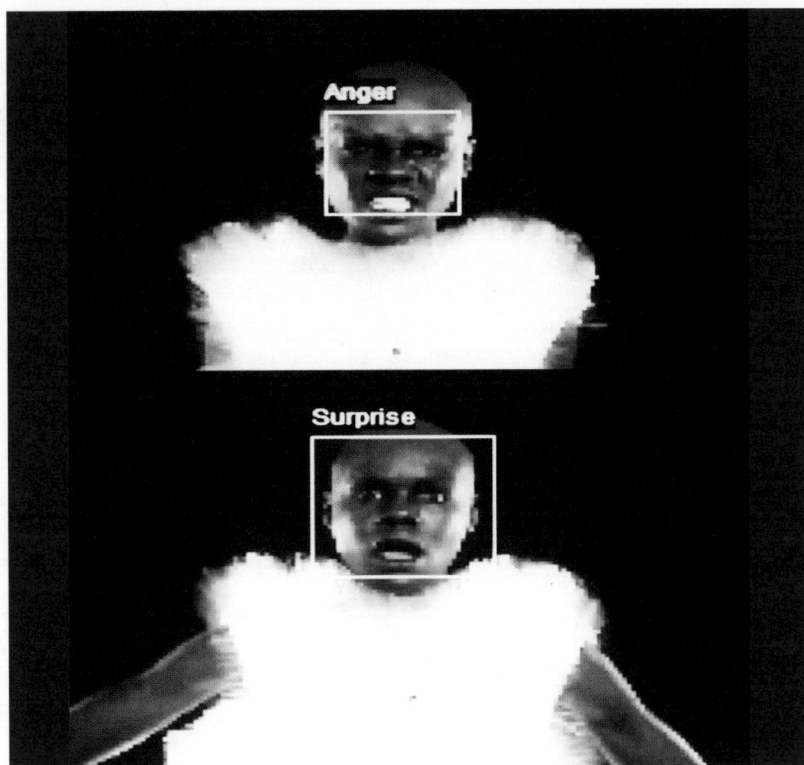

Figure 5. *Opale* (2017): As computational devices are getting smaller and embedded into the substrate of matter, materials become intelligent in tracking our emotions. Image courtesy of Behnaz Farahi.

capacity to communicate during social interaction. Inspired by animal fur, *Opale* is composed of a forest of fiber optics embedded in silicon which bristle when the wearer is under threat. (Figures 4 and 5)

The material development of this project was extremely iterative and experimental. The intention was to control the location and orientation of hair-like elements within a surface so that they could respond to underlying forces not dissimilar to how hair stands up. Known as goose bumps or piloerection, these involuntary responses within the skin are due either to temperature changes or the experience of emotions such as fear, sexual arousal, or other forms of excitement. With hair, each follicle is deeply rooted within and between layers of skin (located beneath the dermis and epidermis). So too with *Opale*, it was important to have a thick flexible surface base in which each fiber could be carefully implanted, to create a hair-like material system. Similar to the procedure for a hair transplant, the project required a unique fabrication process in which every element was carefully placed and oriented in a soft silicon base. Once all fibers were mounted into a temporary surface, silicon rubber was poured into the surface, which later cured and turned into a flexible surface. The location and height of each fiber was based on a study of the surface location of human hair. Data captured from an analysis of the surface curvature of the human body and the underlying contours of the muscles informed the location, density, and height of each fiber

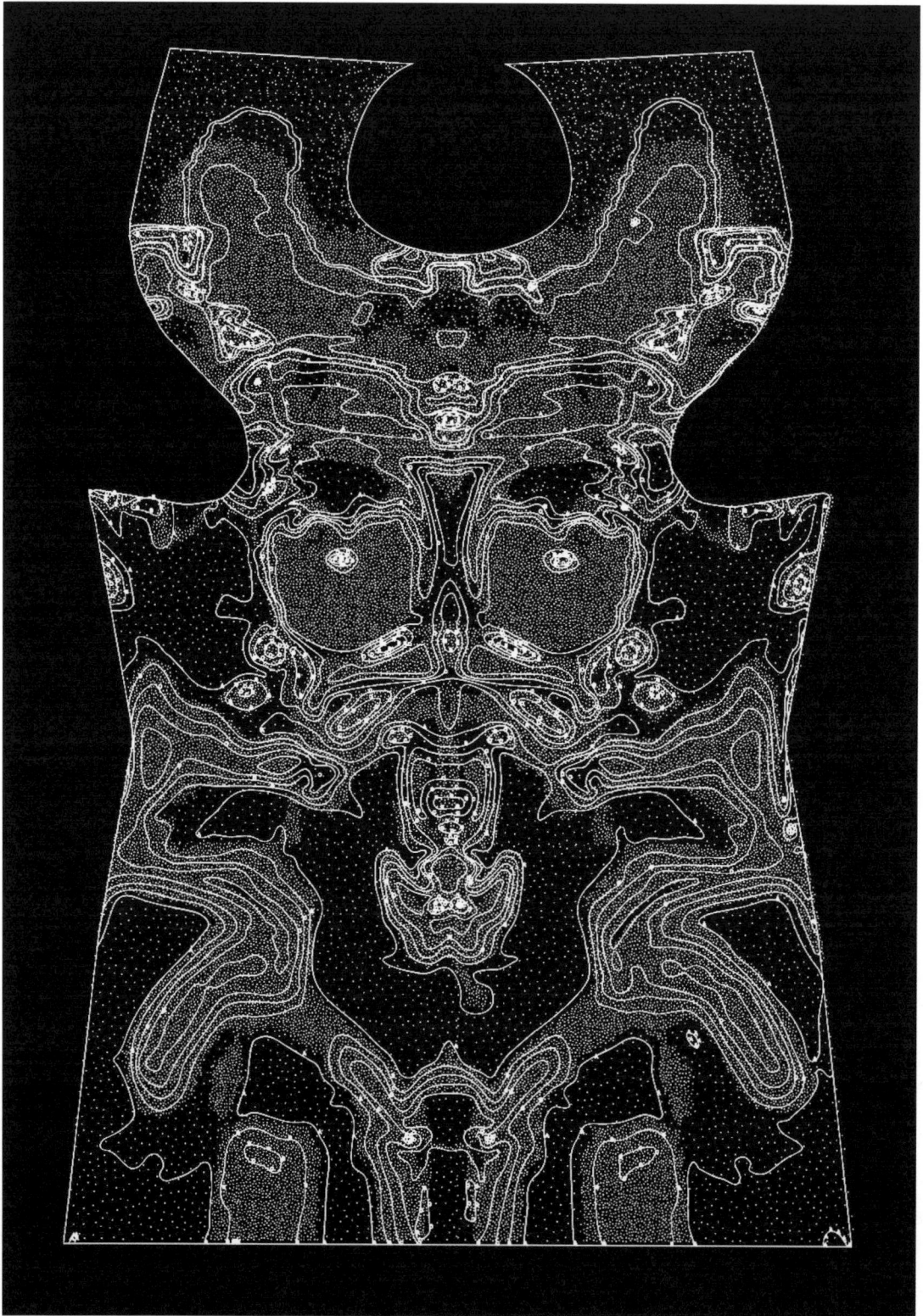

distribution. (Figure 5) The intention was to exaggerate the movement of underlying muscles by having the denser and longer fibers follow the contours of the curvature underneath.

The human physiological response of 'goosebumps' is caused by contraction of tiny muscles that are attached to each hair follicle. Each contracting muscle creates a small bump on the skin surface causing the hair to stand up. To make the fiber optic surface move like bristling fur, six irregular inflatable silicone pockets—following the contours of the human body—were incorporated beneath the skin to generate deformations in the texture and surface volume. This allowed the form to change in a more controlled fashion specifically in various areas such as the wearer's shoulders, chest, and hips. All layers, including the fur-like surface and its inflatable pockets, were then glued together using an additional light coat of two-part silicon rubber to produce the material composite for this wearable. (Figure 6)

The inflatable behavior of the so-called 'goose bumps' was controlled using a custom designed electrical board attached to an Adafruit Feather microcontroller board (M0 with ATSAMD21G18 ARM Cortex M0 processor) capable of controlling an array of six low powered 3-port medical solenoids (LHLA Series). (Figure 7) The Arduino programming environment enabled easy computational control of the air pressure and rapid design of the inflatable interactions. This in turn allowed us to adjust the duty cycle, frequency, and pulse width of square waves which controlled the inflation pattern of each solenoid. For this project, a series of initial experiments on controlling the behavior of the system were conducted in which various combinations of on/off switches for the solenoids were tested. Consequently, various types of motion representing various types of emotions (happy, agitated, etc.) were achieved by adjusting this pneumatic control system.

From the perspective of interactive design, this project looked closely at the dynamics of social interaction. We tend to respond to people around us through our unconscious facial expressions and bodily movements. When surrounded by smiling people, we often smile back; when threatened, we often assume a defensive stance. We mimic each other's emotional expressions via neuron mirroring. Likewise, animals use their skin, fur, and feathers as a means of communicating. Dogs, cats, and mice bristle their fur as a defensive mechanism or as a form of intimidation. Darwin was the first to examine the emotional signals in humans and animals in his 1872 book, *The Expression of the Emotions in Man and Animals*.[25] According to author Dylan Evans in his book *Emotion: A Very Short Introduction*, "Darwin was interested in these expressions as he thought they were good evidence that humans had descended from other animals. He argued the way our hair stands on end when we are scared is a leftover from a time when our ancestors were completely covered in fur": whose role it was to make them look bigger.[26]

The challenge we faced with *Opale* was to explore whether emotions expressed in our social interactions could be represented

Figure 6. Data from surface curvature analysis of the human body informs the location, density, and the height of fibers. Image courtesy of Behnaz Farahi.

Figure 7. Fabrication of fur-line *Opale* wearable by embedding fiber optics inside rubber like silicon. Image courtesy of Behnaz Farahi.

in a non-verbal way through the motion of a garment. Could the garment become an expressive tool or apparatus that responds to the emotions of onlookers? (Figures 8 and 9) For this purpose, the dress was equipped with a facial tracking camera that could detect a range of facial expressions on the onlooker's face: happiness, sadness, surprise, anger, and neutrality. Each detected emotion was sent to a microcontroller capable of activating the solenoids to generate various patterns and inflation speeds in each air pocket. For example, if a small facial tracking camera in the dress were to detect an expression of 'surprise' on an onlooker's face, the wearer's shoulder area would start to inflate. Or, when an onlooker expressed 'anger,' the wearer's shoulder and chest started to inflate and deflate with a frantic, aggressive motion. When people smiled and demonstrated 'happiness,' the dress rippled subtly from top to bottom. In the end, the task of expressing the range of emotions via soft robotic interfaces remains a significant design challenge which requires an interdisciplinary approach uniting design, material science engineering, computer science and Human Computer Interaction (HCI). Nonetheless, it opens up radically new opportunities to address psycho-social issues.

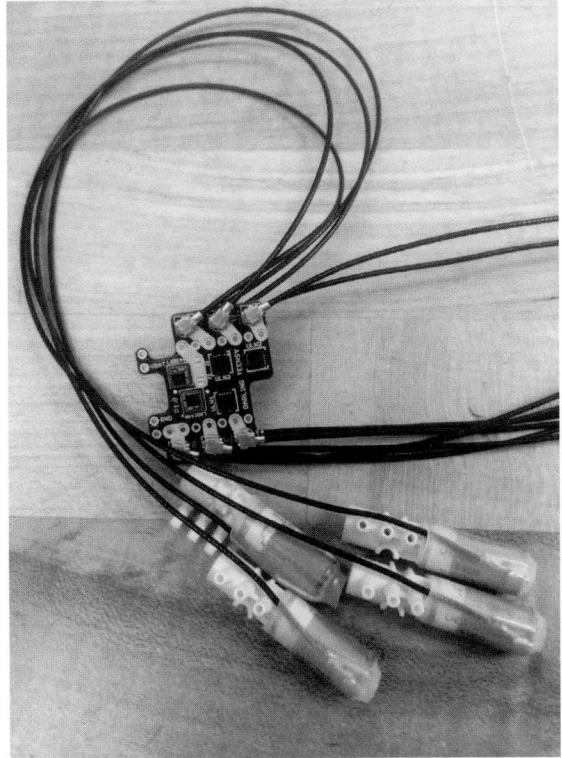

Figure 8. A pneumatic control circuit consisting of six 3-port solenoids valves with coax cable connections. Image courtesy of Behnaz Farahi.

Conclusion

This paper has introduced the research initiative that is, Emotive Matter: an attempt to develop bio-inspired material systems equipped with computational technologies that can interface with human emotions. Through the lens of 'new materialism,' the role of material expressivity in living and non-living systems was examined to develop a framework for the development of active matter. Moreover, through the lens of 'affective computing,' this paper has demonstrated how computational material systems are able to detect, monitor, and decode the emotional state of users. Materials and emotive sensing can be integrated into a single entity inspired by biological systems. As demonstrated in project *Opale*, the research initiative that is Emotive Matter spans several scales from that of fashion to that of architecture. It adopts an interdisciplinary approach, engaging the fields of science, technology, design, and humanities. By embracing the two seemingly distinct topics of 'emotion' and 'material', *Opale* and Emotive Matter are informed by psychology, neuroscience, cognitive science, and social science on the one hand, and by new trends in design, ethology, material science, and artificial intelligence on the other. This hybrid interdisciplinary research is one where theoretical investigations inform design processes, no less than design processes inform theoretical investigations.

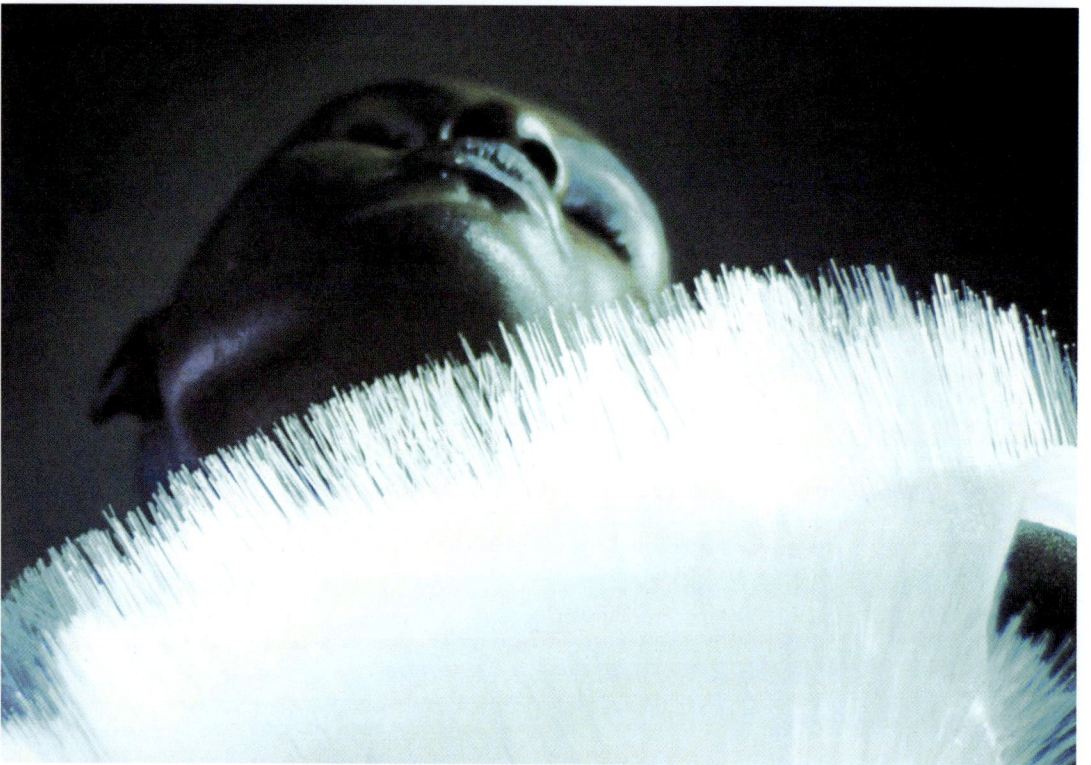

Figure 9. The dress is responding to the onlooker's emotions with various types of dynamic behavior. Image courtesy of Behnaz Farahi.

Figure 10. Facial tracking camera embedded into the silicone dress, able to detect onlookers' facial expressions. (Happiness, Sadness, Surprise, and Anger). Image courtesy of Behnaz Farahi.

Acknowledgments

The research into pneumatic soft robotic systems is part of a broader ongoing collaboration with Paolo Salvagione and Julian Ceipek. I would like to thank them for their advice and helpful contributions to the production of *Opale*. Also, I would like to thank the USC Bridge Art + Ksiense Alliance Research Grant Program that funded *Opale*.

ENDNOTES

1. As noted by Ana Paiva, Iolanda Leite, and Tiago Ribeiro, "Defined by Höök (2009), the Affective Loop is the interactive process in which 'the user [of the system] first expresses her emotions through some physical interaction involving her body, for example, through gestures or manipulations; and the system then responds by generating affective expression, using for example, colours, animations, and haptics' which 'in turn affects the user (mind and body) making the user respond and step-by-step feel more and more involved with the system." Ana Paiva, Iolanda Leite and Tiago Ribeiro, "Emotion Modeling for Social Robots" in Calvo, Rafael A., Sidney D'Mello, Jonathan Gratch, and Arvid Kappas, eds. *The Oxford Handbook of Affective Computing* (Oxford University Press, 2014), 298.

2. Manuel DeLanda, "Matter Matters," *Domus no. 895* (September 2006): 7.

3. Tim Ingold, "The Textility of Making," *Cambridge Journal of Economics*, (2010): 92. http://sed.ucsd.edu/files/2014/05/Ingold-2009-Textility-of-making.pdf.

4. Gilles Deleuze, and Felix Guattari, *A Thousand Plateaus: Capitalism and Schizophrenia*, (Minneapolis: University of Minnesota Press, 1988), 408.

5. Manuel DeLanda, *Deleuze: History and Science*, (New York: Atropos Press, 2010), 68.

6. DeLanda, "Matter Matters," 262-63.

7. Ibid.

8. Deepak Trivedi, Christopher D. Rahn, William M. Kier, and Ian D. Walker, "Soft robotics: Biological inspiration, state of the art, and future research," *Applied Bionics and Biomechanics 5*, no. 3 (2008): 99-117.

9. Carmel Majidi, "Soft Robotics: A Perspective – Current Trends and Prospects for the Future," *Soft Robotics 1*, no. 1 (2013): 5-11.

10. The Soft Robotics Toolkit grew out of research conducted at Harvard University and Trinity College Dublin. https://softroboticstoolkit.com/

11. Jifei Ou, Mélina Skouras, Nikolaos Vlavianos, Felix Hejbeck, Chin-Yi Cheng, Jannik Peters, and Hiroshi Ishii "AeroMorph – Heat-sealing Inflatable Shape-change Materials for Interaction Design," *UIST 16*, (2016): 12.

12. Ursula Hess and Pascale Thibault, "Darwin and Emotion Expression," *American Psychologist 64*, no.2 (February-March 2009): 120-8. doi: 10.1037/a0013386. PMID: 19203144.

13. Tone Roald, Kasper Levin, Simo Køppe, "Affective Incarnations: Maurice Merleau-Ponty's Challenge to Bodily Theories of Emotion," *Journal of Theoretical and Philosophical Psychology 38*, (2018): 205-18.

14. António Damásio, "Feeling Our Emotions," *SA Mind 16*, No 1, (2005): 14-15.

15. Damásio, "Feeling Our Emotions," 14-15.

16. António Damásio, *Descartes' Error: Emotion, Reason, and the Human Brain* (Penguin Publishing Group, 2005), 163.

17. Matthew Nelson, Jan Stepanek, Michael Cevette, Michael Covalciuc, R Todd Hurst, Jamil Tajik, "Noninvasive Measurement of Central Vascular Pressures with Arterial Tonometry: Clinical Revival of the Pulse Pressure Waveform?" *Mayo Clinic 85* (2010): 460-72.

18. Manfred Clynes, *Sentics: The Touch of the Emotions* (Bridport, Dorset: Prism Pr Ltd, 1989), 112.

19. Rosalind W. Picard, "Affective Computing," *MIT Media Laboratory Perceptual Computing Section Technical Report, no. 321* (November 26, 1995): 15. https://affect.media.mit.edu/projectpages/archived/applications.html

20. Picard, "Affective Computing," 15.

21. For information on the *Kinect* devise see, "Watch Your Heartbeat on Xbox One's New Kinect" https://mashable.com/2013/05/22/xbox-one-kinect-heartbeat/#:~:text=One%20of%20Kinect's%20more%20remarkable,the%20pulse%20in%20your%20face.

22. Picard, "Affective Computing," 1-26.

23. Paul Ekman, *Emotions Revealed: Recognizing Faces and feelings to Improve Communication and Emotional Life*, (New York: Holt Paperbacks, 2007), 47.

24. Paul Ekman, Wallace V. Friesen, *Unmasking the Face: A Guide to Recognizing Emotions from Facial Expressions*, (Cambridge, Massachussetts: Malor Books, 2003), 51.

25. Charles Darwin, *The expression of the emotions in man and animals*. (1872) https://doi.org/10.1037/10001-000

26. Dylan Evans, *Emotion: A Very Short Introduction*, (Oxford; New York: Oxford University Press, 2003), 42.

DESIGN SPACE EXPLORATIONS:

ON THE ROLE OF ROBOTIC FABRICATION-INFORMED VISUALIZATIONS IN ARCHITECTURE

STEFANA PARASCHO

ABSTRACT

Robotics and architecture are no longer disparate fields. Digitally controlled machines have been introduced into architectural design and construction, shaping the field like few other technologies have in the last decades. However, their integration did not come without challenges. As popular as robotic technology is today, it remains difficult to find a smooth process that combines architectural design workflows with robotic developments. In particular, the interdisciplinary nature of construction robotics has required all participants in the process to reconsider the hierarchical workflow from design through engineering to fabrication planning, to find ways to interact, communicate, and develop new tools in a collaborative manner. Visualizations that empower designers to fully comprehend and explore the design spaces resulting from different criteria are central to the successful integration of robotics and structural considerations in design. More importantly, by working with design space visualizations, one can combine various criteria and address temporal parameters. By describing the workflow used in two design-build projects that focused on multi-robotic assembly, this article reflects on the implications of visualization for integrating new technologies into architecture, as well as potential and shortcomings of using these tools and techniques.

+ computational design
+ digital fabrication
+ structural form-finding
+ design space visualizations
+ robotics

Figure 1. Detail of *LightVault*, displayed at the *Anatomy of Structure: The Future of Art & Architecture*, London. Copyright CREATE Lab Princeton.

Theorizing Technology

Architectural design is a creative process that results in a design's materialization through a number of rationalizing steps. Traditionally, architectural design was the result of individual, creative acts performed by a lead architect, charged with defining desired outcomes and searching for feasible solutions to given problems. With the use of computation, this process has shifted in allowing designers the ability to extend their capabilities, especially when dealing with complexity.[1] Computation provides designers with tools and methods to increase their capacity for processing and using spatial, material, and fabrication-based information.[2] Rather than aiming to 'rationalize' a pre-determined design, computational tools and methods enable new ways of designing. In his seminal work *The Alphabet and the Algorithm*, architectural historian Mario Carpo argues that the introduction of the digital in architecture has led to more participatory design processes.[3] Architects no longer provide final design outcomes, but rather participate in the creation of digital tools that enable non-specialists to engage with the design process. Similarly, architectural historian Antoine Picon has characterized the current design context as no longer one of authorship but of ownership, similar to how copyrights over tools and templates enable others to generate final design from said devices.

This technological turn risks, however, advancing a design process that disregards the limitations that accompany new tools. Many contemporary digital design processes rely on opaque languages, pre-defined controls, and aesthetics that are already encoded in their parametric definitions, thus indirectly pre-defining the design outcome. While computation enables architects to handle more and more complexity, it also limits available design-space. This shift from design to software development engenders a process in which users get the impression of being actively engaged in design, all the while being limited in the possible solutions. In this context, it is questionable whether the process can still be considered design. What is needed instead is a shift towards an understanding of architects as technically educated practitioners who are versed in handling the complexities of a fast-changing technological world in their designs. This doesn't require tools for pre-defining design characteristics, but a new type of design methodology, based on visualization and exploration.

Some architects have long based their work on a technical education combined with an intuitive understanding of space and construction. However, current advancements in technology surpass the intuitive understanding of space, whose ever more complex processes exceed the scope and capacity of the classically educated architect. For example, today, we see robots employed for construction, using feedback and sensor technologies, while computer algorithms involving optimization and machine-learning are used for generating unprecedented forms. Incorporating and utilizing these new technologies by mere intuition is extremely challenging. We need to develop connections between the designer's world of intuition, sensibility, and creativity and the technological aspects of computation, robotics, and data processing.

At present, architects interested in technology adopt one of two positions: that of end-users who require tools with simple but often very limited user interfaces developed by other specialists, or that of niche specialists in a technical field, far outside the reach of their discipline. The first position is limited by ready-made tools, by pre-existing biases in any algorithm, and by the shift of design control from the designer to the tool developer; the second places architects at risk of losing touch with their discipline and becoming stuck in a narrow technological niche. Missing is a connection between these two extremes, a common language that designers and technical experts can speak and tools that they can use for translating the intricate technical knowledge that these technologies bring to architecture.

Design Space Visualizations, Beyond Intuition

How do we bridge the gap between intuition-based design and technological developments such as algorithmic tools, new machines and their processes, software, and hardware? This article is focused on architectural construction processes that make use of robotic technology and its corresponding software tools, such as path planning and structural simulations. However, the visualization concepts here described are considered applicable to other present and future technologies that may impact architectural design and construction.

A robotic fabrication process is usually governed by the degrees of freedom of the robotic manipulator (usually a result of six rotational axes) and may be informed by additional parameters such as stability, speed, and materials. As opposed to the three-dimensional design space that designers are used to work within, robotic fabrication introduces a multi-dimensional space that is much less intuitive to navigate. Even today, the mapping of these design spaces onto our two-dimensional computer screens simply does not have the capacity to adequately communicate the resulting complexity. What this means is that even if one can numerically define the design space, it is still nearly impossible for a designer to articulate a structural and path planning solution that fits within this space, let alone to intuit how to adjust a design to make it 'better' or to modify a specific design feature.

We can see this problem arising in computational tools that include highly unintuitive information, such as structural form-finding and robotic path-planning software, whose information cannot be easily described or navigated in the architect's design space. The typical design strategy chosen in these cases is a trial-and-error process wherein designers explore, more or less blindly, options until they identify one that feasibly relates to their design intention. This process is often long, cumbersome, and uninformed, far from a creative process that is the basis of architecture. Designers become consumers of existing tools that strongly constrain the creative space and ultimately

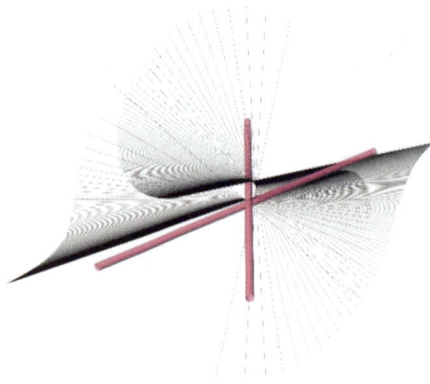 

 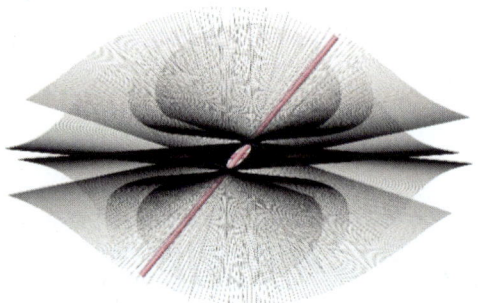

define the outcome before the design process has even started. Recalling Carpo's discussion on authorship we are led to ask: Is the designer but a simple executor of a program developed by someone else? Or can the designer be the creator of such programs and directly translate their design intentions into computational tools? This is not a new discussion, but rather one based on centuries of tool development. The tools that we use inevitably impose constraints on the resulting designs, from a chisel that leaves a specific mark on stone to Building Information Modeling (BIM) software that pre-defines building components.

However, given the speed with which the fields of robotics and computer science are changing, we are moving in a direction where defining the software needed for highly complex digital processes exceeds the capacity of even highly technological enthusiasts. Needed are tools that do not constrain design outcomes through opaque algorithms: designers should not need to devote their careers to understanding complex algorithms and robotic technology. Indeed, needed are processes that make the technology accessible to designers through the rationalization and visualization of design possibilities that such new technologies offer. For example, instead of proposing a tool that automatically generates a brick wall to be automatedly built by robots, why not provide designers with the opportunity to understand the possibilities that exist when building walls with robots? We can find ways of visualizing the potential for differentiation offered by a robotic process while at the same time exposing its physical limitations. Ultimately the fields of computational design and digital fabrication will shift away from developing deterministic, one-click tools for automated processes and move towards investigating the relationship between human actors and computational machines. We seek new ways of communication between these entities that integrate the designer within technical processes instead of pushing them towards the extreme roles of users or developers.

Project 1: *Cooperative Robotic Assembly*

Cooperative Robotic Assembly is the title of the author's PhD thesis completed in 2018 at the ETH Zurich, in the Gramazio Kohler Research unit. The project investigated the complex interrelations between different design parameters in a multi-robotic process. Besides proposing a cooperative assembly technique that utilizes multiple robots to assemble spatial structures without guides and supports, it explored possibilities for defining and exploring the complex design space that results from the robotic assembly method.[4] (Figure 2) A significant part of the work addressed technical challenges of geometric description, sequencing, robotic movement, and structural optimization. However, the project also questioned the role of the designer in processing the complexity emerging from new technological tools, such as robotic fabrication. While solutions to different individual constraints, such as structural efficiency, robotic reach, and geometric

Figure 2. Two robots cooperatively assembling a spatial structure by alternating between acting as placing agent and support. *Cooperative Assembly* Fabrication Process. Copyright Stefana Parascho.

Figure 3. Design space for one bar that fulfills double-tangent constraint to two existing bars. (top: perspective view: one solution vs. two overlapped solutions, bottom: top view: one solution vs. two overlapping solutions). The area not covered by the sweep represents the area without solutions that a designer needs to avoid. *Cooperative Assembly* Double Tangent Bar. Copyright Stefana Parascho.

+Z

Figure 4. Large-scale
demonstrator of cooperative
robotic assembly method.
Cooperative Assembly
Demonstrator. Copyright
Stefana Parascho.

conditions, can be found, it becomes increasingly difficult for a designer to intuitively process and understand their interconnection. Let's consider robotic movement as an example: humans are familiar with thinking and working in three-dimensional Euclidean space, yet robots act in a six-dimensional rotational space which is not intuitively understandable by humans. This means that there is a necessary translation step from one space to the other, between a design and its robotic construction which usually happens via a set of calculations executed in the computational background and not easily accessible to designers. In addition, the complexities of highly constrained geometric systems, non-linear structural behaviors, and robotic reachability issues result in design problems that can no longer be intuitively solved but require the use of computational methods.

Cooperative Robotics Assembly proposed a new assembly method for spatial structures built using robotic fabrication and based on the alternating placement of elements in space. The method allows for one robot to remain connected to the structure at any point in time, acting as a support, while the other places a new item, in this case a steel bar.[5] As soon as the new bar is placed a welded connection is manually executed. The robot's precision and its capacity to dynamically change role during the construction process, means that highly differentiated structures can be constructed that are not limited by prefabricated connectors in number and attachment angles, or by standardized bar dimensions. This method introduces new potentials for differentiation, allowing material to be placed where needed and breaking free from regular space-frame geometries. (Figure 4) However, the assembly method relies on a connection system that requires every new bar to be tangential to two already placed bars, and thus introduces hard constraints in the geometric system. Implications of these constraints go beyond intuitively controllable conditions. Each bar's position in space is determined by a range of possibilities for the tangential connections, a purely deterministic result of discrete decisions which cannot be informed without computational processes. To incorporate these conditions in a design process, it is necessary to identify the feasible solution space for all possible orientations of a given new bar, including areas which could not provide a solution. (Figure 3) As a designer, one can only guess what each of these choices means in terms of geometry, structural behavior, and robotic feasibility, even if one uses simulations to check the effects of any given choice. This remains but a trial-and-error method of design.

In *Cooperative Robotics Assembly*, a crucial component is the sequence of design choices. As opposed to traditional design, here a structure's geometry is based on strong interdependencies between fabrication parameters and time. We can no longer pre-define a final geometry and rationalize it for structural or fabrication efficiency. Rather, in wanting to keep control over the design process, we consider every temporal step of the fabrication process and allow the design process to respond to it. This means that designing is no longer an act of finding one form through a series of parallel decisions; it is a sequence of decisions that build upon previous ones, step by step. At every step of the design process, the parameters of the future bar to be placed, are determined at the time of its placing. Thus, design decisions are focused on choosing a new point in space and a set of bars to connect to, as opposed to high-level information based on aesthetics, spatial relations, and functionality.[6] This process requires the designer to focus on the details of the chosen geometric, structural, and fabrication logic, and hence to confront the technical details of each parameter.

This method forgoes, therefore, the expectation of full control and global optimality in order to provide the designer with more low-level control. It works by offering a breakdown of the entire design task into individual sequential steps, thus limiting the field of view and the number of parameters that need to be considered. While zooming into each particular component increases our understanding of the implications of each design decision, we also run the risk of losing sight of the overall design task. It is imperative that while we engage detailed technical information, we keep in mind high-level implications of every design decision.

Project 2: *LightVault*

LightVault is a project undertaken by the author in collaboration with SOM Engineering and the Form Finding Lab at Princeton University. This project showcased a more high-level approach to multi-constrained interdisciplinary design. The goal of *LightVault* was to design and robotically fabricate a glass vault structure for the *Anatomy of Structure Exhibition* installed in 2020 in London, England. (Figure 1) The constellation of the research team included architects, engineers, material specialist consultants from Delft University, and industry partners, all of whom focused on connecting different knowledge pools and articulating an interdisciplinary strategy for developing a design and fabrication method that ensured the buildability of the structure. Previous research in the field has explored the use of robotic manipulators for masonry construction of brick walls and columns that use glue adhesives,[7] of differentiated facades,[8] of dry-stacked assemblies of large-scale brick walls.[9] In matters of structural design, the project builds on the long-standing tradition of research on form- and sequence-finding for discrete-element structures. Investigations include form–finding of compression-only structures,[10] finding sequences for stable assemblies,[11] and explorations of historic brick-laying patterns (such as herringbone).[12] *LightVault* was challenged with combining form-finding techniques in masonry with robotic fabrication, alongside questions of robotic payload, sequencing, and reachability into a successful process that would enable the construction of efficient masonry vault structures that do not use form- or falsework. In addition to the technical questions solved, tangential questions quickly emerged regarding the role of the human worker

Figure 5. Robots cooperatively assembling central brick arch as a backbone for the vault construction. *LightVault* Central Arch. Copyright CREATE Lab Princeton.

in the fabrication process and the interaction between designer and computational tools in the design process.

While the goal was to find solutions to issues of robotic feasibility and structural stability, for the quick and safe construction of the glass vault, the project raised questions about human agency and machines.[13] Besides the different human actors involved in the process, machines also play a crucial role, since design and fabrication processes must yield a solution that satisfies both the expectations of designers and the constraints of robots. The design process can thus be described as a negotiation between participants. Reacting to problems during construction and being open to adjusting the geometry at any point of the project led to a win-win situation in which designer and machine worked together to achieve a common goal.

Architecturally, the project's main challenge consisted in identifying the intersection of different design spaces with a structurally efficient geometry, a stable brick pattern, and a feasible sequence of collision-free robotic movements. With multiple parties involved in the design process, it became clear that fabrication constraints had to be considered from the very first steps, to reach an optimal, beautiful, and efficient structure. Instead of proposing a fully automated, generative design approach, a process was chosen that served to visualize the many constraints and possible design spaces. This was an iterative process in which one design space was mapped over another and adjusted until it facilitated the intersection of several constraints.

The project began by defining the construction logic of the vault that was needed to ensure its stability without the use of additional scaffolds or supports. Timbrel-vaulting is a construction method for compression-only geometries that does not require supports during construction.[14] As opposed to classical masonry construction, it uses tiles whose building sequence and multi-layer configuration ensure stability during construction, requiring only guides for geometric precision. For *LightVault*, full-sized bricks were used in combination with robotic arms that served as interim supports, instead of form- or falsework. This resulted in a cooperatively assembled central arch where two robots alternated the placing of bricks: thereafter the arch was completed by the robots working independently on both sides of the arch. The resulting robotic setup defined the geometric space of possibilities for the vault since all bricks had to lie within the reach of the robots. In parallel, a structural form-finding approach operating in the geometric volume of the robotic reach space was used to determine feasible geometries that would lead to compression-only structures. This interaction between structural form-finding and robotically constrained spaces led to the vault's unique shape. Moreover, in a separate step, brick patterning and sequencing were computed and visualized to identify an additional range of feasible solutions. Hence, the project's qualities emerged when all factors were considered, both design opportunities and limitations.

Existing software solutions usually focus on visualizing one aspect of the design-fabrication process. For example, there are easy-to use plug-ins that address structural form-finding via simulations,[15] and new libraries and plug-ins that allow for simulating robotic movements and visualizing these in an environment that is familiar to designers.[16] These tools remove crucial barriers in accessing technical information during the design process, but they typically offer only a way of evaluating a given design option instead of representing the space of possibilities of different design criteria. Research in robotics has resulted in new types of visualizations, including that of robotic reach spaces, even if these representations are still inaccessible to designers.[17] (Figure 6)

For *LightVault*, using structural form-finding and robotic path-finding software, the team employed an iterative design process, sharing visualizations of potential design spaces, defined by either the robotic setup or structural constraints. This required the interlaced use of multiple software tools to cover a broad range of requirements including the different workflows of the many parties involved. By combining everything into a visualization based on the COMPAS library and the popular CAD environment of Rhino/Grasshopper, it was possible to easily exchange geometries, but also to communicate spatial requirements of the fabrication method and structural design. This way all parties involved could understand the constraints and, more importantly, the possibilities of different parameters. For the robotic setup, a key visualization was the intersection volume of the robots' overlapping workspaces. So too was the ability to approximate spatial design transformations from the center of the workspace towards the work areas of individual robots, based on assumptions about their reachability. (Figure 6) We were able to identify transitions from lower arches to higher ones which improved structural stability through double curvature but also increased the freedom of

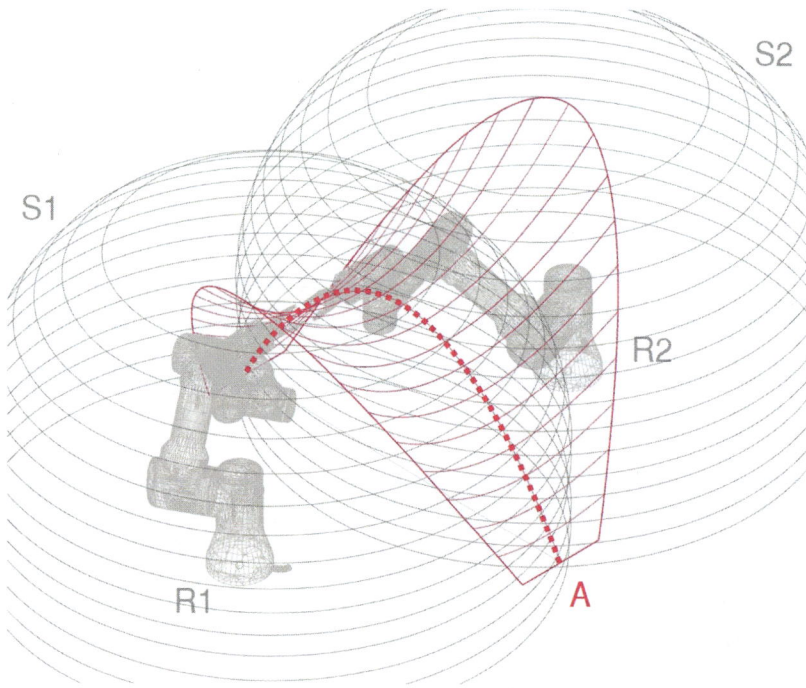

Figure 6. Initial exploration and visualization of the design space resulting from robotic reach constraints on small-scale robots. *LightVault* Robotic Design Space. Copyright CREATE Lab Princeton.

Figure 7. Exploration of different brick patterns and orientations combined with robotic reach spaces. Copyright CREATE Lab Princeton.

movement of the robots at each step of the construction. Structurally, design space visualizations included potential outlines and brick pattern orientations that would ensure stability during construction. (Figure 7) Introducing these parameters within the design, while maintaining the flexibility to adjust it even during construction, was crucial to the successful implementation of the project.

Ultimately, the construction process began with a cooperatively assembled central arch where two robots alternated the placing of bricks. (Figure 5) It continued with the construction of both sides of the arch with the robots working independently. The robotic setup defined a geometric workspace of approximately 9m x 2.5m x 4m with an overlapping area of 2m x 1m x 2.2m. (Figures 8 and 9) By applying the structural form-finding method to this area, it was possible to define a doubly curved surface, which in turn ensured compression only forces. At the same time, the herringbone brick pattern and self-supporting sequencing were computed and visualized to identify possible solutions. By rotating through each of these parameters, we developed the design into a feasible, yet not pre-defined geometry. As opposed to some form-finding approaches, this design was not the deterministic result of an optimization process. Instead, it allowed for change and adjustment even during construction. For example, the shape was recomputed when unpredicted robotic collisions emerged during the construction of the vault's edges. However, given our streamlined workflow that combined robotic space, structural form-finding, glass brick patterning and sequencing, we were able to react quickly and alter the vault's shape according to a collision-free volume.

While many form-finding techniques rely on a static set of conditions and constraints, such as initial shape, supports, or load-case, the design of structures intended for robotic fabrication is always time-dependent. Research does exist on computing ideal sequences for robotic processes and ideal geometries for stability during construction, however the dynamic nature of using different parameters requires new ways of visualizing change that allows for time-variables to 'picture' the implications of every construction step on the design space.[18] We envision a much more interactive non-deterministic design process in which human actors and computational tools find new ways of creatively interacting and exploring that go beyond the problem of the optimum. Ultimately, *LightVault* is the result of a time-based process in which designer, fabricator, computer, and machine work together not solely for the purpose of materializing a design but in a dynamic, non-predefined design exploration. The project not only showcases a glass brick structure built by robots, it is also the embodiment of an interactive, interdisciplinary design process which bridges different fields, languages, and tools. The vault's beauty is not only the result of structural efficiency and material elegance but also a representation of new design approaches based on communication, visualization, and adaptation.

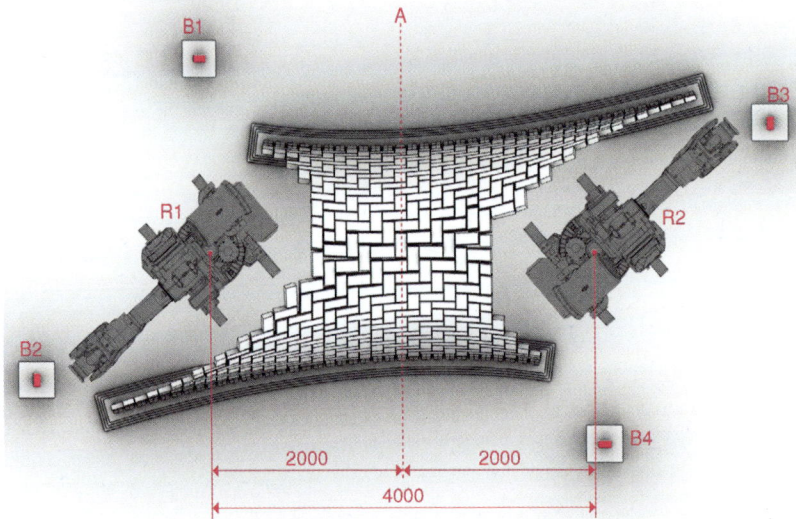

Figure 8. Robotic setup and workspace. *LightVault* Robot Workspace. Copyright CREATE Lab Princeton.

Conclusion

Visualization in architecture has long been used as a means of representation that includes translating a design into construction through the medium of execution drawings. However, with today's tools and computational means we can rethink the role of architectural visualization. Instead of utilizing it to communicate possible designs, visualization can help uncover the complex relations between fabrication, structural performance, and function by picturing new design spaces. For example, the reachability area for a multi-axis robot decreases quickly if obstacles are added. (Figure 10) By visualizing the relationship between placed elements and the robot's reach we offer designers a more in-depth understanding of its constraints that does not require specialist knowledge. This way, architectural visualization can serve as an intermediary language between design and technology and help translate ever-changing and increasingly complex technical knowledge into understandable design constraints. Similarly, by including the temporal aspect of fabrication, visualization can help identify the implications of one sequential step for the following ones. Like in a game of chess, the decisions taken at each step influence the range of possibilities that emerge for subsequent steps. Using the possibilities of computation and visualization to uncover these dependencies, rather than projecting a final design option, offers a powerful tool in today's architectural context.

Figure 9. Final *LightVault* prototype at the *Anatomy of Structure* exhibition in London, 2020. *LightVault* Prototype. Copyright CREATE Lab Princeton.

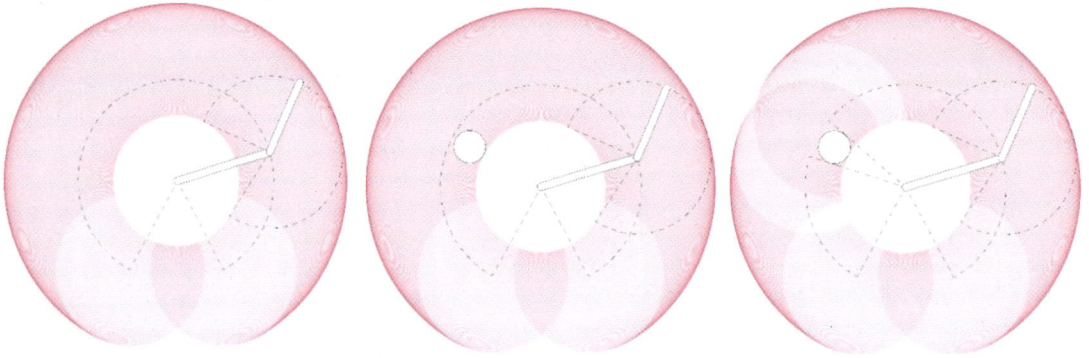

Figure 10. Visualization of reachability for a two-dimensional two-axis system. Left to right: without and with obstacle. Copyright Stefana Parascho.

Collaborators

Cooperative Robotic Assembly was completed in collaboration with D. Thomas Kohlhammer, Augusto Gandia, and Gonzalo Casas. *LightVault* is the result of an interdisciplinary research team including Princeton researchers Sigrid Adriaenssens, Isla Xi Han, Edvard Bruun, Ian Ting, Lisa Ramsburg, Vittorio Paris and Nicola Lepora, with support from Chase Galis, Lukas Fuhrimann, Grey Wartinger and Bill Tansley, and SOM engineers Alessandro Beghini, Samantha Walker, Michael Cascio, David Horos, Mark Sarkisian, Masaaki Miki, Max Cooper, Stuart Marsh, Matteo Tavano, Dmitri Jajich, and Arthur Sauvin.

Funding

Cooperative Robotic Assembly was funded by the National Center for Competence and Research Digital Fabrication (NCCR DFAB). *LightVault* was supported by Princeton's Metropolis Project, and the Princeton Catalysis Initiative.

ENDNOTES

1. Achim Menges and Sean Ahlquist, "Introduction," in *Computational Design Thinking* (Chichester: Wiley, 2011), 10–29.

2. "Scripting Cultures," in *Scripting Cultures* (John Wiley & Sons, Ltd, 2013), 8–12, https://doi.org/10.1002/9781118670538.ch1.

3. Mario Carpo, "Agency," in *The Alphabet and the Algorithm*, (Cambridge, Mass: The MIT Press, 2011), 106–20.

4. Stefana Parascho et al., "Computational Design of Robotically Assembled Spatial Structures: A Sequence Based Method for the Generation and Evaluation of Structures Fabricated with Cooperating Robots," in *AAG 2018: Advances in Architectural Geometry* 2018 (Klein Publishing, 2018), 112–39, https://www.research-collection.ethz.ch/handle/20.500.11850/298876.

5. Stefana Parascho et al., "Cooperative Fabrication of Spatial Metal Structures," *Fabricate*, 2017, 24–29.

6. Parascho et al., "Computational Design of Robotically Assembled Spatial Structures."

7. Tobias Bonwetsch et al., "The Informed Wall, Applying Additive Digital Fabrication Techniques on Architecture," Acadia 2006, 2006; Ralph Bärtschi et al., "Wiggled Brick Bond," in *Advances in Architectural Geometry 2010*, ed. Cristiano Ceccato et al. (Vienna: Springer, 2010), 137–47, https://doi.org/10.1007/978-3-7091-0309-8_10.

8. Tobias Bonwetsch, Bearth & Deplazes, and Gramazio & Kohler, "Gantenbein Vineyard Facade," in *ACM SIGGRAPH 2008 Art Gallery*, SIGGRAPH '08 (New York, NY, USA: Association for Computing Machinery, 2008), 52, https://doi.org/10.1145/1400385.1400413.

9. Luka Piškorec et al., "The Brick Labyrinth," in *Robotic Fabrication in Architecture, Art and Design* (Springer, 2018), 489–500.

10. Diederick Veenendaal and Philippe Block, "An Overview and Comparison of Structural Form Finding Methods for General Networks," *International Journal of Solids and Structures* 49, no. 26 (December 15, 2012): 3741–53, https://doi.org/10.1016/j.ijsolstr.2012.08.008; Philippe Block and Lorenz Lachauer, "Closest-Fit, Compression-Only Solutions for Free Form Shells," in Proceedings of the IABSE-IASS Symposium 2011 (London, UK, 2011).

11. Mario Deuss et al., "Assembling Self-Supporting Structures," *ACM Transactions on Graphics* 33, no. 6 (November 19, 2014): 1–10, https://doi.org/10.1145/2661229.2661266; Gene Kao et al., "Assembly-Aware Design of Masonry Shell Structures: A Computational Approach," in Proceedings of the International Association for Shell and Spatial Structures (IASS) Symposium, 2017.

12. Vittorio Paris, Attilio Pizzigoni, and Sigrid Adriaenssens, "Statics of Self-Balancing Masonry Domes Constructed with a Cross-Herringbone Spiraling Pattern," *Engineering Structures* 215 (July 2020): 110440, https://doi.org/10.1016/j.engstruct.2020.110440.

13. Stefana Parascho et al., "Robotic Vault: A Cooperative Robotic Assembly Method for Brick Vault Construction," *Construction Robotics*, November 2, 2020, 117–27, https://doi.org/10.1007/s41693-020-00041-w.

14. John Allen Ochsendorf and Michael Freeman, *Guastavino Vaulting: The Art of Structural Tile* (New York: Princeton Architectural Press, 2010), http://site.ebrary.com/id/10472727; Philippe Block et al., "Tile Vaulted Systems for Low-Cost Construction in Africa," *Journal of the African Technology Development Forum (ATDF) – Special Issue on Architecture for Development* 7 (January 1, 2010): 4–13.

15. Matthias Rippmann, Lorenz Lachauer, and Philippe Block, "RhinoVault - Interactive Vault Design," *International Journal of Space Structures* 27, no. 4 (December 2012): 219–30, https://doi.org/10.1260/0266-3511.27.4.219; Clemens Preisinger and Moritz Heimrath, "Karamba—A Toolkit for Parametric Structural Design," *Structural Engineering International* 24, no. 2 (May 1, 2014): 217–21, https://doi.org/10.2749/101686614X13830790993483.

16. Romana Rust et al., COMPAS FAB: Robotic Fabrication Package for the *COMPAS Framework*, 2018, https://github.com/compas-dev/compas_fab/; Johannes Braumann, S. Brell-Cokcan, "Parametric Robot Control: Integrated CAD/CAM for Architectural Design," in *ACADIA 11: Integration through Computation* (13-16 October, 2011): 242-51 (CUMINCAD, 2011), http://papers.cumincad.org/cgi-bin/works/paper/acadia11_242; Thibault Schwartz, "HAL," in *Rob | Arch* 2012, ed. Sigrid Brell-Çokcan and Johannes Braumann (Vienna: Springer, 2013), 92–101, https://doi.org/10.1007/978-3-7091-1465-0_8.

17. Oliver Porges et al., "Reachability and Dexterity: Analysis and Applications for Space Robotics," n.d., 7; Wael Suleiman, "On Inverse Kinematics with Inequality Constraints: New Insights into Minimum Jerk Trajectory Generation," *Advanced Robotics* 30, no. 17–18 (September 16, 2016): 1164–72, https://doi.org/10.1080/01691864.2016.1202136; Abhijit Makhal and Alex K. Goins, "Reuleaux: Robot Base Placement by Reachability Analysis," ArXiv:1710.01328 [Cs], October 3, 2017, http://arxiv.org/abs/1710.01328.

18. Yijiang Huang et al., "Robotic Extrusion of Architectural Structures with Nonstandard Topology," in *Robotic Fabrication in Architecture, Art and Design* (Springer, 2018), 377–89; Deuss et al., "Assembling Self-Supporting Structures."

DESIGN AS AN EXTENSION OF OUR SENSES:

ENERGY FLOWS AND THE PROTOTYPING OF ALLIESTHESIA

DORIT AVIV

ABSTRACT

Thermodynamically, architecture is as an active agent: a mediating layer in the exchange between external climatic forces and internal human sensations. What does it mean, therefore, to design with heat? How do we shape spaces to interact with the thermal sensation of people moving through them? While we are visually blind to the flows of thermal energy surrounding us, they are acutely perceived by our skin's thermal receptors. To use architecture to shape and direct heat exchanges between our bodies and the environment, we must be able to measure and represent these interactions. The internet of things (IoT), simulations, and imaging techniques enable detection and representation of invisible physical phenomena. Once characterized, it is possible to actively design the paths of heat flow through buildings, thus reducing our fraught overreliance on mechanical systems and external energy supply. This paper discusses two full scale installations designed, built, and operated in Seoul and Singapore respectively. Using infrared radiation, these spatial constructions create a unique relationship between the human body, energy flows, and the surrounding surfaces, with the goal of enhancing people's thermal sensation and achieving alliesthesia.

+ architectural prototypes
+ energy flows
+ sensors
+ human body
+ alliesthesia

Figure 1. Detail of a thermal image of a person amongst the hot and cold surfaces in *Thermally Alive Space*. Image courtesy of Forrest Meggers.

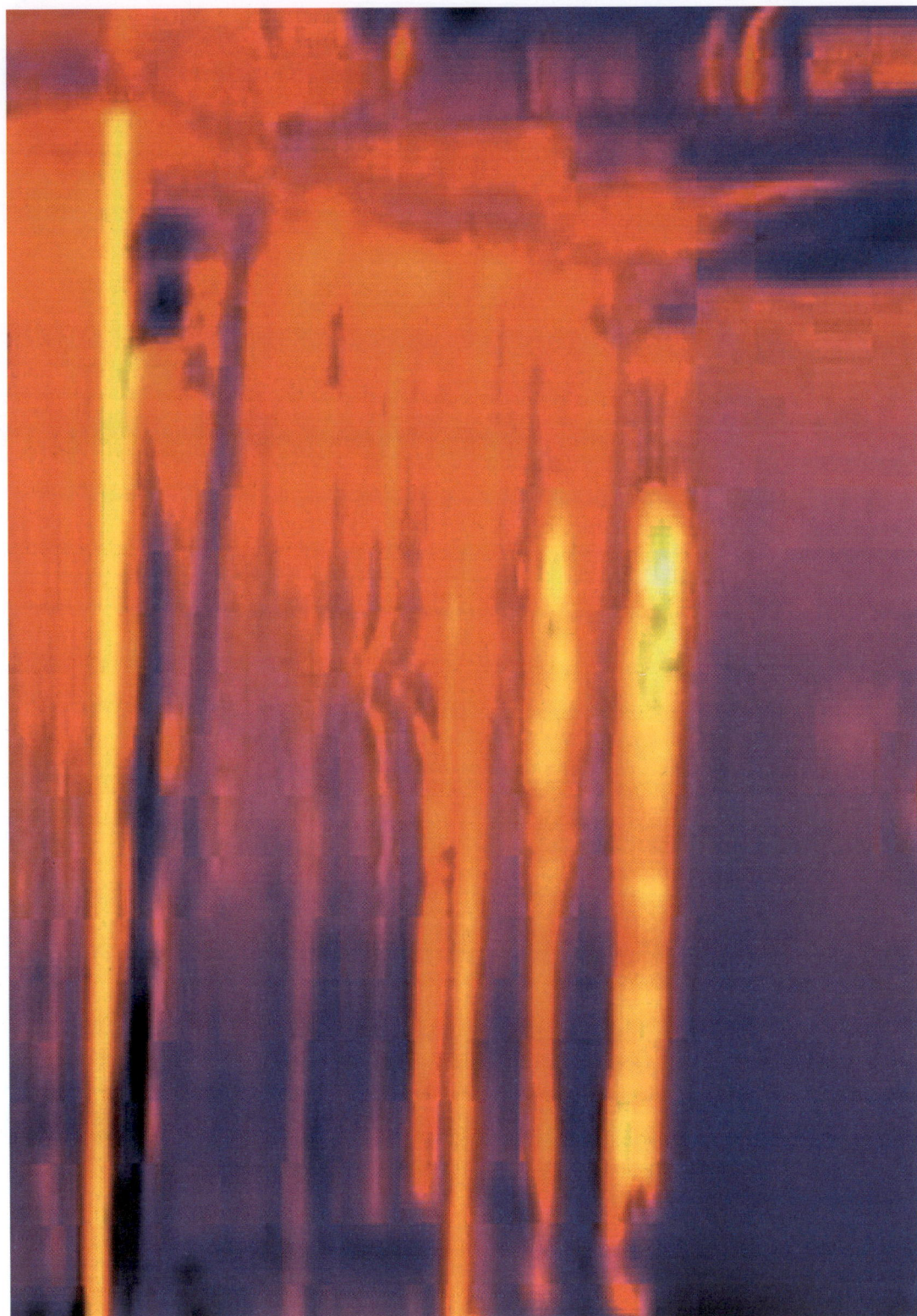

Energy and Architectural Design

Rising global energy consumption continues to increase greenhouse gas emissions, which in turn is responsible for the warming of the planet. Unless we reconsider the fundamentals of thermal control in building interiors, this vicious cycle will only contribute to more energy expenditures.[1] The building sector's share of electricity use has grown dramatically in the past six decades, from twenty five percent of United States annual consumption in the 1950s to more than seventy-six percent by 2012.[2] The majority of energy consumed in buildings is due to mechanical heating and cooling, with spikes in electricity consumption closely correlated to the near ubiquitous spread of air conditioning in buildings.[3] During this period, architects have grown to rely on mechanical systems designed by engineers to provide a ventilated, climatically-controlled interior environment rather than on the architecture itself.

At the same time, technological innovation in architecture, most notably in the field of computational design, has contributed multiple new tools for describing and constructing increasingly complex building surfaces, forms, and volumes. Largely overlooked, however, are the energetic implications of these intricate design decisions on surface radiation and volumetric airflow. Engineers have developed and employed sophisticated internal climatic control systems for buildings, however, these systems most commonly reside within mechanical rooms and hidden spaces that are independent from the building's materiality and form. As such, contemporary buildings are conceived as mere containers of environmental systems, disconnected from the exterior climate within which they reside. How did we arrive at this moment? Architects as 'master-builders' meticulously orchestrate the design and construction details of buildings yet maintain almost no agency in determining the way energy is consumed in buildings. As a result, they inadvertently facilitate buildings' contribution to greenhouse gas emissions and the warming of our climate.

In the second half of the twentieth century newly available air conditioning technologies were described by Reyner Banham in *Architecture of the Well-Tempered Environment* (1969) as powerful partners at the architect's disposal to aid in environmental management.[4] Banham pointed with excitement at the prospect of having "full control" over the design of one's environment.[5] Decades later, the result of this "control" has been that to this day mechanical systems of chillers, ducts, and vents are treated by architects as black-boxes, untouched and unseen. Robert Venturi argued in *Complexity and Contradiction* (1977) that their spatial expression in modern architecture had been contained in the building's "*poché* space"—within wall cavities, shafts, and dropped ceilings. [6]

To challenge the perception that thermal exchanges are separate from a building's form and materiality, what is proposed in this paper is neither an exposing of the mechanical space previously hidden in the *poché*, nor a continuation of conventional architectural engagement

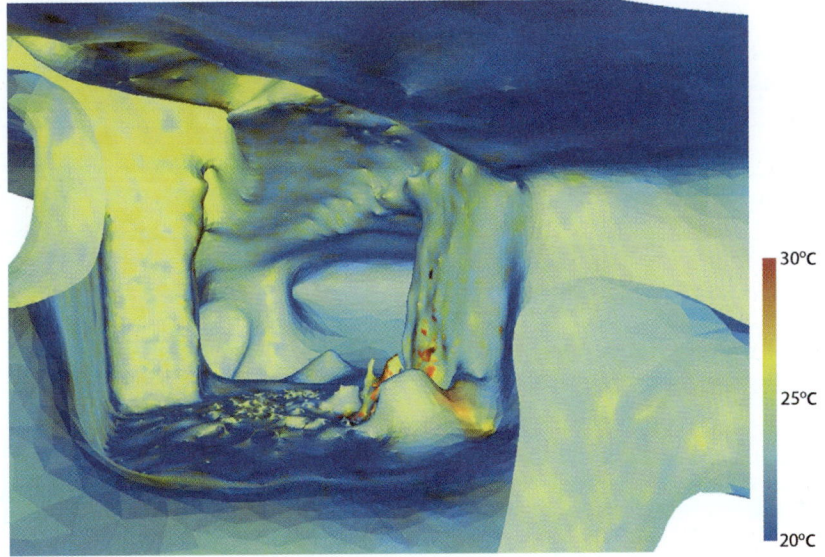

Figure 2. Building interior as a thermally heterogenous environment. Image generated with data from a thermal surface scan, detecting both the surface geometry and temperature. Image credit Nicholas Houchois and Dorit Aviv.

30°C

25°C

20°C

with heat transfer as that limited to the R-value of walls. Instead, this paper reconsiders architectural surfaces and interior volumes as active players in a dynamic energy field that exists between the human body and the spaces surrounding it.

Space, Heat and Architecture: Filling the Void

Siegfried Gideon in *Space, Time, and Architecture* (1941) insisted that "architectural development" during any given historical period depended upon that period's conception of space.[7] Gideon claimed that modern architecture could be defined by the "plastic volumes" it produced and by its resulting interior spaces that were "flowing" rather than rigid.[8] This concept of flow was used by Gideon to describe the formal manifestation of interior spaces, conceived as the inverse of matter—as voids.

However, as argued in this paper, the flows of energy and matter that take place across the interior volumes of any building are very much physical realities. Spatial voids are, in fact, filled with heterogenous material layers and thermal interactions. And yet, airflows and energy exchanges which constitute this interior space have been largely ignored by architectural discourse, even though they are intimately defined and conditioned by the geometric and material properties of such spaces.

This is clearly visible in a 3D thermal scan of a building's interior, where using infrared sensing, surface geometries and temperatures are captured and represented with a color gradient.[9] (Figure 2) The hottest spot in the space, represented by a red region, is due to the presence of a person sitting and typing on a laptop. Both the person and the computer not only heat up the air flowing around them via convection but also emit heat to the surfaces around them via radiation. Light sources and exposure to the outdoor environment contribute to thermal differences amongst interior zones and further influence thermal interactions that occur within buildings.

If we view our discipline from the perspective of thermodynamics, architecture can be understood as a mediator in the transfer of heat between the human body and the larger environment which extends from its immediate surroundings to the cosmic scale of the sky and stars.[10] Heat transfer physics connects these different scales directly; while conduction occurs only between objects which touch each other, convection transports heat through air motion affected by atmospheric conditions. Thermal radiation—part of the spectrum of light—exchanges heat at all distances, as close as a hand's reach and as far as the sun, at various wavelengths. To shape and direct energy flows architecturally, to and from the human body, we must begin to understand and describe architectural elements in relation to the extents of their thermodynamic capacity and influence, rather than as autonomous isolated objects.

The Human Body, Thermoregulation, and Alliesthesia

The human body is the living epicenter of these thermodynamic interactions. Buildings do not care if they are hot or cold, but humans do. Our bodies have complex sensorial systems to detect and manage heat with multiple sophisticated mechanisms for exchanging energy with the environment. Our core temperature is regulated by our skin that acts as a medium for heat transfer, in the body's attempt to maintain a state of homeostasis.[11] The hypothalamus in the brain is the regulator of this process: when the core temperature rises above the homeostasis state, it initiates heat-releasing mechanisms such as dilating superficial arteries, increasing blood flow to increase the rate of heat loss to the air, initiating sweating, and decreasing heat production through metabolism.[12] Conversely, when the core temperature is too low the hypothalamus signals superficial arteries to constrict while it initiates shivering and increases the metabolic rate.[13] This thermoregulation process controls the temperature and sweat level of the skin's surface and thus alters the rate of heat exchange. This exchange is dominated by evaporation, radiation, and convection, while conduction is only responsible for a small percentage of energy transfer from the body.[14]

Architectural elements can enhance or diminish the transfer of heat from the human body to its colder surroundings (cooling), or vice versa, from a hot environment to the body (heating). The material properties and geometric configuration of walls, floors, and ceilings define the boundaries of this exchange. While this dynamic process is dependent on multiple variables, we have gotten used to relying on thermostats that can only measure air temperature as the sole source of data in the thermal environment. The conspicuousness of thermostats codifies the ubiquitous use of air conditioning as temperature control system in buildings. The neglect of other variables (environmental and

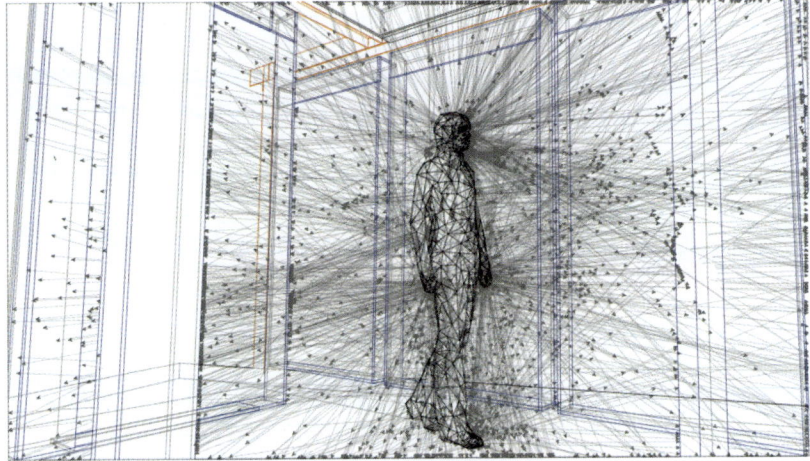

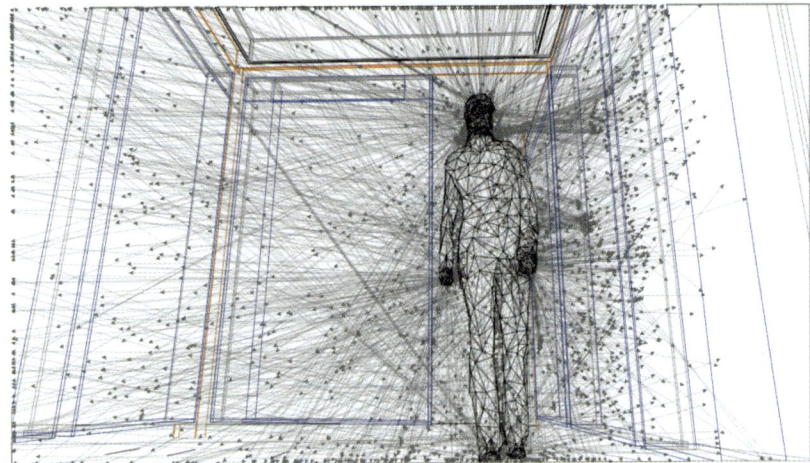

Figure 3. Visualization of heat fluxes from human skin to the environment. Each line is a vector representing heat fluxes leaving the surface of the body once exposed to colder building surfaces. Image credit Dorit Aviv.

personal) which impact heat transfer from the body, has disassociated architectural elements and the human body from the control of thermal sensation in buildings. And yet, we know that energy exchanges closely link human bodies with the architectural space that surrounds them. This interaction is illustrated in a vector representation of the direction of heat fluxes leaving the skin and being absorbed by the surrounding surfaces. (Figure 3) This illustration is part of a simulation process that helps determine the amount of energy exchanged due to the temperature difference between the skin and the surface it faces. This temperature difference—the delta between objects, spaces, or materials—is what drives all thermodynamic processes in nature. The geometry of our body's surface and that of spaces surrounding the body dictate the amount of 'optical exposure', of electromagnetic radiation that is transferred between them, and result in different rates of heat exchange from different areas of the skin. Like the rays of visible light emitted by a light source, the lines in the illustration trace the path of thermal energy emitted by our body and absorbed by the surfaces surrounding it. (Figures 4a and 4b) Additionally, the amount of clothing, airflow, and sweat level also contribute to the multiple interactions that are possible between surfaces and humans.[15]

The field of study broadly termed 'Thermal Comfort' has been preoccupied for most of its history with seeking the measure of equilibrium and neutrality in these interactions. Scientist and engineer Povl Ole Fanger (1934-2006), whose thermal comfort model has dominated the field for decades, defined thermal comfort in terms of achieving a heat balance between our metabolic rate and the rate of heat exchange of our body with the environment, arriving at a neutral state in which we neither feel too hot nor too cold.[16] It is the lack of discomfort, or lack of excessive heat loss or gain, which defines the ideal thermal condition for our body as a steady-state, both internally and externally.

This thesis does not consider, however, that for a large fraction of the time both the external climate and our own internal metabolism constantly undergo dynamic transformations. In fact, an excess of heat gain or heat loss can be advantageous in arousing sensorial responses to thermal changes. This concept of 'alliesthesia,' explored in publications by building scientist Richard de Dear, investigates the benefits of thermal pleasure that arise from the combination of environmental and metabolic transient conditions.[17] Imagine the experience of entering a building after a bike ride or a morning jog, when exhaling and sweating at peak metabolic rate a strong gust of cool air on your skin is most welcome (even as the same type of chilled air flow is drafty

	30.00<
	29.00
	28.00
	27.00
	26.00
	25.00
	24.00
	23.00
	22.00
	21.00
	<20.00

Figure 4a. Simulation of the exposure of different regions on the surface of the body to fluxes from the various surfaces surrounding it. Mapping of different levels of exposure of the body surfaces to surrounding heat fluxes. Image credit Dorit Aviv.

and unpleasant when sedentary or resting in bed). Conversely, the experience of entering a warm bath or sauna is pleasurable precisely because of the concentrated excessive heat gain it provides. As Lisa Heschong wrote in her doctoral thesis *Thermal Delight in Architecture* (1978), "The more sensory input we experience, and the more varied the contrasts, the richer is the experience and its associated feelings of delight." [18]

Sensing Technologies and the IoT

For buildings to respond more effectively to external climatic conditions, as well as to internal dynamics of contained spaces, we need live, timely, and communicable information to make these conditions legible. Just like our brain processes sensorial data from our skin and responds in a feedback loop, the immense growth of novel sensing devices provides us with real-time data that describes invisible physical processes, extending our sensory register to the environment around us. Low-cost sensors and microcontrollers, broadly categorized as the "Internet of Things" (IoT), with open-source code and mass availability have changed the way we construct and test physical prototypes. At present, building occupants can purchase environmental sensors for a fraction of the cost of a decade ago. Equally, inexpensive microcontrollers are available with open-source code that can be used without the need to buy expensive dataloggers and software licenses to process the data. Employing a variety of thermal imaging techniques, 3D scanning, and environmental sensors

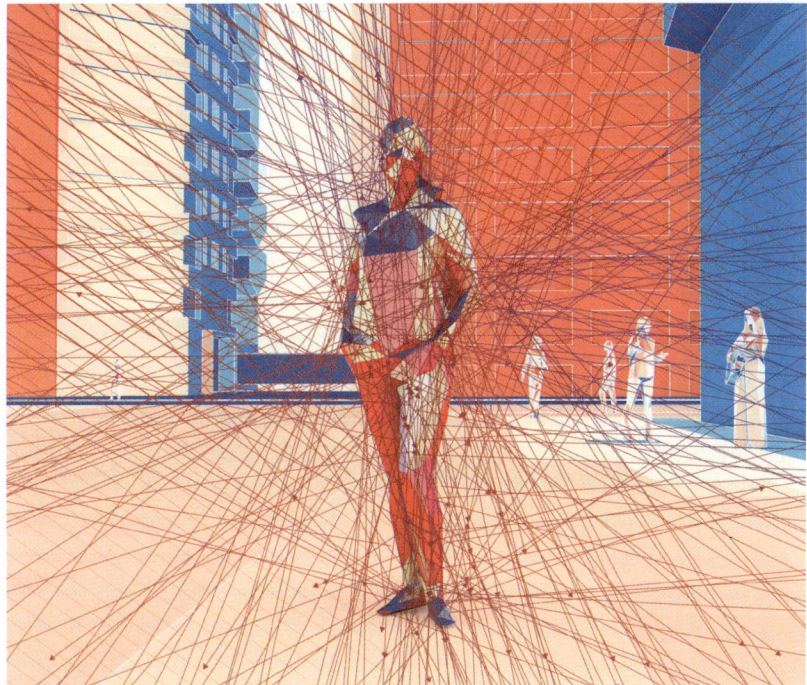

Figure 4b. Simulation of surface temperature change of different areas of the body based on the degree of exposure to heat fluxes from buildings' longwave radiation and solar shortwave radiation. Image credit Dorit Aviv and Jiewei Li.

to gather data on the ever-changing dynamic flows of heat and matter through space enables the design of architectural elements that respond to temporal changes in measurable environmental forces such as wind speed and direction, solar radiation, relative humidity, and air temperature.

Once this data is collected, the question remains: How does it become legible to designers who would like to use it? Visualizing these thermodynamic processes is essential for interpreting them. Two experimental projects are presented in the following pages which aim to investigate how architectural design methods can leverage the many potential trajectories of energy flow in buildings. These pavilion-sized prototypes were built and deployed at full scale to test and illustrate interactions of heat with architectural form and materials. Both pavilions were designed to magnify the influence of radiant heat fluxes on the human occupants who entered them. Sensor measurements were taken from the prototypes and used to analyze their environmental performance.

Thermally Alive Space, Seoul Biennale (2017)

The first project, *Thermally Alive Space*, was a full-scale installation for the 2017 Seoul Biennale for Architecture and Urbanism.[19] Its aim was to explore the constant thermal exchange that occurs between the human body and an architectural interior. This exchange, resulting in a gradient of thermal sensations, was used as means for architectural space-making, defining the spatial perception of people moving through the installation. Infrared cameras captured thermal data, recorded thermal events, and provided means for visualizing both.

Architecturally, the installation was conceived as a darkened tunnel, inserted within the ground floor of a small building. Its walls and ceilings were constructed of reflective aluminum surfaces shaped into parabolic troughs. Upon entering, the darkness helped subdue one's vision, which conventionally dominates our sensorial response to architectural spaces. (Figures 5a and 5b) In this way, darkness was designed to heighten the thermal sensations of visitors who were now more attuned to the temperature variations sensed by their skin.

The vertical and horizontal concave parabolic troughs had water pipes placed at their focal points.[20] (Figure 6) The water temperature of the pipes was regulated by a central heat pump to vary by thirty degrees Celsius, between cold and hot pipes. (Figure 8) To optimize their impact, the pipes were covered with matt black paint which is highly emissive in the infrared range and like parabolic light fixtures, here too, the parabolic shape of the room's surfaces created a special reflection effect. The multi-directional fluxes from the pipes placed at the focal point of each parabola were turned into parallel rays facing the person in the plane ahead. (Figure 9) Thus, the installation acted like a series of parabolic mirrors, orienting the heat emitted from the pipes and directing it towards pavilion visitors walking through it. Opposing radiant fluxes from the surfaces created a perception of an

Figure 5a. Thermally Alive Space, view of entrance into the installation. Image credit Dorit Aviv.

Figure 5b. Thermally Alive Space, view of installation interior showing the dark space, the reflective surfaces of the aluminum parabolic troughs, and the thermal images captured at the end of the tunnel. Image credit Jeon Byung-Cheol.

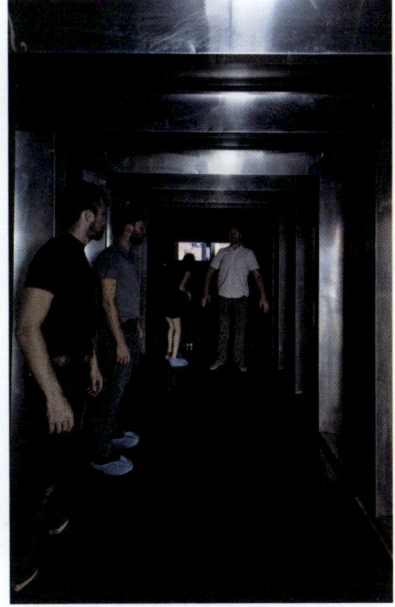

intense heat gradient throughout the space. This gradient became the differentiating element in a space that visually looked dark and repetitive. (Figure 7) The thermal image captured in Figure 7, which shows the resultant gradients, can be considered as the defining 'thermal plan' of the installation.

While walking through the tunnel, these gradients produced an acutely perceptible transition between the sensation of concentrated heat to that of concentrated cooling. This invisible yet experienced spatial contrast provided a sensory immersion for the body's thermal receptors, similar to what Heschong playfully describes as the "delicious" delight "of a balmy spring day as I walk beneath a row of trees and sense the alternating warmth and coolness of sun and shade."[21] In addition to the skin's sensorial response, infrared cameras provided a visual recording of the varying temperatures of both the installation's surface elements and those of people's skin and clothing while moving through the space. (Figure 10) Different colors visualizing infrared radiation were used to reveal the hidden play of radiative heat transfer otherwise invisible to the human eye.

The Cold Tube, Singapore (2018)

The second full-scale prototype project was an experimental cooling pavilion that displayed a groundbreaking development in radiant cooling for hot-humid climates.[22] The Cold Tube was designed, built, and operated as part of a collaborative venture between multiple research institutions and installed at the United World College of Southeast Asia in Singapore in 2018.[23] The pavilion was defined by a spatial envelope of approximately six by eighteen feet and constructed

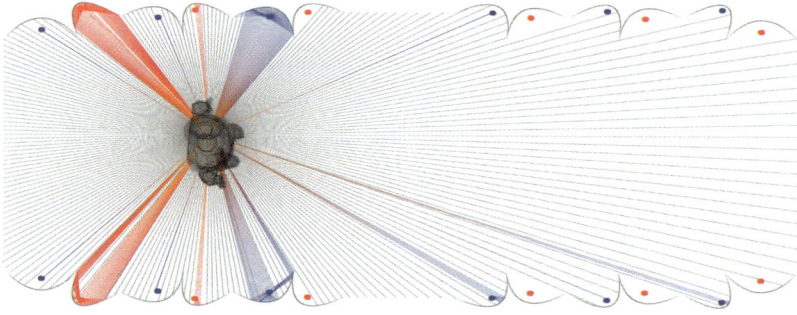

Figure 6. Visualization diagram of radiant heat fluxes reflected by the surrounding curved surfaces onto the surface of the body of a person traversing the space. Image credit Dorit Aviv and Tyler Kvochick.

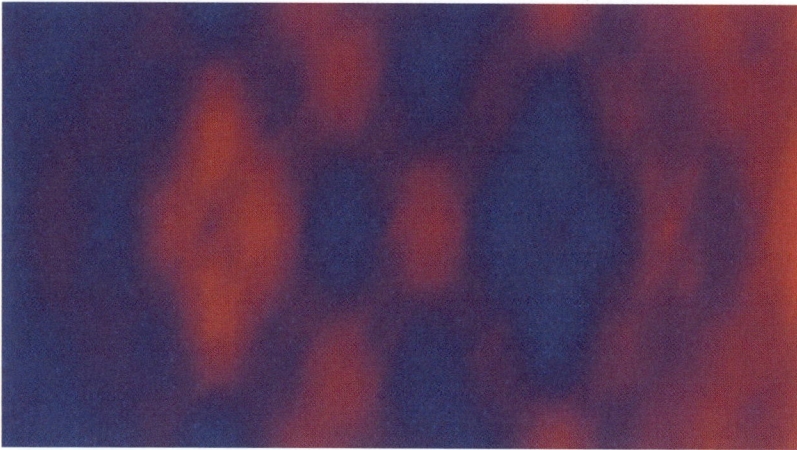

Figure 7. Visualization of the resultant hot and cold regions throughout the installation. The space was defined by zones of different thermal sensations rather than by physical barriers. Image credit Tyler Kvochick.

using ten, four by eight foot panels; two horizontal panels at the top and eight vertical panels, with north and south facing entrances. (Figure 11)

The goal of this structure was to provide radiant cooling by panels with embedded chilled water tubing, while avoiding the occurrence of condensation in the humid climate of Singapore. Custom variable speed chillers were used to cool down the water that ran through blue polymer capillary mats. Despite their very low temperature the surfaces of the cold mats were protected from condensation by a polyethylene membrane placed between them and the space occupied by visitors walking through the pavilion. (Figure 13a) This ultra-thin membrane acted as a barrier between the hot humid air, which was freely flowing through the space, and the cold surfaces behind the membrane cooled to below the dewpoint temperature. The membrane was not, however, a barrier to the flow of heat emitted by people via infrared radiation. This was transmitted through the membrane to the cold mats just like visible light is transmitted through window-glass. (Figure 13b) Thus, by playing with the material properties of the surfaces, it was possible to avoid condensation by preventing the humid air from reaching the cold surfaces while at the same time allowing radiant energy to flow between people and adjoining cold surfaces. Like the curved reflective parabolas used in the previous installation, here too material and geometric properties of architectural elements were used to modify the thermal sensation of people in space.

The spatial arrangement of the pavilion was configured to further optimize its cooling impact: it enveloped the people inside, isolating them from the hot surrounding environment and exposing them to thermal exchange with the cold panels. The resultant thermal gradient field through the pavilion's space was captured representationally. (Figure 12) When standing deep inside the pavilion, a person would exchange heat mainly with the panels, perceiving a sensation of cooling, but when walking towards the opening, the rest of the environment's impact would increase, resulting in a surging sensation of heat. Multiple sensors were placed in different locations inside the pavilion to assess its overall performance and to tune the system controls to achieve the intended cooling impact.

Thus, without the introduction of any form of forced air conditioning, the pavilion provided much-needed thermal shelter in an incessantly hot-humid climate. This is an important achievement in a constantly warming planet: the creation of a unique thermal sensation in which the skin is cooled through its exposure to surfaces. Many visitors who walked through the pavilion, witnessed firsthand the surprising, pleasant sensation of being immersed in a thermal state of alliesthesia, able to spend time in an open-air structure in the sweltering climate of Singapore, yet surprised with a cooling sensation.

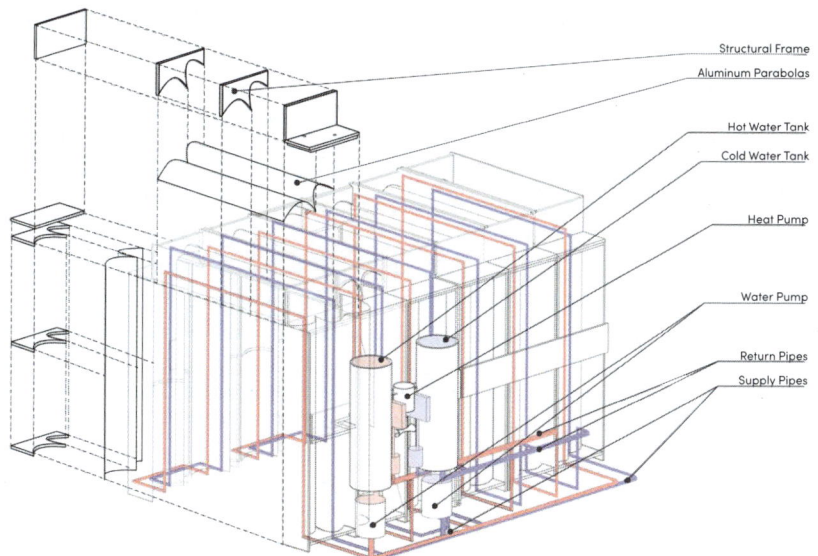

Figure 8. Diagram of the hydronic system providing radiant heating and cooling within the installation. Image credit Dorit Aviv.

Structural Frame
Aluminum Parabolas

Hot Water Tank
Cold Water Tank

Heat Pump

Water Pump

Return Pipes
Supply Pipes

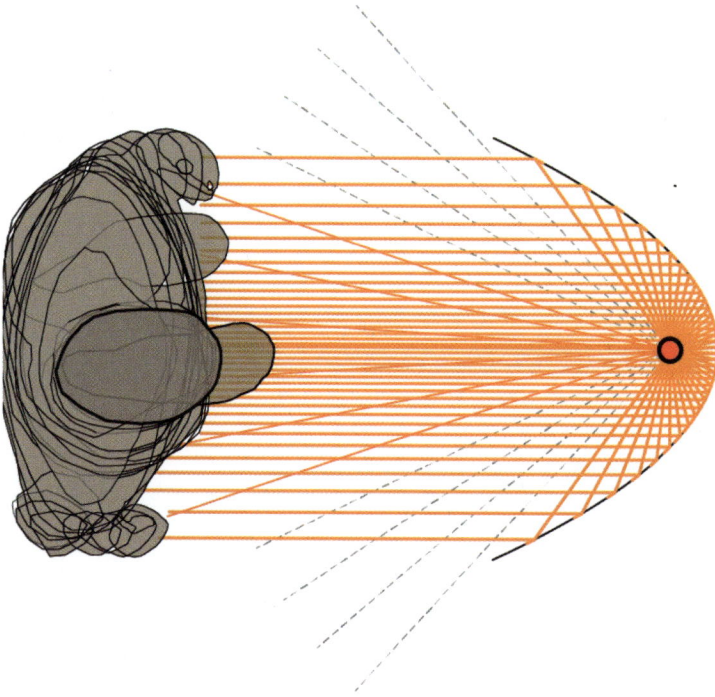

Figure 9. Illustration of heat fluxes from the hot pipe placed in the focal point of a parabola, reflected by the parabolic surface onto a person facing it. Image credit Dorit Aviv.

Concluding Reflections

Both installations demonstrate how design can contribute solutions to the pressing need for energy conservation, and reciprocally, how the inclusion of thermodynamic interactions within the design process can enrich the architect's palette with generative formal and material strategies. The vectors, lines, meshes, and color gradients found in the analytical representations were an integral part of the research projects. They provided the means for visualizing all manner of data associated with heat fluxes, energy flows, and temperature differences between objects and people in space. The color representation of surface temperatures is, of course, a fiction since infrared radiation occurs outside of the visible range, beyond the RGB palette. Yet it does serve to delineate something deeply significant: that the potential of heat transfer is dependent on gradients—deltas between the energy state of exterior climate and interior climate, as between people and architectural elements. With this awareness, designers can begin to exploit these naturally occurring gradients within their designs.

On a more fundamental level, these modes of representation encourage a shift from an architecture which is dependent on black-box mechanical systems for its climatic adaptation to an architecture which integrates energy as a guiding element for building form itself, connecting specific geometric and material arrangements to specific embodied sensations of people in space.

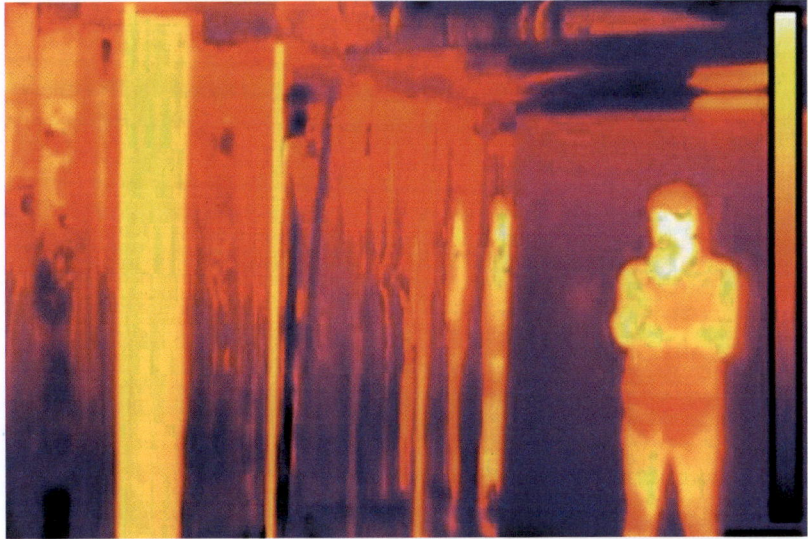

Figure 10. Thermal image of a person amongst the hot and cold surfaces in the pavilion. Image courtesy of Forrest Meggers.

Figure 11. View of the constructed pavilion, *The Cold Tube*. Image credit Lea Ruefenacht.

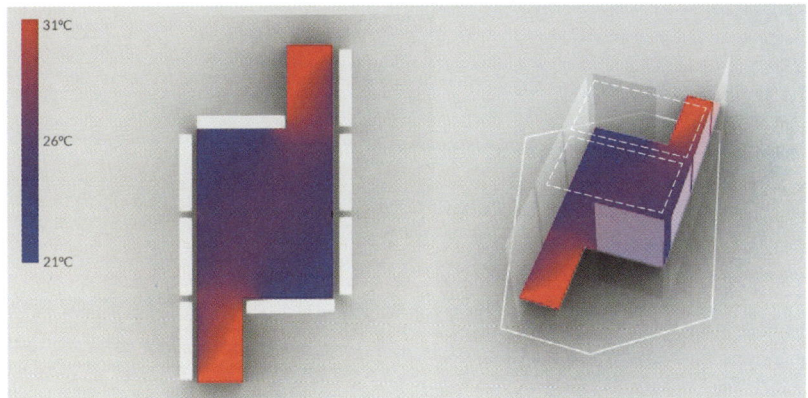

Figure 12. Visualization of the thermal zones created by the pavilion's geometry. Image credit Dorit Aviv.

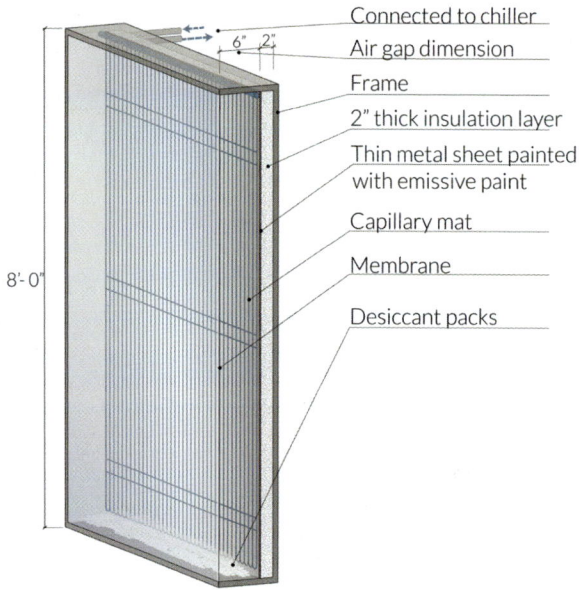

Connected to chiller
Air gap dimension
Frame
2" thick insulation layer
Thin metal sheet painted with emissive paint
Capillary mat
Membrane
Desiccant packs

8'- 0"

Figure 13a. Diagram of membrane–assisted radiant cooling panel components. Image credit Dorit Aviv.

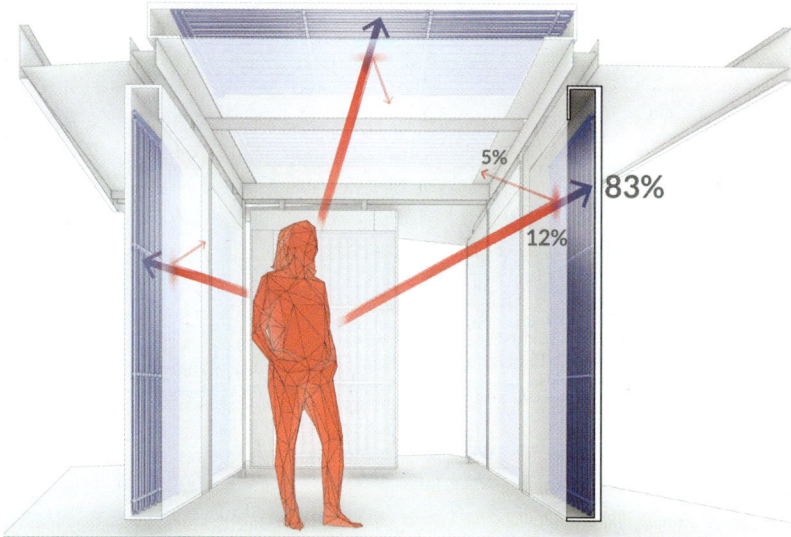

5%

83%

12%

Figure 13b. Diagram of person's heat loss from the warm surface of the skin and clothing to the surrounding cold panels through the transmissive membrane. Image credit Dorit Aviv and Jiewei Li.

Collaborators

Kipp Bradford, Kian Wee Chen, Hongshan Guo, Nicholas Houchois, Tyler Kvochick, Forrest Meggers, Jovan Pantelic, Lea Ruefenacht, Adam Rysanek, and Eric Teitelbaum.

ENDNOTES

1. Lucas W Davis and Paul J Gertler, "Contribution of Air Conditioning Adoption to Future Energy Use under Global Warming," *Proceedings of the National Academy of Sciences* 112, no. 19 (2015): 5962–67.

2. US Department of Energy, "Quadrennial Technology Review: An Assessment of Energy Technologies and Research Opportunities," *Manufacturing Energy Consumption Survey.* (Washington DC: EIA, 2013).

3. Liu Yang, Haiyan Yan, and Joseph C Lam, "Thermal Comfort and Building Energy Consumption Implications–a Review," Applied Energy 115 (2014): 164–73; Jeff E Biddle, "Making Consumers Comfortable: The Early Decades of Air Conditioning in the United States," *The Journal of Economic History* 71, no. 4 (2011): 1078–94.

4. Reyner Banham, *Architecture of the Well-tempered Environment* (Architecture Press, 1969).

5. Ibid, Chapter 2: "Environmental Management."

6. Robert Venturi, *Complexity and contradiction in architecture,* Vol. 1., (New York: The Museum of modern art, 1977).

7. Sigfried Giedion, Space, *Time and Architecture: the growth of a new tradition* (Harvard University Press, 1967): xlviii.

8. Ibid, 460.

9. Image previously published here: Dorit Aviv, Nicholas Houchois, and Forrest Meggers, "Thermal Reality Capture: Merging Heat-Sensing with 3D Scanning and Modeling to Characterize the Thermal Environment," *Proceedings of the 39th Annual Conference of the Association for Computer Aided Design in Architecture* (2019): 338–45.

10. Kiel Moe, *Insulating Modernism: Isolated and Non-isolated Thermodynamics in Architecture* (Birkhäuser, 2014), https://doi.org/10.1515/9783038213215; William W Braham, *Architecture and Systems Ecology: Thermodynamic Principles of Environmental Building Design, in Three Parts* (Routledge, 2015).

11. J Gordon Betts et al., "Introduction to the Human Body: Homeostasis," in *Anatomy and Physiology* (Houston, Texas: OpenStax, 2013), https://openstax.org/books/anatomy-and-physiology/pages/1-5-homeostasis; Lindsay M Biga et al., "Anatomy & Physiology," (OpenStax /Oregon State University, 2020).

12. Connie Rye et al., "Metabolism," in *Biology* (Houston, Texas: OpenStax, 2016), https://openstax.org/books/biology/pages/6-1-energy-and-metabolism.

13. J Gordon Betts et al., "Introduction to the Human Body: Homeostasis," in *Anatomy and Physiology* (Houston, Texas: OpenStax, 2013), https://openstax.org/books/anatomy-and-physiology/pages/1-5-homeostasis.

14. A review of the forms of heat transfer from the body and the thermal potential of active building surfaces is provided in Kiel Moe, *Thermally active surfaces in architecture* (Princeton Architectural Press, 2010).

15. Maohui Luo et al., "High-Density Thermal Sensitivity Maps of the Human Body," *Building and Environment* 167 (2020): 106435.

16. Povl O Fanger, *Thermal comfort. Analysis and applications in environmental engineering* (Copenhagen: Danish Technical Press, 1970).

17. Richard De Dear, "Revisiting an Old Hypothesis of Human Thermal Perception: Alliesthesia," *Building Research & Information* 39, no. 2 (2011): 108–10; Thomas Parkinson and Richard De Dear, "Thermal Pleasure in Built Environments: Physiology of Alliesthesia," *Building Research & Information* 43, no. 3 (2015): 288–301; Sijie Liu et al., "Dynamic Thermal Pleasure in Outdoor Environments-Temporal Alliesthesia," *Science of The Total Environment* 771 (2021): 144910.

18. Lisa Heschong, *Thermal delight in architecture*, PhD diss. (Massachusetts Institute of Technology, 1978): 2.

19. The installation was designed and constructed in collaboration with Forrest Meggers, Eric Teitelbaum, and Tyler Kvochick.

20. Dorit Aviv, Eric Teitelbaum, Tyler Kvochick, Kipp Bradford and Forrest Meggers, "Generation and Simulation of Indoor Thermal Gradients: MRT for Asymmetric Radiant Heat Fluxes," Proceedings of the International *Building Performance Simulation Association*, Vol. 16 (2019): 381–88. https://doi.org/10.26868/25222708.2019.210702

21. Lisa Heschong, T*hermal delight in architecture*, 30.

22. Eric Teitelbaum et al., "Membrane-Assisted Radiant Cooling for Expanding Thermal Comfort Zones Globally without Air Conditioning," *Proceedings of the National Academy of Sciences* 117, no. 35 (2020): 21162–69.

23. Amongst the participating institutions: Princeton University, University of British Columbia, National University of Singapore, Eidgenössische Technische Hochschule Zürich (ETH), and University of California, Berkeley.

CODE, GRAPHICS, FORM:
FORM AN AESTHETIC INQUIRY OF MECHANICAL MOTION

VIOLA AGO

ABSTRACT

Form and aesthetics have been central to architectural discourse for decades. More recently, advances in technology, tools, and production processes have fueled new methods of inquiry in architectural form and its role in the larger discipline. This article positions 'form' as a state of becoming rather than a static entity by focusing on the creative and perceptive event that produces the work. In looking at the 'event of form,' one more carefully examines the forces and motions that give rise to form and subsequently, to the agency that it has in architectural thinking. The first part of this essay introduces theoretical and historical concepts of motion in drawings produced by artists, scientists, and experts in workplace efficiency. The second part describes and examines four drawing projects produced by the author which capture motion in its transition from hand production to machine production. This design research work argues that the project of 'indeterminate form' can further eradicate disciplinary divides and introduce undiscovered territories for aesthetic inquiry in architecture.

+ motion
+ formless
+ aesthetics
+ becoming
+ drawings

Figure 1. Systems and Geometry Drawing *House 3*, 2021. Credit Viola Ago.

290

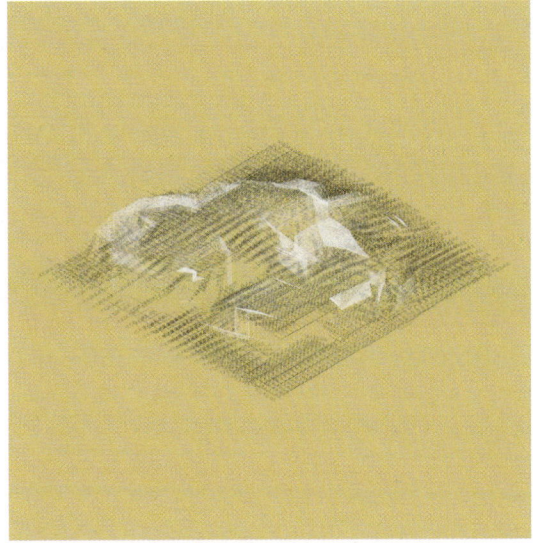

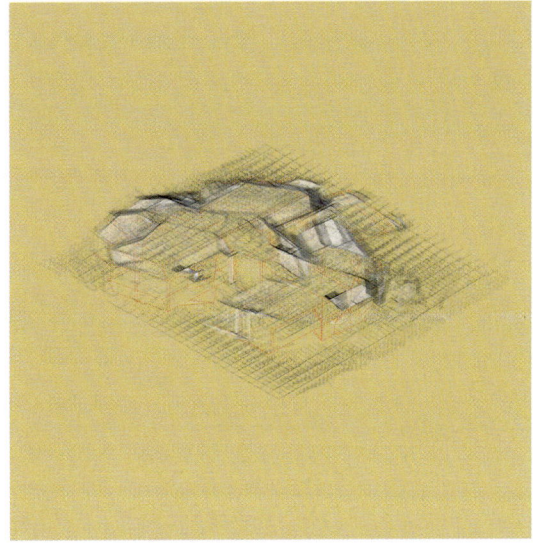

Becoming and Motion

In her dissertation titled *From Grid to Matrix: (Im)material Events and the Emergence of Smooth Space at the Limits of Contemporary Architecture*, architect, educator, and Dean Ila Berman described architectural form as a continually changing effect dependent on two distinct moments: one of confrontation (further defined as an action) and one of experience (or rather event).[1]

> [T]he capacity to think architecture as either action or event, dislocation or becoming, demands that one approach architecture's ever-changing relationship with time. All architectural relations, actions, experiences and interpretations, occur either across this interval between states, or intensively within a becoming that inheres in, and is endured through time. It is the event of architecture which inspires its living relation, and the movement of such an event which inevitably results in its dis-location. [2]

This quote unravels a convincing description of the architectural event as a condition that is simultaneously material and force (immaterial). In alignment with this idea, this essay asserts that architecture can be formless and indeterminate, can resist static definition, can be spatially non-conforming, and can occupy ever more fictitious formal or formless narratives. It can, at the same time, be an altered state that depends on the observer or the occupier of space. In other words, architecture can be thought of as constantly in a state of becoming rather than being.[3] Concepts of forms that emerge from an event (or process) and of forms that resist decisive (static) states are not new or peripheral to architecture, as is evident from Berman's 1994 dissertation. In architecture, late modernists, and thereafter deconstructivists, operated within similar frameworks. Adjacent disciplines such as painting, sculpture, and film, offered their own discourses on the different manifestations of becoming including the formless, the indeterminate, and other aesthetics of anti-rigid forms.[4] Perhaps most decisively, philosophical thought and anthropological inquiries have pursued the concept of becoming, and by extension 'process' in these creative disciplines.

Before turning to examples of work that effectively illustrate alternative formal inquiries, it's important to introduce one common denominator that binds (and simultaneously creates) the work: motion. In the presence of becoming, transition, transfer, repetition, action, and communication, there is motion. In his edited volume *Redrawing Anthropology: Materials, Movements, Lines*, anthropologist Tim Ingold, a long-time pursuer of the study of lines and of beings in the world—at times as simultaneous events—introduces us to a collection of essays by invited authors that speculate on the various binaries that frame the book's thesis.[5] Some of these binaries include object/material, repetition/habit, meshwork/network, linearity/correspondence, and discrete/flow, among others. They are intertwined, existing within the grasp of each other.[6] In his introductory chapter, Ingold offers the reader a lens with which to read the book that includes his interpretation of the analytical binary that is visible and invisible motion.[7] For Ingold, motion manifests transition, be it in terms of material, movement, or lines. In this, Ingold asserts a continuum (a meshing, or 'meshworks') that favors states of transition where objects are the consequence of other things (a material in motion for example). Ingold expands his idea of motion alongside his other major concept of "correspondence".[8] Instead of reinforcing the typical delamination of event and perception, he asserts that the event and the receiving of said event occur simultaneously because the two are in "correspondence" with one another.[9] Another way to think of this is as the simultaneity of moments in the creative 'manifestation' of an event and in its reception. Simply put, these phenomena occur 'alongside' one another.[10] Returning to the question of becoming through the lens of correspondence and continuum, indeterminacy in relation to motion, notwithstanding the author/user (creator/observer) dilemma, can be an incredibly powerful tool for creating an architecture that exists in the zone where action and result are not separated, but rather at their limit.

Aiming to release form from the constraints of static boundaries and definitions, let's begin the conversation from a contemporary context with two artists whose sculptural installations engage concepts of liminal finalities, these being Ann Veronica Janssens and Lara Favaretto.[11]

In the first example, Janssens's *Untitled (blue glitter)* (2015) consists of a thick layer of sand glitter that rests on the gallery floor. This layer, which was originally an undisturbed mound, has been literally kicked, repeatedly, into place by the artist while dressed in protective coveralls. The artist's physiognomy exists within the work and once the viewer is confronted with the work, the artist's motions that shaped the piece recur in real time with onlookers observing. Similarly, in the second work by Favaretto *Village of the Damned* (2009), part of the installation includes art collector Bob Rennie jumping and propelling confetti particles with his feet while standing on what was originally a cubic form. In these examples, the becoming of the work is determined by the synchronization of actions (kicking, shoving), by the material (glitter, confetti) and its volumetric containers (layer, cube), and ultimately, by the formal event (in correspondence with the observer). This three-part system, which suggests recurring movement and changes in one's physical state rather than static form, renders their work spatial and architectural in two ways: systematically (with a procedural apparatus) and affectively (with an element of chance). These two works are not designed to belong to any one state, as captured in any one photograph, but rather each state is one of many possible outcomes arising from given formal circumstances. And the element of chance in the synchronization of actions, materials, and event asks of us to consider the following question: How does motion, mechanized or kinesthetic, create form? More specifically, what is the relationship between the motion of an action and the indeterminate attributes of form?

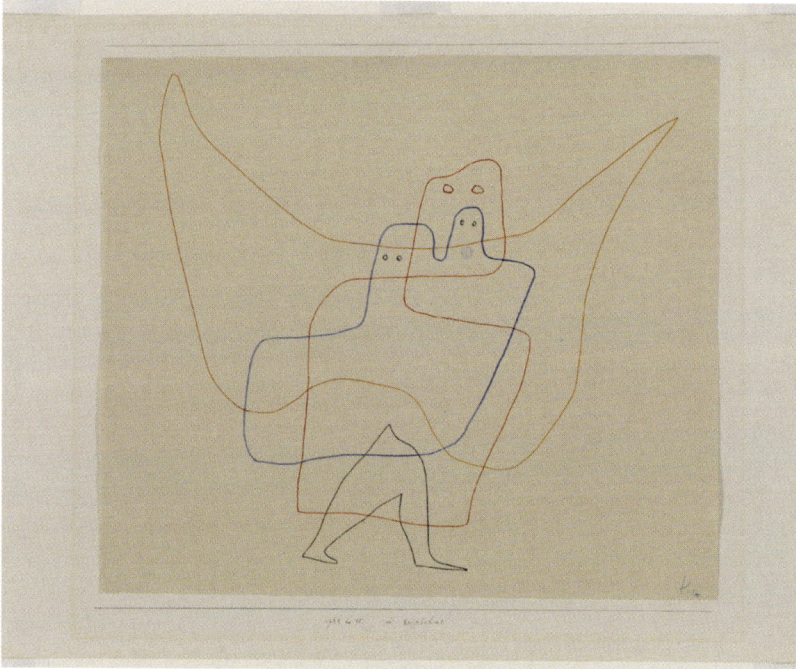

Figure 2. In Angel's Care, Paul Klee, 1931. Solomon R. Guggenheim Museum, New York Estate of Karl Nierendorf, By purchase, © 2018 Artists Rights Society (ARS), New York / VG Bild-Kunst, Bonn.

Tracing Motion, Becoming Form: A Brief History of the Mechanics of Motion

Mechanics as the Study of Motion

To understand action as motion, let's turn to a brief historical account of movement in thought. Aristotle, philosopher of the fourth century BCE and a father of Western thought, theorized extensively on natural systems.[12] In his text on mechanical problems titled *Mechanica*, Aristotle analyzed multiple scenarios of objects in motion, both in action and reaction. His observations gave rise to the origins of mechanics in his discussion of levers, pulleys, and balancing structures.[13] Although complex, Aristotle's descriptions were observant, poetic, and analytical. In the text's second problem, Aristotle questioned:

> If the cord supporting a balance is fixed from above, when after the beam has inclined the weight is removed, the balance returns to its original position. If, however, it is supported from below, then it does not return to its original position. Why is this? It is because, when the support is from above (when the weight is applied) the larger portion of the beam is above the perpendicular. [14]

The study of mechanics is broadly defined by motion and forces on parts that act together for an event to take place. Aristotle's understanding of mechanics may be commonplace in the modern world, yet having originated in antiquity, it persisted, unchanged and unchallenged for almost two millennia. It wasn't until the eighteenth century, with the advent of the industrial revolution, that mechanization developed into more complex systems.[15] In his 1948 book *Mechanization Takes Command*, historian and architecture critic Sigfried Giedion (1888-1968), a seminal figure in the history of formal analysis, discussed the transition from mechanical to industrial systems in terms of methods of production.[16] He argued that during this evolution an important event took place in the transition from handmade to machine-made: a phenomena he interpreted through the relationship which people had with the objects and materials they produced. This relationship transitioned from one where there was a one-to-one bond between producer and product, to one where the producer created the parts that made a machine that made the products.[17] The assembly line, which according to Gideon was first deployed in slaughterhouses, exemplified the transformation.[18]

Tracing and Registering Motion

With industrialization (and the subsequent rise of capitalism), corporations developed an obsession with production; speed meant more products, and in turn, more profits. As a result, the study of efficiency in human motion in service to production gained an unprecedented momentum. Frank (1868-1924) and Lillian Gilbreth (1878-1972), industrial engineers, efficiency experts, and pioneers in the field of scientific management, set out during the early 1900s to determine the laws of efficiency.[19] The Gilbreths recorded the hand motions of labourers to propose more efficient movements while working. This was not an easy task at the time; the technology for such recordings was rudimentary at best. The method they developed included attaching light sources to the worker's hands to record (through photographic long exposures) their motions. They then studied the film stills and identified movements that could be omitted or repeated in a different pattern to increase efficiency. They categorized these motions into lists of essential movements that workers should perform to improve production methods and to maximize value.[20]

The history of motion and its visual traces has been studied and explored in multiple disciplines. In visualization, alignment with the scientific work of the Gilbreths can be found in the artistic work of Albanian-born photographer Gjon Mili (1904-1984), and in that of Eadweard Muybridge (1830-1904), Étienne-Jules Marey (1830-1904), and Harold Eugene Edgerton (1903-1990). In *Talking Pictures: Clement Greenberg's Pollock*, art historian Caroline A. Jones theorized these new explorations as works that moved towards the internalization of mechanized motions. These systems of recording the world in motion, entered the realm of visual studies as equal parts, art and technology.[21] In films such as Robert Bresson's 1959 *Pickpocket*, the camera is obsessed with the human hand and its stealth movements as it removes personal belongings from people's pockets. The film reveals yet another method of recording and exposing the human hand, which

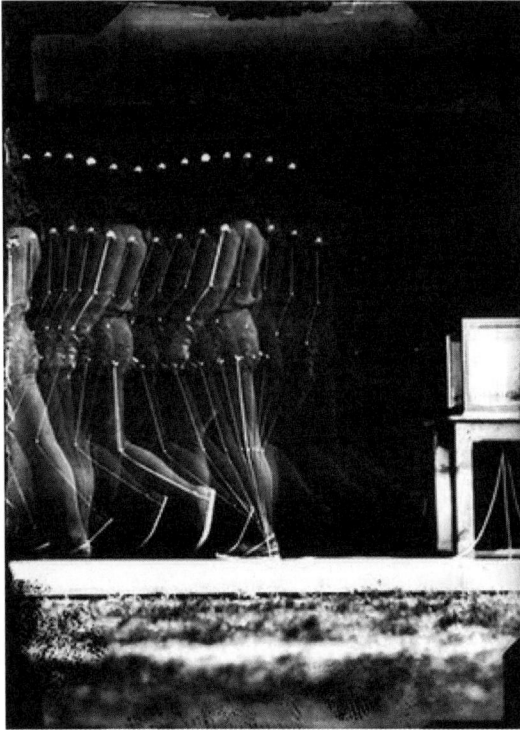

Figure 3. Man Walking (Chronophotography), Jules Etienne Marey, 1890-91. Wikimedia Commons. https://en.wikipedia.org/wiki/File:Marey-_Man_walking,_1890-91.jpg

albeit not diagrammed, forces the viewer's eyes to follow insistently its motions in its visual and verbal narrative. This tracing of the hand is experiential and subjective, and it belongs to both the visual sphere and perceptual apparatus of the viewer.

Hence, during the early part of the twentieth century, motion in humans, animals, and machines evolved from scientific recordings to artistic interpretations of lines of motion: a transition from diagrams of precise machine-production (as in the Gilbreth's chronocyclegraphs) to figures as emergent subjects (as in Pablo Picasso's light drawings photographed by Gjon Mili). In art, Paul Klee (1879-1940), Wassily Kandinsky (1866-1944), Marcel Duchamp (1887-1968), and Pablo Picasso (1881-1973) abstracted the figure in their respective works using linear traces. The emergence of the figure—or subject—resulted from the motion that the line traced. Klee's modernist paintings provided a toolset for abstraction, a newly emerging genre at the time. The first lesson in his *Pedagogical Sketchbook* began with defining drawing as "an active line on a walk, moving freely, without goal."[22] This was part of his pedagogical contribution while teaching at the Bauhaus.[23] For Klee, free-hand drawing was liberating; the result or the final figure was not predetermined, the element of chance was fundamental. Klee was interested in the unknown figure that would eventually emerge from walking a line on paper and he was fascinated by the impossibility of redrawing the same identical figure more than once when drawing freehand. The walked line that resembles a body could not be drawn in precisely the same way again. (Figure 2)

Surely, it can be argued that lines as traces of motion have gained aesthetic and cultural agency. In the pursuit of abstraction and of figures in art and technology, Muybridge, Marey, and Edgerton contributed ideas and works that considered the temporal aspect of motion. Traces were not simply a continuous line registered during one long exposure, but rather singular, layered records that depicted a subject in motion. These experiments produced a sense of cinematic temporality that unapologetically favoured the tracing of motion rather than the subject itself. In these works, the figure is always blurred, muted, and rendered almost irrelevant, without limits. Emphasis is placed on the compositional qualities that emerge between each of the tracked actions of the subject in motion, that is, on the relationship between self-similar units in a system. (Figure 3)

Indexing Motion and Two-sided Painting

Picasso and Jackson Pollock (1912-1956) were crucial in revolutionizing two aspects of painting: painting as index and painting with double-sided picture planes. Picasso's continuous line drawings can be thought of as indexes of motion rather than artistic attempts at best representing a subject, as in his drawing with light series from 1949.[24] As modernism operated between the virtual and the real, between illusion and the actual, the line became an activator of tensions between ambiguity and precision. And like in earlier examples by Muybridge, Pollock's drip paintings of lines are also the material transcription of motions and paths.

Unique to Picasso were his light drawings, 'paintings' he completed using light and a double-sided picture plane. As described by Ben Cosgrove, these 'drawings' assumed the placement of the observer and author on either side of the canvas, with the latter a transparent sheet of glass.[25] So drawing, the actant (person drawing), and figure (observer) were conflated. Whereas in the long history of art the picture plane was always looked at in one direction, in these works the drawing's/painting's frontal orientation is no longer a known truth.

Hence, with drawing as a form of double sidedness and alternatively as an index of motion, one cannot help but see an analogue in the production machinery that was the punch card system.[26] Enter the era of automation. The original logic of this manufacturing system, invented in the 1800s, was to feed information to a machine to complete a repetitive task, first used in the production of textiles.[27] On a punch card, numbers are arranged on a predetermined two-dimensional grid. The predefined holes on the 2D grid indicate that something needs to occur, and the void/solid binary (punched/unpunched) can be thought of as a system of zeros and ones—true or false. This marks the beginning of the logic of executable files, later associated with computer programming—from numerical codes (NC) that were fed through the punch card system to the computer numerical code (CNC) fed through the digital interface.[28]

Transitions from hand motion to mechanical motion to data motion are important when considering that automation is widespread in both technical industries and creative fields. I have argued that it is undeniable that there is an aesthetic dimension to these developments. It is, therefore, important to question the aesthetic qualities resulting from mechanized and digitized systems of production. We occupy a landscape inundated with digital fabrication and optimization processes that promise (and sometimes deliver) precision, fidelity, and the democratization of contemporary tools and methodologies. In the second half of this paper, I discuss what our field can gain by aligning its aesthetic inquiries with current advanced technologies for which motion and line drawings are central.

Drawing Exercises that Register Motion

Motion, mechanized systems, and aesthetics are paramount in my work on computational processes, formal compositions, and fabrication approaches. An extensive part of my design and research endeavors are focused on drawing and representation, not as alternative ways to communicate a design proposal, but rather, as methods by which we can understand and assimilate the simultaneity of both process and product (author and observer). Earlier in this essay, in my discussion of *Mechanization Takes Command*, I pointed to the transition from hand-to-product to hand-to-machine-to-product in the human history of making. With the proliferation of advanced software and hardware technologies we can now think of these relationships (hand/product, hand/machine, and computer/product) as flattened and nonlinear. In other words, this is a condition wherein producing the work and experiencing the work can happen simultaneously, and in which a representational tool, such as a drawing, is not an afterthought, but one that exists throughout the entire process. The drawings I discuss below capture immaterial and material events as they celebrate motion in the manner of—returning to an earlier proposition by Tim Ingold—"correspondence." These drawing exercises capitalize on mechanized movements and demonstrate the working through of concepts of simultaneity, chance, and indeterminacy in architectural production.

Drawing Exercise 1: Excerpts from *House 2*

Operating in the landscape of computation broadly speaking, this first drawing exercise uses algorithmic loops that draw parallel lines in parceled out areas on a 2D plane. Though the process, and the overall work, is based on the use of 3D geometries, the drawing exercise is not interested in volumetric representation proper. (Figure 4) Instead, the composition of lines alludes to a geometrical assembly by using lines in motion. Here, instead of taking one line for a walk, the overall figure

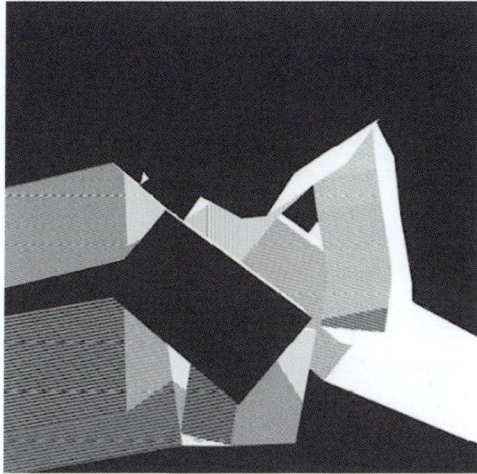

Figure 4. Drawing Exercise *House 2*, 2021. Credit Viola Ago.

Figure 5. Drawing Exercise Wall Series, 2016. Credit Viola Ago.

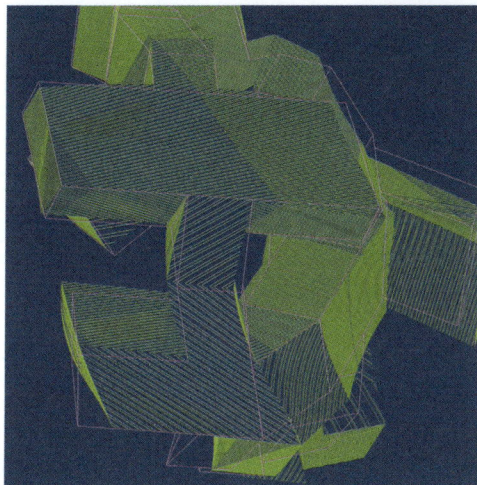

Figure 6. Drawing Composition *House 3*, 2021. Credit Viola Ago.

emerges from taking between fifty and a hundred lines for several different walks. Originally, the lines were coded to 'walk' equidistant from one another. However, the in-between distance of the parallel lines was altered as a function of the slope of the surfaces of the underlying 3D planar geometry. Another way to describe this spatial relationship is that the coefficient that alters the distance between adjacent parallel lines was measured from the XY plane and then remapped onto a mathematical domain from 0 to 1, which is the reason that the distance between the lines never increases, but only decreases (since the domain ends at one, and multiplying any value by a number in the domain of 0 - 1 will always result in something equal to 1 or less than 1).[29] Once the lines—drawn in white—were so close together that they overlapped, they created what appeared to be a solid fill, although drawn using vector data alone. Vectors are pivotal to the work for two main reasons; they are scaleless and therefore their fidelity is maintained throughout the process, and (as opposed to rasters) they function explicitly on coordinate data—a vector from 0,0,0 to 0,1,0 displays that information directly.[30] Using this framework, a sphere would be drawn with line geometry rather than rendered with shaders.[31]

Drawing Exercise 2: Wall Studies from Models are Drawings that Believed in Miracles

Where Klee referred to using the hand to walk a line, I look to computational and numerical processes (further divorced from hand and author) to walk lines.[32] The transition from hand-to-product is now from hand-to-computer-program-to-machine interface program-to-machine tool-to-product. Computer languages that support the translation of information from digital files to physical artifacts are quite advanced and have gained autonomy in architecture and peripheral design disciplines.[33] The use of vector lines as motion paths enables design agency to rest somewhere in between author and machine, avoiding the privileging of one over the other. This second drawing exercise is a physical artifact: an object generated from the carving operation of a 3D drawing. (Figure 5)

The artifact, milled using a 5-axis computer numerically controlled (CNC) machine, uses vector lines as its only geometry (no surface data was used as input in this executable file). This drawing is the literal registration of parallel lines in 3D digital space carved out of material stock through a tool that is coded to follow predetermined paths. In this example, one could say that the CNC machine is taking these lines for a walk. The 5-axis CNC machine has an end mill attached to a tool turret which is then attached to the spindle motor that moves in the YZ and XZ planes while the gantry which holds the material stock moves in the 2D XY plane. Taking advantage of these possible movements of the end mill, this exercise capitalizes on the impressions that remain on the material stock after the end mill has moved and spun through it. With this exercise, however, there is an additional layer of visual phenomena that emerges from the process. The traced motions in the material create additional geometry that is visible in the grooves along with other small-scale indentations that are left on the surfaces of the artifact. Whereas previously, we strictly looked at linear and vector-based data, here shading, or in digital terms, surface-based geometrical attributes (embossing, indentations, bump maps) create a visual phenomenon that is the result of precise machine motion paths. With this, the work has unintentionally entered the realm of the figural—or even the painterly, as the relationship between the input geometry and the code-based fabrication process becomes more intricate, interrelated, and even codependent. In other words, though the work is produced from input vector data (more specifically, line-segments) devoid of conventional visual properties, the material remnants produced by the otherwise hyper precise CNC machine gives rise to the emergence of added visual qualities. And while this refuse material remnants are typically considered a hindrance to computer-controlled machine processes, this series of wall studies celebrates, and by extension argues, that the residue of machine motion possesses aesthetic agency that lurks somewhere between the realm of mathematical space (vector data) and perception (pictorial and figural effects).

Drawing Exercise 3: *House 3*, Graphic Motion Studies

This third drawing exercise continues to explore the agency of parallel lines that are taken for a walk with an end mill in the production of a drawing. (Figures 8 and 9) The drawing traces the paths that a tool would travel in order to mill graphic lines upon the cladding of a house.[34] More conventionally, architectural drawings represent geometry, visual phenomena, or instructions for assembly, fabrication, and construction. This drawing exercise instead, conflates these aspects of drawing production, from the conceptualization to the fabrication phase.[35] More importantly, the drawing resists being placed within the traditional definition of architectural design phases (concept, schematic, design development (DD), construction drawings (CD), or presentation). Surely, *House 3* does not propose that the panels of the cladding system be milled on site at full scale. Rather, this drawing represents the agency of the design process as one that encompasses both concept and production. The graphic lines—or in other words, the vector-based geometry—in the design of this house are central to the overall architectural design ambition. The house's design starts and ends with motions, paths, and lines (and in this particular example the relationship between lines as graphic and lines as emergent form). The original line composition defines the formal strategy, structural system, as well as the assembly and production of *House 3*. Not unlike Muybridge's long exposures and the Gilbreths' chronocyclographs, this drawing study is primarily concerned with displaying mechanized production rather than emphasizing the subject (which, in this case, is the house's geometry). The compositional and resultant aesthetic

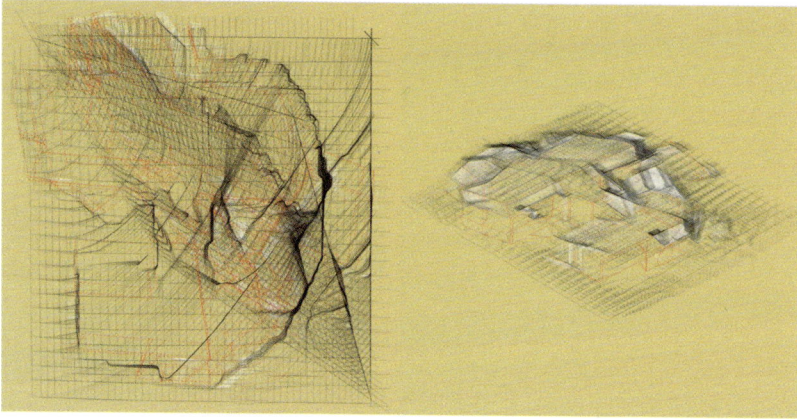

Figure 7. Excerpt, Systems and Geometry Drawing *House 3*, 2021. Credit Viola Ago.

conditions of the traces of mechanized movements produce a figure that vaguely registers some aspect of the underlying form with no regard for precision, explicitness, or fidelity. As such, the ambition of this drawing is to celebrate the event that occurs between the enactment of predetermined, precise toolpaths and the resultant figure in real time.

Drawing Exercise 4: Motion in the Imaginary, from House 3

Registering motion in visual and technical projects highlights a new territory of exploration and experimentation in contemporary visual practice. A number of current architectural firms, whose work is characterized by a similar shift from object to system, have produced work in the spirit of choreographed operations and mechanical movements.[36] As argued, drawings can host design processes by focusing on the traces of movements that register the immaterial effects of an animated act. Yet this final drawing pushes the boundaries of criteria used in previous exercises. (Figure 6) In addition to tracing motions in favor of indeterminate figure or form, this drawing adds an additional layer of information of the kind typically found in production oriented architectural drawings: numerical data. The legibility of annotations in architectural drawings, when included, takes precedence over other kinds of information (for example, a dimension line displaying code compliance values for a staircase takes precedence over the lines drawn to represent the stair treads and risers). These annotations, though they capture spatial information in three dimensions, are typically projected onto a 2D drawing plane for visibility and clarity and it can be argued that in so doing, they participate in their own niche aesthetics in the long history of architectural drawings.[37] My drawing from Exercise 4, however, resists the conventional aesthetics of annotations in order to produce new systems of value for numerical data in architectural drawings. Although illegible, the numbers convey factual information, even as their illegibility is a direct result of their spatial and 3D conditions. The position that the geometry and its

Figure 8. Drawing Study *House 3*, 2021. Credit Viola Ago.

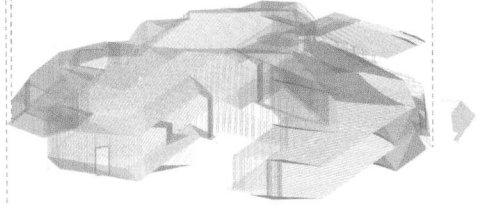

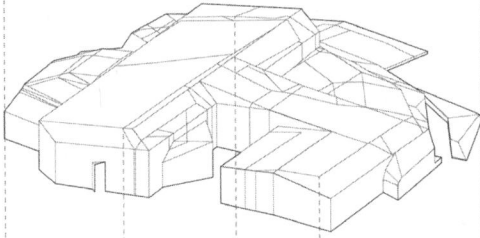

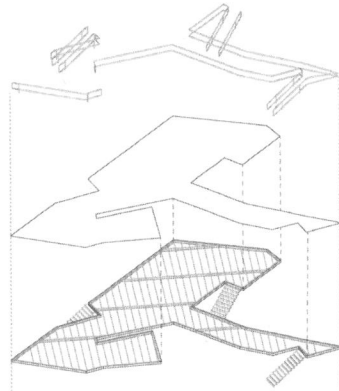

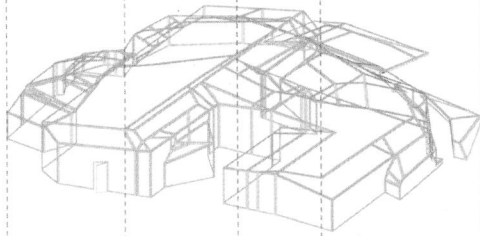

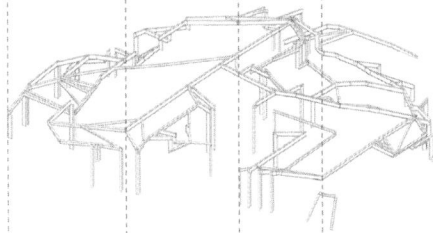

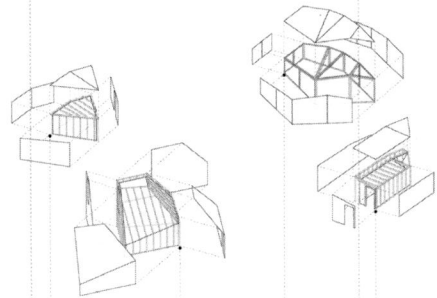

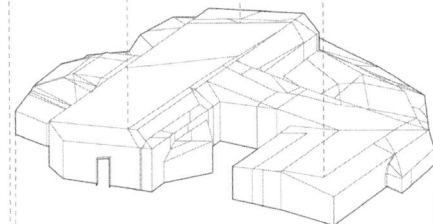

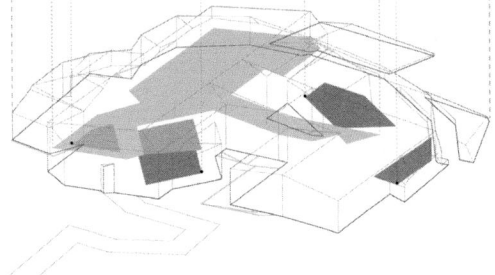

numerical information occupy in space have not been altered to serve the human eye. Maintaining the accuracy of the data challenges what we currently accept as drawing norms and hence the question this position elicits is: As building construction and construction management inch closer to eradicating the use of printed 2D CD sets, in favor of communicating directly with the 3D Building Information Model (BIM), what will happen to the drawing?

Conclusion

This essay on the role of motion in architectural representations serves to assist in synthesizing and evaluating the ways in which the discipline associates and translates motion and movement in drawings traces. There have been many debates surrounding the question of the architectural drawing in recent years in symposia, exhibitions, and themed publications. This article and the work displayed here seeks to insert yet another trajectory into the conversation, one in which the question of drawing in architecture is understood as an event rather than static form. By looking at scientific management, visual studies, and the fine arts, the objective of this work is to engage more pluralistic methods for finding more productive and intersectional territories for our discipline.

Acknowledgments

I would also like to thank the Yessios Fellowship provided by the Knowlton School of Architecture at the Ohio State University for their generous funding in support of the design, research, and experimentation work captured in this essay. In addition, I would like to thank Satoru Sugihara who taught me how to code and how to push the boundaries of conventional drawing techniques.

Funding

This work was partially supported by funding provided during my Muschenheim Fellowship at the Taubman College of Architecture and Urban Planning at the University of Michigan. Some of this work was also completed with support from the Scholarly and Creative Works Subvention Fund from the Office of Research at Rice University.

ENDNOTES

1. Ila Leslie Berman, "From grid to matrix:(Im) material events and the emergence of smooth space at the limits of contemporary architecture." (D.Des diss., Harvard University, Cambridge, MA, 1994), 36–43.

2. Ibid., 41.

3. 'Becoming' has been theorized heavily in disciplines such as philosophy and critical theory from antiquity to present times. The ambition here is not to continue that lineage of thought in which 'becoming' is understood as part of the human condition, in opposition to being, or as an isolated ontological problem; understood through the linear history of philosophers from antiquity (Heraclitus) to the realist (Hegel), early existentialists (Nietzsche), and the post-structuralists (Deleuze). Rather, 'becoming' here is an idea that considers the human, machine, action, and atmosphere in its totality. More specifically, 'becoming' is seen through the lens of multiple agencies: the hand, eye, machine, and numerical systems.

4. For a survey of these types of projects, see Viola Ago, "Compositional Physics and Other Diagrams of Force," *Log 46* (Summer 2019):33–43.

5. Tim Ingold, ed. *Redrawing anthropology: Materials, movements, lines* (New York: Routledge, 2011), 1–20.

6. Ibid., 5–18.

7. Ibid., 4–5.

8. For a more detailed explanation of Ingold's correspondence, see Tim Ingold, "On Human Correspondence." *Journal of the Royal Anthropological Institute.* 23, no. 1 (2017): 9–27. For an earlier writing on his concept of "correspondence," see Tim Ingold, "Anthropology is Not Ethnography," (2009).

9. Tim Ingold, "Anthropology is Not Ethnography," 87–88.

10. Ingold, "On Human Correspondence,"14.

11. Two analogous post-war artists whose artworks identify with this framework are Alice Aycock's "Sand/fans" (1971) installations and Lynda Benglis' polyurethane and paint drip sculptural works.

12. Aristotle, and Walter Stanley HETT. *Aristotle: Minor Works: On Colours; on Things HEARD; Physiognomics; On Plants; On Marvellous Things Heard; MECHANICAL Problems; On INDIVISIBLE LINES; Situations and Names of Winds; On MELISSUS, Xenophanes, and Gorgias.* (Cambridge, MA: Harvard University Press, 1936).

13. Ibid., 327–411.

14. Ibid., 347–349.

15. Sigfried Giedion, *Mechanization takes command: a contribution to anonymous history.* (Minneapolis: University of Minnesota Press, 2014): 14–15, 31–32.

16. Ibid., 46–50.

17. Ibid., 46.

18. Ibid., 86–96.

19. Andre Baumgart and Duncan Neuhauser, "Frank and Lillian Gilbreth: scientific management in the operating room." *BMJ Quality & Safety* 18, no. 5 (2009): 413–415, http://dx.doi.org/10.1136/qshc.2009.032409.

20. Giedion, *Mechanization takes command: a contribution to anonymous history*, 114–15.

21. Caroline A. Jones, "Talking pictures: Clement Greenberg's Pollock." in *Things That Talk: Object Lessons From Art And Science*, ed. Lorraine Daston (New York: Zone Books, 2004), 329–73.

22. Paul Klee and Sibyl Moholy-Nagy. *Pedagogical sketchbook.* (London: Faber & Faber, 1953), 16.

23. Paul Klee and Will Grohmann, Paul Klee, (New York: H. N. Abrams, 1969).

24. For Picasso's light drawings see Ben Cosgrove, "Behind the Picture: Picasso Draws With Light," *Life.* https://www.life.com/arts-entertainment/behind-the-picture-picasso-draws-with-light/

25. Ibid.

26. Leo Steinberg, "The flatbed picture plane," in *Other Criteria: Confrontations with Twentieth-Century Art* (Chicago: University of Chicago Press, 2007): 61–98.

27. M. Mitchell Waldrop, *The dream machine: JCR Licklider and the revolution that made computing personal,* (New York: Viking Penguin, 2002).

28. In computer programming, an executable file is a file type that completes a set of commands. Unlike recordings, which are meant to be diagrammed or represented in legible manners, an executable file is not meant to be read but rather to initiate a series of operations.

29. Any number multiplied by a number between zero and one will be either equal to itself (multiplied by 1), less than itself (for example 0.5), or zero (multiplied by 0).

30. As opposed to raster data which is based on calculations that are not explicitly available upon observation.

31. The term 'shader' is borrowed from computer graphics and refers to the calculations that a rendering engine processes to determine the amount of shading of a 2D pixel that represents an incredibly small region of a 3D surface geometry.

32. I have previously written about working in simulation engines to generate design processes that are based on chance, indeterminacy, and imprecision. In that article, I make a case that the use of the hand and the computer mouse is implicated in the decision-making process of running a simulation. See: Viola Ago, "Fusion Axis: A Phenomenological Inquiry on Process" in *PLAT Journal*, (Houston: Rice University, 2020): 48–57.

33. Here, one thinks of projects especially designed with advanced fabrication tools in mind that are strictly and entirely in service to the possibilities afforded by the new and advanced novelty of the tool, machine, or process.

34. This drawing belongs to *House 2* from the three-house series by Miracles Architecture. *House 2* is a house for a sculptor. https://www.miraclesarchitecture.com/#/house-2/

35. Another way to think about this is through the lens of Morphosis's *House 2-4-6-8* assembly axonometric drawing if the assembly were to be automated using numerically controlled systems.

36. Contemporary practices include First Office, the LADG, Nemestudio, Curtis Roth, Office CA, Hans Tursack, Stock A Studio, MR Studio, and A/P Practice.

37. Drafting software packages such as Autodesk's Autocad, Autodesk's Revit, Bentley's Microstation, and Graphisoft's Archicad have forged their own standardized annotation aesthetic over the years. The trained eye can recognize the software package by simply looking at a drawing that it produced.

ARTIFICIAL ARBORETUM:

HARVESTING PHOTOGRAMMETREES AND CULTIVATING SPECULATIVE REALITIES

JACQUELINE WU

ABSTRACT

Artificial Arboretum is a speculative project whose collection preserves, studies, and cultivates "photogrammetrees" found on Google Earth™. *Artificial Arboretum* highlights the vulnerability of these trees that—for all their wonderful deformities—are endangered by the speed and vision of urban development and technological progress. Their data lives, but at the mercy and politics of each software update. In a society that strives for pixel-perfect digital commodities, these photogrammetrees will soon be phased out for ones increasingly indistinguishable from reality. This paper and the work described illuminates the structure and ownership of data collection. It also questions the conventional production of digital aesthetics that is typically aimed at the 'usefulness' of commercial technologies and their allied industries. To this end, it proposes an alternative world-view associated with the process and politics of 3D imaging and of reconstruction technologies that operate at the scale of our planet. The project *Artificial Arboretum* embodies values and beliefs critical of the mechanisms subtending Google Earth™: it favors visual and aesthetic idiosyncrasies over the high fidelity of representation and presents photogrammetrees not as proprietary data but as data belonging to, and nurtured by, the public domain.

+ photogrammetry
+ speculative design
+ archive
+ digital
   representation
+ Google Earth™

Figure 1. Photogrammetree mesh detail. Image by Jacqueline Wu.

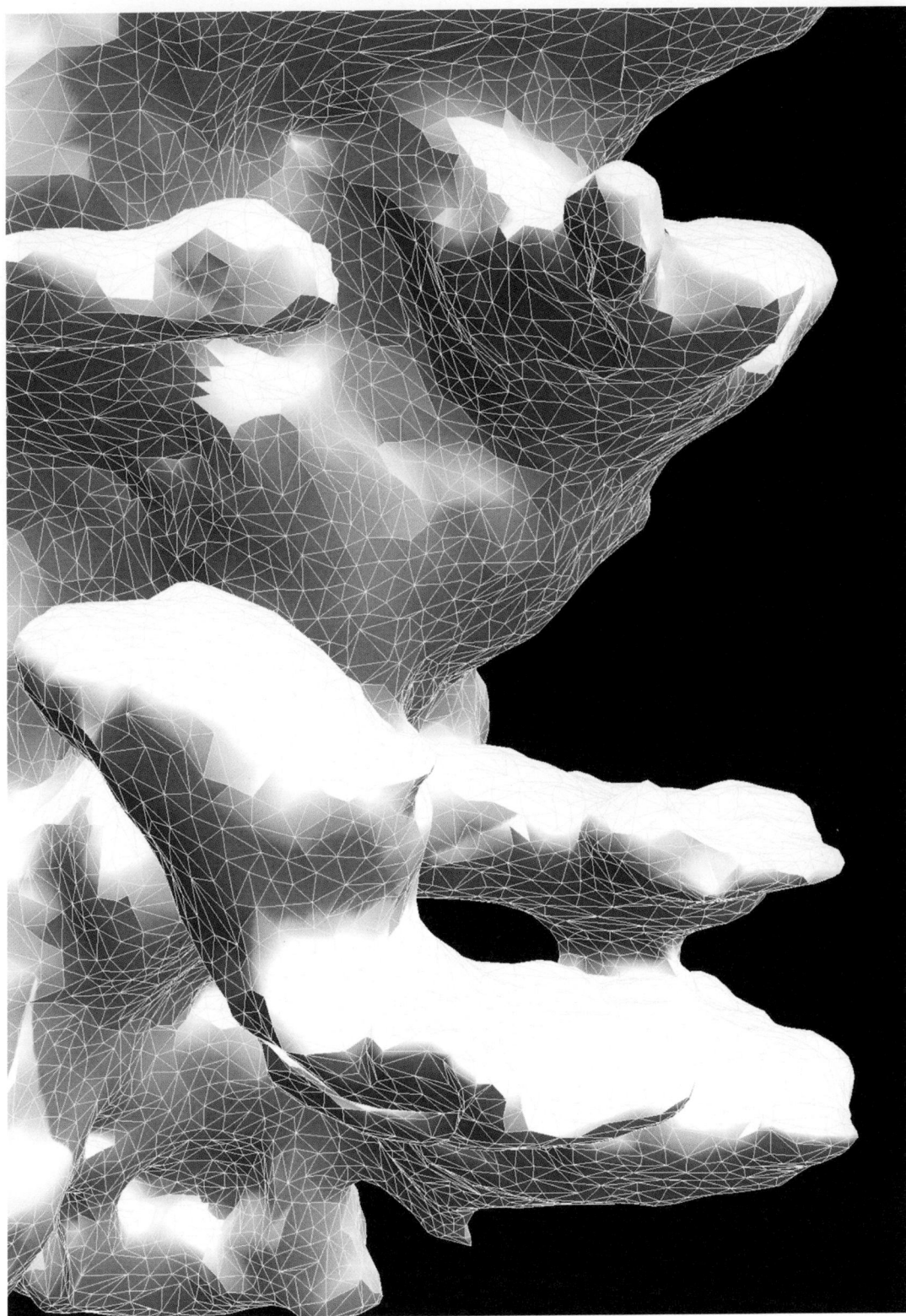

The Gap

Photogrammetrees exist at the interstices of Google Earth™. They sprout across the virtual equivalents of our municipal parks, city parking lots, sidewalks, traffic islands, residential backyards, and commercial courtyards. However, they do not translate into software pixels as easily as the smooth and broad strokes of skyscrapers and highways. They are familiar and close to home, yet they read and reflect our physical world in ways that we may not fully grasp. They often defy gravity, hovering above the surface of the ground plane or are disjointed at mid-branch. (Figure 2) Some subjects of photo-grammetrees merge with their neighbors and others fuse with the surfaces of buildings. They do not cast shadows themselves, but the shadows of their true form are baked onto the surface of Google Earth's digitized virtual mesh. Their wonderful deformities remind us that organic matter remains elusive to the virtual world.

I started collecting photogrammetrees because of their visually compelling and peculiar, often contradictory, qualities. They became of interest, however, for far more than how they looked. Indeed, the very nature of their forms began to tell a story about how they came to be and what they could become. Artist and writer James Bridle uses the term "New Aesthetic" to describe this emerging practice of computational sense-making, which draws attention to visual incon-gruities at the convergence of the physical and digital.[1] As machines are increasingly used to process, format, organize, and understand the world, there is warrant to catalogue, categorize, connect, and to make known the differences and the gaps in these overlapping but distant realities. As such, this paper, and the work it discusses are not merely about the objects themselves, but equally a commentary on the systems—sometimes technological, spatial, legal, and political—which permit, shape, and produce them.[2]

The uncanny typologies of photogrammetrees offer a glimpse into a far greater system of data collection and assembly. Their forms are representative of a distinct digital process built upon algorithms, computers, storage systems, automated cameras, and custom-built aircrafts that generate the virtual interface that is 3D Google Earth™. As peripheral artifacts resulting from a process that is meant to map urban buildings and infrastructure, photogrammetrees help us visu-alize and understand how an algorithm written to reproduce cities reconstructs nature. With seemingly hard surfaces and hollow interi-ors, photogrammetrees are structured more like buildings than their 'nature'. The uninterrupted mesh of photogrammetrees is similarly sharp edged, flat, abstract, and geometrically configured to quickly load and run on a browser.

For Google™, the photogrammetrees' representational gaps are merely technical problems to be resolved. The former Vice President of Engineering at Google Maps™, Brian McClendon, has called it "The never-ending quest for the perfect map" whose vision is largely reflective of a society that strives for pixel-perfect digital commodities.[3]

Figure 2. Sample of Google Earth™ photogrammetree, harvested by Jacqueline Wu. Image by Google EARTH ©2022. Map Data SIO, NOAA, U.S. Navy, NGA GEBCO / Image Landsat/Copernicus / Data LDEO-Columbia, NSF, NOAA

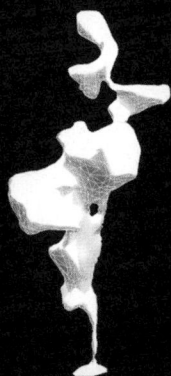
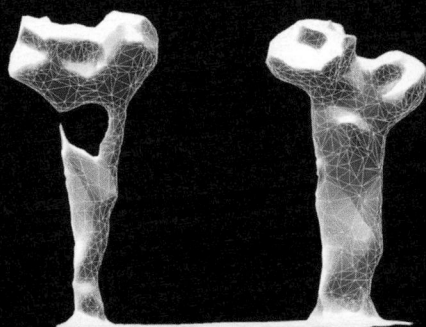
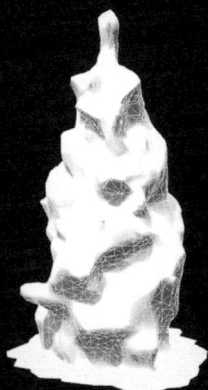

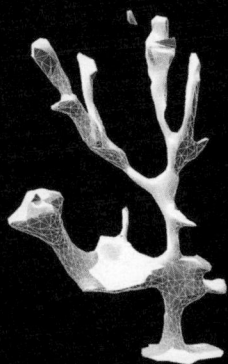
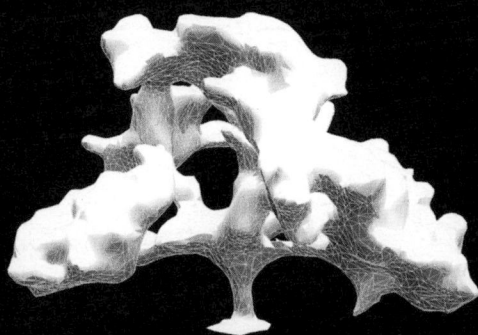

With evermore data at higher resolutions, increased processing power, and scalable solutions, all limitations to the creation of perfect maps will eventually be overcome. However, according to this view of the world, today's photogrammetrees are endangered by the speed and appetite for perfect pixilation. Their data lives but at the mercy and politics of each software update, a cadence that is extremely volatile and opaque. As a temporal species—emerging, evolving, and disappearing every few months or years—it is uncertain whether these photogrammetrees will ever again be reconstituted in their current forms. (Figure 3)

Artificial Arboretum, a speculative project

The concept for *Artificial Arboretum* materialized out of a fascination for photogrammetrees. Galvanized by the existential threat to their survival, I have been dedicated to the collective preservation, study, and cultivation of photogrammetrees. The *Arboretum* I've imagined establishes a social fiction that embodies values and beliefs vastly disassociated from those promoted by Google Earth™. It favors representational idiosyncrasies over the search for high fidelity, and in this, it presents photogrammetrees not as proprietary data but as data belonging to, and nurtured by, the public domain. The underlying commentary in *Artificial Arboretum* is meant to illuminate structures of data collection and ownership as well as question conventional digital aesthetics directed at the 'usefulness' of commercial technologies and their allied industries. Ultimately, the project's motivation is to propose a different reality with a different set of ideals where photogrammetrees are celebrated and thrive. *Artificial Arboretum* expands the practice of computational sense-making beyond critical thought. It engages critical materiality by creating and expressing an alternative worldview for the production and politics of 3D imaging and for reconstruction technologies that operate at the planetary scale.

    Artificial Arboretum builds upon and borrows from the speculative design practice of the London based studio of Dunne & Raby, whose principals use design as a medium to stimulate discussion and debate amongst their peers, industry, and the public about the social, cultural, and ethical implications of existing and emerging technologies.[4] Their approach emphasizes thinking through design rather than through words, communicating ideas across form and function. It uses the language and structure of design to engage people, to suggest how everyday life as we know it could be different, and to invite the audience to make believe. While there is an assumption with speculative work that it is oriented towards the future, it can simply be somewhere else, a parallel world contemporaneous to our own.[5] In their A/B text—or manifesto of sorts—Dunne & Raby outline how the primary shift within speculative design is ideological: the role of the designed object is to catalyze thought rather than fulfill utilitarian, aesthetic, or commercial motives.[6] This was the program for *Artificial Arboretum*, a work which exists in three parts, as Archive, Virtual/Physical Exhibition

Figure 3. Photogrammetree mesh typologies. Image by Jacqueline Wu.

space, and Open-Source Manual, all initiatives built around research, cultivation, and education, respectively. The project is a work of speculative design that is brought to life with everyday design artifacts or props, such as digital renders, physical mockups, and digital and physical models. These invented yet functional objects and images are situated within the familiar typology of the arboretum (a place for collecting and showcasing trees and other botanical specimens). Yet, they have added layers of estrangement and absurdity, and perhaps even humor, that signal their ambiguous status as simultaneously real and unreal. This series of artifacts, described in the next three sections, are prompts to a much larger speculative world. *Artificial Arboretum* is at once a commentary, a proposal, a repository, a recipe, and an invitation to collectively examine and shape narratives around the systems that produce Google Earth™.

The Arboretum as Archive

The core program of the *Arboretum* is that of research arm responsible for harvesting, classifying, and developing methods of preservation and storage of photogrammetrees. The collection serves as a living record of our planet, as photographed on the ground by digitally operated satellite cameras. The scans capture not only the geometry of the trees, but also their age, seasons, shadows, and surrounding environments. As residual artifacts in a process meant to map buildings and infrastructure, they remind us that organic matter remains elusive to the virtual world. Google Earth™ comes with a promise of unlimited access to a universe of images—its website states "The whole world, in your hands."[7] And yet, its 3D data is not accessible to the public. The process of harvesting photogrammetrees is not so simple. A subterfuge method is required to extract and reconstruct the application's data points.

In principle, what is available via the Google Earth™ application is available to everyone with a computer and access to the internet. Anyone can virtually harvest overhead 2D views and images of photogrammetrees using the same techniques that created them. Google Earth™'s flyover feature allows users to simulate the image collection process, typically performed by aircrafts and custom-built cameras, with a toolset comprised of a computer mouse and screen recording software. (Figure 4) The only difference being that the 3D world captured by Google Earth™ is physical and that from which my photogrammetrees are taken is virtual. In creating the objects housed in the Archive, the 2D images of Google Earth™ are thereafter stitched together into a (new) 3D model in photogrammetry using software such as Autodesk ReCap or Agisoft Metashape. This operation recreates the object-image of what was once but a slice of Google Earth™ into a virtual model whose 3D coordinates are a duplicate of the photogrammetree. Once harvested and re-created, the photogrammetrees are catalogued along with a record of their native digital characteristics, including geolocation, timestamp, file size, texture maps,

Figure 4. Sample of aerial screenshots from Google Earth™ prepared for the photogrammetry process. Image by Jacqueline Wu.

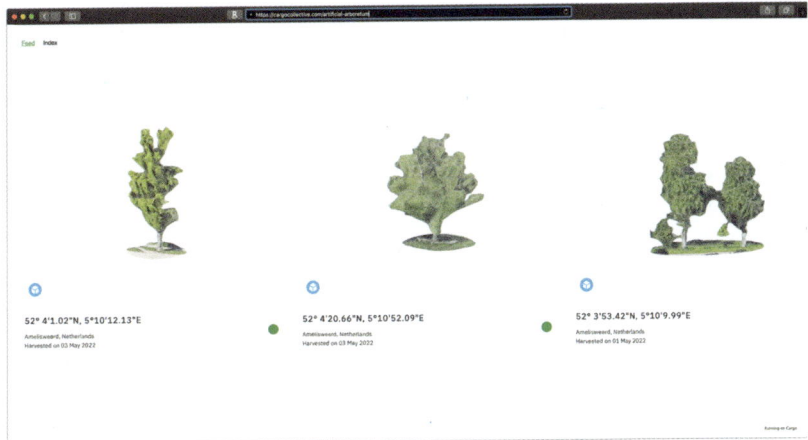

Figure 5. *Arboretum* as public space. Image by Jacqueline Wu.

Artificial Arboretum

1.

30°16'5.04"N—97°45'58.74"W
917 KB
19901 polygons
15346 vertices

2.

33°47'58.29"N—116°31'42.64"W
241 KB
2156 polygons
1649 vertices

3.

33°47'58.29"N—116°31'42.64"W
241 KB
2652 polygons
1892 vertices

4.

35°41'5.58"N—139°42'39.23"E
265 KB
3422 polygons
2758 vertices

5.

35°41'4.97"N—139°42'39.97"E
354 KB
4836 polygons
3837 vertices

Notes — *1.* Accessed 09 Mar 2019; *2.* Accessed 21 Feb 2019; *3.* Accessed 21 Feb 2019; *4.* Accessed 23 Feb 2019; *5.* Accessed 23 Feb 2019

and mesh data. Due to variability in web applications, these data points are crucial in documenting the lineage of photogrammetrees and in tracking the history of changes between each Google Earth™ product update.

The process of collecting and cataloguing these discrete moments in time mirrors the fascination and urgency of artist Clement Valla's *Postcards from Google Earth*.[8] Valla casts himself as a tourist in the virtual space of browsers, capturing screenshots of uncanny 3D-generated forms to export them from Google EarthTM's update cycle. *Artificial Arboretum* is also focused on registering technical artifacts like photogrammetrees to make sure there is a record that captures particular images and the 3D model they produce. (Figure 5) Alongside other artists faced with the 'endless archive' of the Internet, my work in creating the *Artificial Arboretum* Archive is an act of curation, not in the reductive sense of the word as simply an act of selection but as a form of caretaking—a purposeful act of preservation that ensures these fragments are seen.[9]

The *Arboretum*'s Archive is modeled loosely after seed banks common in agriculture and gardening, which typically use an 'ex situ' strategy to conserve biodiversity and to offer imperiled plant species protection from predation and mortality. Indeed, photogrammetrees at present exist primarily in cities and urban areas because this is what Google Earth™ prioritizes when mapping. The Amazon, for example, does not have a 3D layer on Google Earth™, but only a 2D satellite view. Moreover, the Archive is structured according to digital attributes such as timestamp and location coordinates in support of its survival. By duplicating and storing photogrammetree data in new locations "off-site," uncoupled from Google Earth™'s network, it is no longer subject to the volatile dynamics and technological pressures of the update cycle. Here, it is not the seed that is stored, but the 3D CAD model of the photogrammetree. Yet, like the seed, in its hybrid state of being and non-being, the CAD model bridges an ontological gap between presence and disappearance.[10] (Figure 6) The digital format allows it to multiply and move into the hands of many. Not only does it safeguard the data for future access and restoration, but it holds the genetic information to unlock new and distributed modes of making, learning, sharing, and revising alternate realities.

The Arboretum as Physical/Virtual Exhibition Space

The most ambitious aspect of the project is the *Arboretum*'s Physical/Virtual Exhibition space, a speculative facility for the cultivation of photogrammetrees. (Figure 7) Digital models from the Archive are transmuted into the material world, moving from simulation to reality and into human scale impressions and interactions. This idea gestures to the work of #Additivism—a portmanteau of 'additive' and 'activism' conceived by artists Moreshin Allahyari and Daniel Rourke—whose work captures the possibilities that emerge when CAD models and 3D printers are used in critical, poetic, and disruptive new ways.[11]

Figure 6. *Artificial Arboretum* photogrammetree classification poster. Image by Jacqueline Wu.

Reconstructing photogrammetrees as physical matter adds new dimensions to the *Arboretum*, whose presence, touch, materiality, and scale allows them to be experienced viscerally. Unmediated by a screen, these 3D printed objects help narrow the gap between fiction and reality. Their physical presence locates them in our world while their meaning, embodied values, beliefs, ethics, dreams, hopes, and fears belong to both physical and virtual worlds—each with their set of rules, structure, and goals[12] Materially the collection becomes a hard copy of the network, circulating photogrammetrees within both information space and physical space.

As a complement to the archival seed bank, the Exhibition space of the *Arboretum* expands the mission of the Archive into a civic center and, more generally, into a community space. It is a site for public engagement, hosting educational, recreational, and horticultural programs. Whereas trees in a typical arboretum are generally grouped according to climate biomes (deserts and tropics, for example) or geographically by continent, photogrammetrees in *Artificial Arboretum* are grouped, archived, and exhibited according to the following domains: the Grove of Gravitational Defiance, the Forest of False Positives, the Island of Inconsistent Existences. Their names hint at the unconventional physics, timescales, and tectonics under which Google Earth™ operates. These fictional territories, accompanied by the familiar use of tree label placards, are used to organize the visitor map and the experience of wandering through the space of the arboretum. (Figure 8)

In this way, the Physical/Virtual Exhibition space of *Artificial Arboretum* is reminiscent of the speculative *Map of Eneropa* that AMO, the research branch of Rem Koolhaas' Office for Metropolitan Architecture, created as part of a study for the European Climate

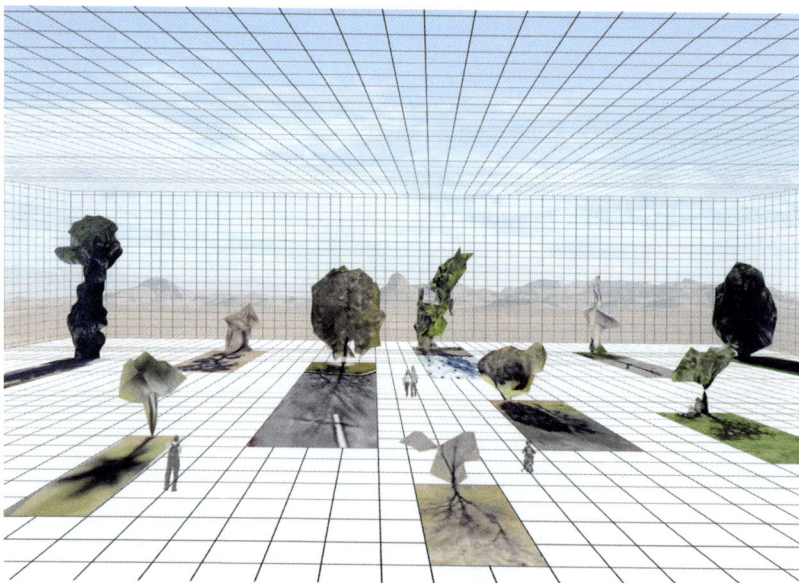

Figure 7. *Artificial Arboretum,* Physical/Virtual Exhibition Space. Image by Jacqueline Wu .

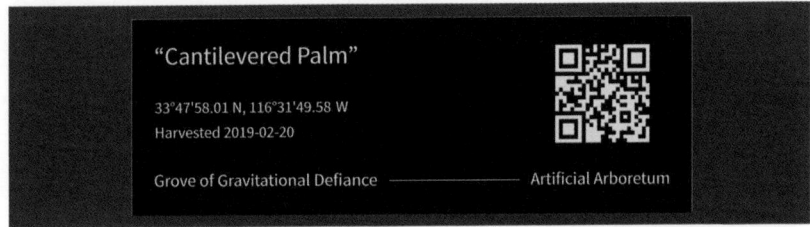

Figure 8. "Cantilevered Palm" tree label located in Grove of Gravitational Defiance. Image by Jacqueline Wu.

Foundation's *Roadmap 2050*.[13] The map charts an alternative Europe with regions renamed according to their main source of renewable energy—Isles of Winds, Tidal States, Solaria, Geothermalia, and Biomassburg. The two graphic narratives are similarly able to condense complex ideas into playful, provocative artifacts that elicit discussion and debate around the alternative worldview they represent.

The Grove of Gravitational Defiance, for example, houses photogrammetrees that suspend, cantilever, bend, fuse, and warp in ways that disregard real world physics, but which resolutely stand in a virtual one. In the Forest of False Positives, visitors find a collection of low-fidelity, shrink-wrapped telephone poles, streetlights, cell towers, and other urban structures that resemble photogrammetrees in silhouette and stature but are not. These False Positives reveal how the photogrammetry algorithm lacks semantic understanding of the world, sometimes mistaking a maple tree for a telephone pole, seeing no difference in the objects it captures. Living or manmade, it recreates all things equally in form. Whereas the Island of Inconsistent Existences transports visitors to the liminal space between our present physical reality and what is represented in Google Earth™. Here photogrammetrees are in limbo: their real-world counterparts no longer exist—whether naturally uprooted, removed for new development, or as a result of something else—yet they remain in Google Earth™ due to the time lag between update cycles. (Figure 9)

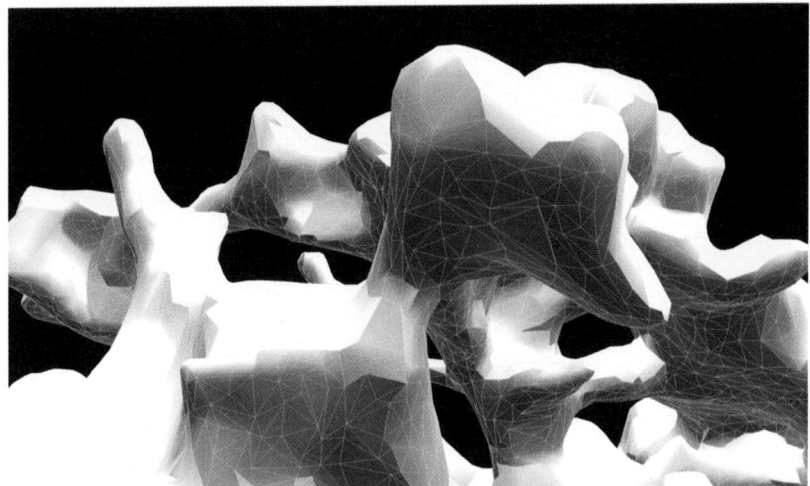

Figure 9. Photogrammetree mesh detail. Image by Jacqueline Wu.

The Arboretum as Open-Source Manual

The third component of the *Arboretum* is a call to action. It consists of an open-source training manual that equips the reader with step-by-step instructions on the technical process of harvesting photogrammetrees. (Figure 10) You too can create your own *Artificial Arboretum*. Indeed, the work of the *Arboretum* will likely never reach completion, the ideas and methods that subtend it will persist as long as they remain in circulation. As software updates continue to accelerate towards the perfect map, this collection of photogrammetrees will be the only accessible evidence of paused 'branches' within Google Earth™'s version of history. The open-source training manual acknowledges the urgency and immensity of the task and thus aims to engage and recruit global participation towards saving these species. By democratizing, decentralizing, and distributing both the data and the techniques, the Arboretum can grow into a collective and sustainable practice. There is no ownership of photogrammetrees as there is no one model of the *Artificial Arboretum*. It exists in a plurality of forms and like the nature of the network it exists anywhere and everywhere. The hope is that shared materials and knowledge will fuel the manufacture of other narratives, alternative landscapes, and new realities for photogrammetrees beyond the one presented here.

Acknowledgments

I would like to thank Anthony Dunne and Fiona Raby for their support in the development of Artificial Arboretum during my time in their Designed Realities Studio and for fostering a space that encourages ideas and projects like this one.

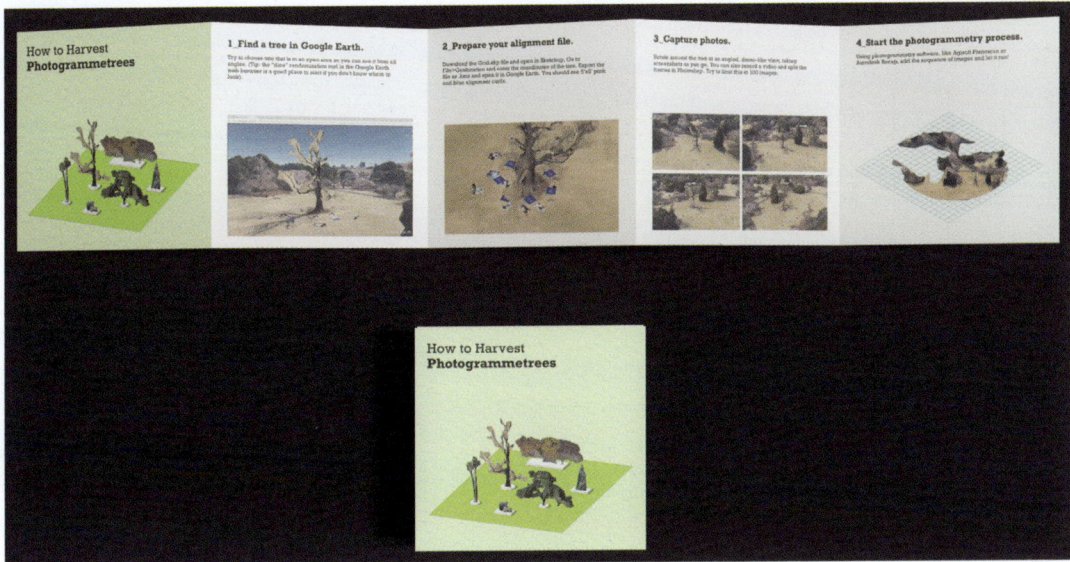

Figure 10. Open-source manual, *How to Harvest Photogrammetrees*. Image by Jacqueline Wu.

ENDNOTES

1. James Bridle, The New Aesthetic. https://new-aesthetic.tumblr.com/

2. James Bridle, "The New Aesthetic and its Politics," Booktwo.org, June 12, 2013, https://booktwo.org/notebook/new-aesthetic-politics/.

3. Brian McClendon, "The never-ending quest for the perfect map," Official Google Blog, June 6, 2012, https://googleblog.blogspot.com/2012/06/never-ending-quest-for-perfect-map.html.

4. Dunne & Raby. http://dunneandraby.co.uk/content/biography.

5. Anthony Dunne and Fiona Raby, *Speculative Everything: Design, Fiction, and Social Dreaming* (Cambridge: MIT Press, 2013), 129.

6. Ibid., vii.

7. Google Earth. https://www.google.com/earth/versions/#earth-for-mobile

8. Valla, Clement. Postcards from Google Earth. http://www.postcards-from-google-earth.com/

9. Joanne McNeil, "Endless Archive", in *Collect the WWWorld: The Artist as Archivist in the Internet Age*, ed. Domenico Quaranta (Brescia: LINK Editions, 2011), 42–43.

10. Paul Soulellis, "The Distributed Monument," in *The 3D Additivist Cookbook*, ed. Morehshin Allahyari and Daniel Rourke (Amsterdam: Institute of Network Cultures, 2017), 130.

11. Morehshin Allahyari and Daniel Rourke, "Introduction," *The 3D Additivist Cookbook*, (Amsterdam: Institute of Network Cultures, 2017), 4.

12. Dunne and Raby, *Speculative Everything*, 44.

13. OMA/AMO, "Map of Eneropa" in *Roadmap 2050: A practical guide to a prosperous, low-carbon Europe*, The European Climate Foundation, 2010, 174–75.

BIO
MATTER
TECHNO
SYNTHETICS
MULTI-DISCIPLINARY PRACTICES
BEYOND THE NATURAL WORLD

AMBER FARROW

To be synthetic is to be artificial, where 'synthesis' is the art of combining multiple parts into a coherent whole. It is imperative within design practices and academia that we 'synthesize' our thinking, to move beyond siloed disciplines and towards a new fluid understanding of the world, that is irreducible to discrete and separate components. The explorations discussed in this final section, Synthetics, are boundless in nature; they challenge the mechanistic view of the world by examining a future relationship between humans, material systems, and technologies. Synthetics contradicts commonplace notions of dualism by exploring how concepts of hybridization in design reshape our everyday experiences. Its authors articulate a new, emergent design language focused on non-deterministic design, new materialism, worlding, authorship, interspecies rationalities and making visible the invisible.

Firstly, implicit perspectives (and biases) are explored by **S.E. Eisterer** and **Clarissa Tossin** whose conversation surfaces two distinct points of view, that of historian and artist. In a discussion of Tossin's artwork, nature, technology, and indigenous culture are examined as related to questions of scale, time, and production methods. Technology as an accelerant is questioned; as is that which accelerates the speed of production, consumption, waste generation, and potentially of our time on earth. Their dialogue asks: Is it possible to decelerate the traumas of extractive technology or is reaching the breaking point a necessary moment in our evolution?

Jenny Sabin challenges the disconnect between humans and technology, highlighted in Hochhäusl and Tossin's conversation. Sabin also explores the question of production, however, doing so through the lens of cyberfeminist theory in order to undermine commonplace binaries such as man/women, human/machine and mind/matter. She speaks to the very tensions revealed by her goal to 'transcend' boundaries and binaries within design. Her work—a hybridization of biology, material studies, technology, and environment—relies at its core on the binary system she critiques. With the result being, that it is the inherent binary logic of 'weaving' that is used to construct her resultant hybrid/cyborg.

Rachel Armstrong's research in biodesign provides an alternative perspective on materials and what we consider to be living systems. Armstrong reflects upon her practice whose key projects have narrated the origins of life, including our (mis)understanding of nature and our role within it. Questioning our definition and interpretation of *bios*, and reclaiming the concept of *zoe*, her work highlights the tension between artifice and nature as a factor of time. Her practice formulates an ecological ethics that establishes a more-than-human approach to design. As biodesign grows as a field, ethical questions around our obligation of care towards the agents we construct and use, become imperative.

In making tangible the invisible, **Patricia Olynyk** discusses immersive artworks that use projective media to affect the viewer's sense of self. Using shadow play, her work occupies a 'lost space' where synthesis occurs in the interaction of the viewer's shadows and the constructed environment. Technology is key in these 'shadow' works which conjure notions of the Frankenstein body and address themes of hybridization, replication, and unpredictability. Using intangible materials such as light—or its absence—as medium, Olynyk highlights that the seemingly invisible can be understood as an operative tool and a means of representation.

Kathy Velikov asks how material and technological constructs cultivate environmental sensibilities in sympathy with the inaccessible and the incomprehensible. Influenced by Walter Benjamin's 1916 essay "On Language as Such..." which argues that artistic practice is founded on "languages issuing from matter," Velikov situates her own practice in a philosophy of material thinking that allows for things themselves to offer us thoughts. In this way material thinking offers us a new language within which to position her work discursively, as it relates to boundaries, environmental mediation, matter, and aesthetics. Through this, Velikov opens alternative trajectories and frameworks by which to discuss and interrogate experimental design work grounded in building technologies and architectural material systems.

Synthetics showcases the design work of a diverse group of artists, historians, researchers, and practitioners whose work challenges our current understanding of the world. Convening a range of alternative approaches and unrestrained by a singular discipline it highlights ethical human actions whose products move beyond the so-called 'natural' world.

FUTURE FOSSIL:

ART, ECO-FEMINISM, AND DECOLONIZING HISTORIES OF ARCHITECTURE - A CONVERSATION

S.E. EISTERER AND CLARISSA TOSSIN

ABSTRACT

The artwork of Los Angeles-based Brazilian artist Clarissa Tossin addresses themes of ecology, material extraction, and neo-colonial power relationships around the world. It reflects upon post-human futures that are visual and material while it considers how critical insights from fourth wave feminism and queer theory, environmental history, and post-colonial theory might allow envisioning these futures. In dialogue with architectural historian and theorist S.E. Eisterer, Tossin discusses the role of artistic works that challenge historical narratives, particularly those which have long been steeped in extractive paradigms and settler colonial violence. To this end, the conversation considers how we might write more equitable narratives of the built environment, especially against the backdrop of the impending climate catastrophe.

+ eco-feminism
+ post-extractive
  paradigms
+ amazon
+ Octavia E. Butler
+ future fossil

Figure 1. *Future Fossil*, 2018, (detail). cedar tree trunk, rocks, roots, leaves, bark, soil, sand, plaster, cement, silicone, foam, resin, aluminum foil, electronic waste, recycled melted plastic from artist's own waste, 20 ft. x 14 in. x 17 in. Commissioned by Harvard Radcliffe Institute. Image courtesy of Clarissa Tossin and the Harvard Radcliffe Institute. Photo by Stewart Clements.

Exhibition

What I've noticed about the human species over my lifetime is that we tend to go over the edge, more often than we ought to. We go to the edge and then we realize 'my God, that's a precipice, we could fall over, we could die,' and we draw back. The problem is, with something like global warming, you can't just draw back and make it okay ... after you spent over a century messing things up, by the time you decide to fix it, your grandchildren might see some results, but chances are, you won't.

Octavia E. Butler, *Fast Forward: Contemporary Science Fiction* [1]

The centerpiece of the exhibition *Future Fossil* is a striated drill core. A sleeping column of sorts, it exposes layers of soil and earth, cement, electronic waste, and recycled plastics. Over a speaker, the voice of African American award-winning science-fiction author Octavia E. Butler is audible to visitors, with excerpts from interviews given over the years. Siri, the software deemed an "intelligent assistant" by its maker Apple, responds to prompts by Butler's recorded voice and talks about the future.[2] The artworks in the exhibition by Brazilian-American artist Clarissa Tossin address themes of ecology, material extraction, and neo-colonial power relationships. They reflect upon post-human futures that are visual and material, but they also consider critical insights from fourth wave feminism and queer theory, environmental history, and post-colonial theory. Tossin's artwork contributes to the writing of critical histories and theories of architecture, in the context of an impending climate crisis. It does so by taking a close look at the past and the future of the built environment.

Conversation

SEE: Clarissa, *Future Fossil* is an artistic and spatial speculation about post-human futures, but one that insists it is not enough to celebrate bio-cybernetic dreams. In your work you take up three fundamental lessons imparted by Butler's groundbreaking trilogy *Xenogenesis*.[3] First, your work in *Future Fossil* suggests that while capitalist technological imaginaries have become reality, this has unfolded alongside all forms of injustices including colonial and neo-colonial violence that underpin the logic of material extraction worldwide. Second, the show reminds us that there is an urgent need to envision our "environment-worlds"—as feminist architectural theorist Helene Frichot would call it—as nature-cultural hybrids.[4] Thirdly, your work is a call to care for such environment-worlds. The exhibition also calls for synthetic and social proposals that stand against the often empty and dystopian architectural images that have dominated the discourse on post-humanism.

CT: Butler's late 1980s trilogy, *Xenogenesis* is set in the Amazon Forest. I have an ongoing body of work that looks at the history of extraction in the region and the legacies of industrial exploitation in the forest. Butler's trilogy is also an eco-feminist work. She created an alien world that operates as a living ecosystem with hybrid living things such as animal-cars. The Ooloi, Onakali's non-binary being, can decode, store, and replicate genetic material by ingesting material samples. They also have the capacity to heal others, correct the genetic predisposition of humans to disease, and to build offspring from their mate's genetic material.[5] The aliens' spaceship or planet is a giant forest, a structure inside of a living membrane. The Oankali can communicate with this environment—or building if you wish—directly through touch. Its membrane is like a wall that separates our interior and exterior, yet it communicates with both the outside and inside through tactile input. Human beings, however, cannot communicate with the membrane; only the Oankali are able to navigate their world, in this way. This form of communication is possible because aliens and the world are one and the same thing, and this is contrary to our understanding of the natural world as an order separated from human beings. All these concepts were inspiring to me in conceiving *Future Fossil*. Architectonic structures as living things have yet to be addressed, however.

SEE: Butler's ideas metaphorically and literally inform the piece titled *Future Fossil*, which is the centerpiece of the show. (Figure 1) Can you discuss the layering or the striation of the geological material of the drill core and how it relates to the idea of thinking beyond the opposition of nature and technology/culture? In addition, how does it allow us to think about post-human environments?

CT: My idea was to make a core sample of the future as if extracting a few meters off the earth's strata, thousands of years from now. In doing so, I thought about how our material lives are organized in a system that generates excessive waste, what happens with this waste, what kind of footprint it leaves on the earth, and what this means materially—beyond the life of any one human being—at both geological and temporal scales. When you think about being part of an ecosystem, you must consider all kinds of different perceptions of time: the time of a tree, a mosquito, or a human being.

What I did was also personal. I tried to bring these larger ideas in line with a singular 'action' particular to my life. I kept track of the amount of trash that I was consuming—especially plastic—and melted it into this piece, creating a stratum of earth with my plastic waste. *Future Fossil* thus became a record of my consumption alongside the geological stratum which is a material record of time in the earth's crust. In this way, I projected a future geology using materials of my own consumption.

Future Fossil, the exhibition, is the first in which I address, more directly, my concern with excessive waste in the production of the artwork. The question being: What would it mean to have a sustainable sculpture practice in the twenty-first century? If you work materially as an artist, there is always a lot of waste in testing and transportation. In this exhibition I also made weavings using Amazon.com boxes, which are ubiquitous today, especially during the pandemic. (Figures 2 and 3) The piece speaks to the many contradictions in recycling, especially what gets recycled and what does not. Recycling is a business, and it must be profitable. A few months ago, I learned that Los Angeles declared they would stop recycling cardboard, simply because it is no longer profitable. The mindset that everything must be profitable is where the problems begin. The question for me then becomes: what can I do at the scale of my practice? Using Amazon.com boxes and recycling my own plastic waste are just two gestures that I have incorporated in my work. The work of art, however, is about posing larger global questions rather than believing to have all the answers.

Figure 2. *#AmazonisPlanitia1*, 2019. Image Courtesy of Clarissa Tossin. Suspended circular weaving: Amazon.com boxes and archival inkjet prints on glossy photo paper, 23 in. (diameter) and wall work: archival inkjet print on glossy photo paper with 1/16" acrylic facemount, application of recycled melted plastic from artist's own waste over facemount, walnut frame, 20.7" x 29.9" and 19.6" x 11.2". Image courtesy of Clarissa Tossin.

Figure 3. Detail of *#AmazonisPlanitia1*. Image Courtesy of Clarissa Tossin.

SEE: For more than a decade, your work has consistently raised related questions about resource extraction and many of your pieces reveal these concerns on a vast scale. A number of your artworks build up to *Future Fossil*. Concerns around resource extraction and exploitative labor practices appear prolifically in *Fordlândia Fieldwork*, for example. (Figures 4 and 5) In this work you highlight how cities and regions that are thought to be far apart, or whose histories may seem unrelated, are linked through extractive paradigms. Can you discuss this 2012 piece and how it reflects critically on global capitalist production, as well as on the encroachment and dispossession of indigenous lands? It seems to me that the work points to a chain of relationships that connect production, pollution, and decay. These processes are all integral to the logic of extraction that define the places you study in *Fordlândia Fieldwork*, from Brazil to the U.S.

CT: Fordlândia was a rubber plantation village owned by Ford Motor Company, along the Tapajós river in the Amazon Forest, from 1928 to the early 1940s. In the 1920s, the Brazilian government gave Ford Motor Company a concession of land to create this rubber plantation, with the idea that it would bring development to the region.[6] However, the project was a failure, mainly because it was, and still is, impossible in the Amazon Forest to create a plantation where the *Hevea Brasiliensis* (Pará rubber tree) is native. When planted close together, as required for a plantation, the Pará rubber tree becomes vulnerable to blight and other factors that ultimately kill the tree. In this way, the forest fought back against the logic of capitalism—the logic of 'cleaning up' the environment to grow one specific crop in a vast stretch of land.

In an installation made with double sided satellite images, I addressed the story of Fordlândia, which was only one piece in a body of work dedicated to this industrial village. *Fordlândia Fieldwork* consists of eight double-sided inkjet prints, four feet by four feet each, depicting satellite images of the plantation on one side, and on the other side images of locations in the US that helped bolster the automotive industry, including Detroit, Dallas, and Los Angeles. I folded the paper following an origami pattern to make a car. Folding the paper with this technique made the piece a topography of sorts. The idea was to show the contrast and interconnections between all kinds of landscapes created by this industry, their relationship to the natural materials they extracted, and how these places are affected by the industrialization process. Every time the work is shown the eight prints are organized in a different way. Their arrangement is decided by the installation crew in order to reinforce the idea that the landscape is constantly changing.

Figure 4. *Fordlândia Fieldwork*, 2012, eight double-sided archival inkjet prints on cotton paper (8'x8' each), dimensions variable. Image courtesy of Clarissa Tossin.

Figure 5. Detail of *Fordlândia Fieldwork*. Image Courtesy of Clarissa Tossin.

SEE: Many of the landscapes and topographies that you create in your work, or that you ask us to imagine, are deep reflections on the continued exploitation of indigenous land and labor in the Amazon. One of the critical through lines in your work is how thinking with ancestors could be a way out of the impending environmental predicament. Feminist scientists and anthropologists such as Donna Haraway, Deborah Bird Rose, and Beth Povinelli have suggested that it is time to make "quiet country."[7] Rose illustrates the idea of "quiet country" in the introduction to her book *Reports from a Wild Country: Ethics for Decolonization* which is based on decades of scholarly work living with Aboriginal communities in Australia. Here she writes, that "quiet country" is "the country in which all the care of generations of people is evident to those who know how to see it."[8] She also highlights the "critique of White people's efforts to find their own redemption in Aboriginal people's culture, Law, and teaching as if those we had conquered should now save us."[9] She adds, "yet it is also true that there is much to be learned. At the very least, consider gratitude." I believe that her ideas for decolonization are helpful in thinking about how to restructure the history of architecture and I see your practice aligning with some of these ideas. Your work *Meeting of Waters* follows this path. Do you regard it as a form of decolonizing art practice? (Figure 6)

CT: *Meeting of Waters* reflects upon the presence of an industrial park in the city of Manaus, the capital of the state of Amazonas in Brazil. Manaus was extremely wealthy during the rubber boom

Figure 6. *Encontro das Águas (Meeting of Waters)*, 2016–18. Woven archival inkjet print on vinyl (4 1/2 ft x 50 ft), terra-cotta objects, fishnet, thread, and woven baskets and backpack made out of Amazon.com boxes. Installation View: *Encontro das Águas*, Courtesy of the Blanton Museum of Art at The University of Texas at Austin (2018). Photo by Colin Doyle.

of the late nineteenth century. After 1912, when the British turned to alternative rubber sources within their colonies, the city's economy was greatly affected. Today, it is the port of Manaus which facilitates the movement of capital. Following a decade of deregulation that began in 1957, the city became a Free Trade Zone and now hosts manufacturing plants for such companies as Apple, Sony, LG, Coca-Cola, Dell, Harley Davidson, and Honda Motorcycles. The commodities made at these factories enter the global and domestic market on cargo ships that exit the Port of Manaus. A large percentage of TV sets consumed in Brazil are produced there. Electronics, soda, and motorcycles are the three dominant industries in Manaus, and electronics receive substantial subsidies from the government.[10] In *Meeting of Waters*, the terra cotta objects I crafted are all replicas of these odd objects manufactured in Manaus.

The Manaus Free Trade Zone is very well-known in Brazil, less so outside the country. My work comes from the desire to make this known to more people. Hidden in the forest, if you are not particularly knowledgeable about the region or live in Brazil, you would not think of an industrial park of this scale when you think of the Amazon Forest. It does, however, follow the capitalist logic of bringing 'progress' to the forest. Thankfully, we are entering a different moment with greater understanding that this is not only damaging the natural environment, but also poses questions about what kind of progress is really achieved here, and how these developments affect the economic livelihood of local communities.

For *Meeting of Waters*, I was interested in working with labor-intensive material practices connected to the environment, like weaving and ceramics, and at the same time addressing aspects of my work related to landscape and satellite imagery. Traditionally, weaving and ceramics used local resources such as natural fibers and soil, and its centrality within indigenous aesthetic traditions prejudiced the way they were incorporated in the European canon of art history. I embraced these hand-made, time-consuming processes in my art practice as a way of counteracting mass production and consumption. The idea was to divest myself of immediate production and incorporate a slower way of production. While I make art in a world that is structured around capitalism, I try to engage with another form of timing—that of weaving. For the past few years, I have recycled Amazon.com delivery boxes in weavings loosely inspired by indigenous Amazonian traditions. The slow, laborious process of flattening boxes, cutting them into strips, and weaving them together stands in contrast to wasteful cycles of mass resource extraction, production, and consumption upon which our lives depend. A disposable container made into something to be contemplated signals a broader invitation to stop, look, and reflect.

I recognize that ceramics and weavings do not represent the full scope of indigenous aesthetic practices present today, either in the Amazon or elsewhere. In Brazil, there is no shortage of contemporary indigenous artists working in all media (video, painting, performance, and photography, for example). The choice of teaching myself how to weave has more to do with my desire to engage with a tradition I was deprived of in art school, and that is traditionally related to locally sourced natural resources. I wanted to bring weaving in conversation with a different materiality in my art practice while also engaging with a different scale of time.

SEE: Following in the footsteps of feminist artist Mierle Ukeles and others, in one of your early works *White Marble Everyday* (2009), you engage the question of maintenance.[11] Here you expose a relationship between body and stone, or the exploitative relationships between government and service, through an analysis of a canonic piece of modern architecture. What constitutes this work and how does it speak to questions of maintenance and care?

CT: *White Marble Everyday* was filmed at Brasília's Federal Supreme Court building, designed by Brazilian architect Oscar Niemeyer in 1957. For me, the work's images capture the contradiction between the discourse that surrounds this famous piece of architecture and the actualities of the real building. When I first saw the maintenance crew cleaning the floors outside the Federal Supreme Court, I approached them and started a conversation. I was shocked to learn that cleaning the floors was a daily routine. Every day from Monday to Saturday, they start at six in the morning and spend more than four hours cleaning the white marble exterior floors. The Federal Supreme Court is a heavily trafficked public building on the major plaza in Brasília, the Plaza of the Three Powers. When I made the film, you could step on the white marble floors around the building and take pictures as a tourist. A few years later, when there were protests in Brazil against increases in the cost of public transit, barricades were put around the building and were never taken down. The more protests there were, the more layers of barricades were added, and people were pushed further away from this public building. According to Niemeyer, all the floors around the building should have been public space. Thankfully, they were still accessible to me when I made the film in Brasília.

SEE: An aspect I find striking in *White Marble Everyday*, and which relates to your later work, is that the whiteness of the marble is more than a metaphor. The marble had to be quarried and transported for thousands of miles to Brasília to create this white building, all of which required the application of immense extractive forces. And then, this white marble, this dead, compressed fossil must be maintained every day, by the maintenance crew that completes the invisible, labor-intensive service work for a building conceived to mark the democratization of public services.

CT: Yes, the work surely holds all these contradictions. Public buildings in Brasília probably consumed a whole quarry of white marble, since all the public edifices are covered in it. Even the floors inside the buildings are made of marble. However, the soil in Brasília has an iron and clay rich subsoil, which makes it rather reddish. As a result, it is almost impossible to keep the marble white, because once people walk in, they bring the reddish soil onto those surfaces with their shoes. It is ironic that one would want to use white marble floors in this region.

SEE: The soil of the city finds its way back to resist the whiteness of the design.

In 2018 you ended an interview with curator Meg Rotzel at Harvard University with the quotation by Butler, which we introduced at the outset of this conversation. It seems there is an arc one can trace from Butler's quote to a statement you made about the contemporary political climate in Brazil. You warned of the extractive, destructive, and racist policies of then newly elected far-right president Jair Bolsonaro noting that it would bring "our present teeters precariously on a brink. Bolsonaro's proposed policies threaten both the environment and the livelihood of indigenous Brazilians' communities."[12] In 2019, fires in the Amazon increased by 84% percent and we have seen some of the gravest human rights violations. 2019 was one of the deadliest years with over three hundred murders of human rights activist world-wide, twenty-three of them in Brazil alone.[13] You have since been to the Amazon, can you talk about your experience there?

CT: I was there in August 2019 to study the Amazonian biome. It was a ten-day residency organized for artists and scientists. For me there was a kind of disjunction between my personal experience and what was happening in the forest. On the one hand, I was on an easy and pleasant trip in the Amazon Forest. Everything was organized. There were people in the group who are knowledgeable about the environment, including a local guide and scientists from different fields. We were visiting natural reserves, watching birds, or studying different ant colonies. It was all about observing the life of the forest in close detail. Yet at the same time, the fires were swallowing the land, burning big sections of the forest to 'prepare the land' to raise cattle or for soy plantations. That is how the forest is turned into 'productive land'. It was very disturbing.

Figure 7. Installation view of *White Marble Everyday* (2009) in the Ezra and Cecile Zilkha Gallery, Wesaleyan University, 2017. Image courtesy of Wesleyan University's Center for the Arts and Clarissa Tossin. Photo by John Groo.

I think existing power structures benefit from promoting the idea that the Amazon is a 'wild jungle'. Doing so is very debatable since it erases the fact that the forest is a result of land management practiced by indigenous communities who have lived on the land for thousands of years. Amazonian farmers between 450 BCE and 950 CE fertilized their soil and developed *terra preta*, or black soil, which is the most fertile soil in the Amazon Basin. *Terra preta* is a product of indigenous soil management and slash-and-char agriculture; the charcoal is stable and remains in the soil for thousands of years, binding and retaining minerals and nutrients.

The region's economic problems are complicated, and since the election of Bolsonaro and COVID-19, the situation is deteriorating. The government delayed sending help to the region which greatly affected the ability of indigenous communities to contain and recover from virus infections. This led to the loss of many indigenous leaders, such as Messías Kokama from Manaus, and the death of a leader further weakens the community's structure and organization. Things do not look good in the region right now.

Back in 2019, the trip reinforced just how global the Amazon is. It is connected to global cycles of production, which in many ways put it at the center of the world, and this, not only because of its key role in regulating the earth's climate. Initiatives such as the Fundo Amazônia (Amazonian Fund) raise money from nations around the world in efforts to prevent, monitor, and combat deforestation. They also sought to promote different economies within the Amazon, so that local communities could foster livelihoods without damaging the environment. We visited one of those communities which relies on funds by the Swedish government. Unfortunately, the Swedish government had just announced that week that it was withdrawing its support because of the actions of the Brazilian government. Thus, you see how complex things are and how they are played out in the political arena.

SEE: The discipline of architectural design especially in schools of architecture has been firmly entrenched and infatuated with notions of newness and histories of progress. As such in professional architecture the work of feminist thinkers, such as Donna Haraway and the ideas conveyed in "A Cyborg Manifesto," have at times been emptied of their serious considerations about gender and labor, to evoke and produce dystopian post-humanist imagery.[14] Your work, while drawing on Haraway, takes a fundamentally different approach that is much more aligned with the ethics of "A Cyborg Manifesto."

CT: I have been interested in Donna Haraway's ideas on posthumanism. In *Future Fossil*, there is a sound piece titled *You Got to Make Your Own Worlds (for when Siri is long gone)*. This piece is a 'conversation' between the words of Octavia E. Butler and Apple Inc.'s iOS virtual assistant Siri. The idea was to create a dialogue about the future that showcased each of their anxieties about it. I searched multiple interviews with Butler on YouTube and consulted her archives at the Huntington Library in Los Angeles. I was looking for quotes that talked about the future, or specifically about the future of the environment. After listening to a number of her interviews, I tried to convey the way she spoke, and to incorporate her tone in the piece. For Siri, I really tried to talk with 'her' and to put some of her responses in the dialogue. In this sense, Siri's parts are also quotations. But as a very simplified form of artificial intelligence (AI), Siri has a standard way of answering questions. For example, whenever you ask for her opinion, she offers the 'vacant' response: "What I think is not important. What else can I help you with?" Alternatively, "your opinion is what really matters." This makes it very difficult to personify Siri. Nonetheless, I incorporated her specific way of talking into the work.

Butler's text, however, seeks to touch upon the resilience of human beings, who are empathetic, who try to engage with the difficulties, and hope to make things better. Her writing was abruptly contrasted by the machine's response. At the very end of the dialogue Butler says, "I'm hopeful." And Siri says, "I give up."

SEE: Who has the last word?

CT: I would like to think that the optimism of Butler prevails, but likely it will be AI.

Acknowledgements

The authors would like to thank Qiran Shang for help with initial transcription and editing of this conversation.

Transcription of *You Got to Make Your Own Worlds (for when Siri is long gone)* (2019)

looped sound, voices of author Octavia E. Butler and Apple Inc.'s iOS
virtual assistant, Siri
Listen at: https://vimeo.com/324844931

Siri: *Go ahead. I'm listening...*

Octavia E. Butler: It's all nonsense, nonsense.

S: *Go ahead.*

OB: ...my own feeling of powerlessness

S: *I'm sorry. I don't understand.*

OB: I know that we can do better than we have but I don't know that we will. I know that we can be better human beings than that.

S: *Is there something I can help you with?*

OB: We have this conflict within ourselves. All too often our intelligence serves our hierarchical tendencies so that we tend to one-up ourselves to death.

S: *Sure. How can I help?*

OB: Oh my! Send us off to other worlds.

S: *That may be beyond my abilities at the moment.*

OB: No. No. It's not a matter of deliberately doing it. It's just a matter of behaving as though the world were a little bit more as you wish it were in some way.

S: *I don't know what that means. If you like, I can search the web for "behaving as though the world were a little bit more as you wish"*

OB: No. No. There are a lot of reasons for this ...the environmental situation, ... lots of other things that have simply... they're problems now, they become disasters because, you know, they are not attended to.

S: *Take a deep breath. Now exhale. And repeat.*

S: *Take a deep breath.*

OB: uhmmm....

S: *Now exhale.*

OB: Ohh! You gotta make your own worlds. You got to write yourself in.

S: *I'm looking forward to new beginnings.*

OB: Explain what that is.

S: *Sorry for being obtuse. I'm afraid I do not have an answer to that.*

OB: hummm... but ahhh...

S: *This is about you, Octavia, not me.*

OB: It's dangerous to assume that whatever we've been doing we're going to keep doing that, you know, the future is more of the same only more advanced.

S: *Well, it's your opinion that counts, Octavia.*

OB: Oh my! hummm.. I have no idea on how to answer that.

S: *That's what I thought.*

OB: I think that the one thing we can be sure of is that we won't have, you know, straight line prophecy coming true that whatever technological things we're doing now will just do more of that and better. I think we'll get surprises. It's dangerous to assume that we can actually see the future by only looking at the advancements we've made so far.

S: *I try to be satisfied with what I have. I have everything I need in the cloud.*

OB: It's not about the nasty future of Earth. It's about the people who decide to leave.

S: *I really couldn't say. I give up.*

OB: I'm hopeful. I'm hopeful. I'm... I'm hopeful.

Figure 8. Video stills from
White Marble Everyday, 2009.
Image courtesy of Clarissa
Tossin.

ENDNOTES

1. *Fast Forward: Contemporary Science Fiction*, "Octavia Butler interview - transcending barriers," June 22, 2018, video, 19:54, https://www.youtube.com/watch?v=KG68v0RGHsY&t=967s.

2. Clarissa Tossin, *You Got to Make Your Own Worlds (for when Siri is long gone)*, 2019, looped sound, voices of author Octavia E. Butler and Apple Inc.'s iOS virtual assistant, Siri, https://vimeo.com/324844931.

3. Octavia E. Butler, *Xenogenesis: Dawn; Adulthood Rites; Imago* (New York: Warner, 1989).

4. Hélène Frichot, "Step Four: Chart Your Environment-World," in *How to Make Yourself a Feminist Design Power Tool* (Braunach: Deutscher Spurbuchverlag, 2016), 83–108.

5. Octavia E. Butler, *Xenogenesis: Dawn; Adulthood Rites; Imago* (New York: Warner, 1989). The Oankali have three sexes: male, female, and Ooloi. Oankali have the ability to perceive genetic biochemistry, but the Ooloi manipulate genetic material to mutate other beings and build offspring from their mates' genetic material.

6. Greg Grandin, *Fordlandia: The Rise and Fall of Henry Ford's Forgotten Jungle City* (Prince Frederick, MD: Recorded Books, 2010).

7. Deborah Bird Rose, "Introduction: Into the Wild," in *Notes from A Wild Country* (Sydney: University of New South Wales Press, 2004), 1–8; Donna Jeanne Haraway, *Staying with the Trouble: Making Kin in the Chthulucene* (Durham: Duke University Press, 2016); Donna Haraway, "Making Oddkin: Story Telling for Earthly Survival," October 26, 2017, video, 1:34:02, https://www.youtube.com/watch?v=z-iEnSztKu8; Elizabeth A. Povinelli,

"Transgender Creeks and the Three Figures of Power in Late Liberalism," *differences* 26, no.1 (May 2015): 168–87.

8. D. Rose, "Introduction: Into the Wild," 4.

9. Ibid.

10. "Histórico: A história da Zona Franca de Manaus, em resumo," gov.br, published August 28, last edited July 14, 2020, https://www.gov.br/suframa/pt-br/zfm/o-que-e-o-projeto-zfm

11. Frichot, "Introduction: Taking Instructions," in *How to Make Yourself a Feminist Design Power Tool*, 24–26; Toby Perl Freilich and Mierle Ukeles, "Blazing Epiphany: Maintenance Art Manifesto 1969! An Interview with Mierle Laderman Ukeles," *Cultural Politics* 16, no.1 (March 2020): 14–23.

12. Rachel Vogel, "Conversations Across Space and Time: Clarissa Tossin's *Future Fossil*," print publication for *Future Fossil* (January 31–March 16, 2019, Radcliffe Institute for Advanced Study), https://www.clarissatossin.net/Future-Fossil.

13. Jonathan Watts, "Amazon fires: what is happening and is there anything we can do?" *The Guardian*, August 23, 2019, https://www.theguardian.com/environment/2019/aug/23/amazon-fires-what-is-happening-anything-we-can-do; Nina Lakhani, "More than 300 human rights activists were killed in 2019, report reveals," *The Guardian*, Jan 14, 2020, https://www.theguardian.com/law/2020/jan/14/300-human-rights-activists-killed-2019-report.

14. Donna Haraway, "A Cyborg Manifesto: Science, Technology, and Socialist-Feminism in the late Twentieth Century," *Simians, Cyborgs and Women: The Reinvention of Nature* (London and New York, Routledge, 1991), 149-81.

TRANSINDIVIDUAL AND TRANSDISCIPLINARY DESIGN RESEARCH:

WEAVING, AI, ENERGY FUTURES, AND BIO-BRICKS

JENNY E. SABIN

ABSTRACT

This article presents ongoing transdisciplinary research and design projects that span the fields of cell biology, material science, physics, fiber science, fashion, mechanical and structural engineering, and architecture. The work it discusses and theorizes is focused on contextual, material, and formal intersections between architecture, science, and emerging technologies. The material world this type of research interrogates reveals examples of nonlinear fabrication and self-assembly occurring at a surface and at a deeper, structural level. It also offers novel possibilities for questioning and redefining the scope of architecture within the greater extents of generative design and fabrication, alongside development of biosynthetic theories and transdisciplinary design methodologies. This article deepens our understanding of transdisciplinarity and transindividualism through its coupling of gender theory and computational design. Unearthing the post-gender dimension of the "nature and culture" dialectic, it elucidates research methods, prototypes, and architectural projects completed by the author and her collaborators, focused on three transdisciplinary projects that (re)configure their performances according to local criteria and human interaction. The projects discussed are *Ada* by Jenny Sabin Studio for Microsoft Research, *Sustainable Architecture & Aesthetics: Agrivoltaic Pavilion*, funded by the National Academy for Engineering and the Grainger Foundation, and *PolyBrick/ PolyTile*, funded jointly by the Sabin Lab and Luo Labs.

+ transdisciplinary
+ computational design
+ biosynthesis
+ data-driven design
+ artificial intelligence

Figure 1. Detail of project *Ada*. Photo by Jake Knapp for Microsoft.

My collaborative work and teaching investigate the intersections of architecture and science and apply insights from biology and mathematics to the design of adaptive material structures and spatial interventions. This work encompasses expertise in digital fabrication, algorithmic processes, and transdisciplinary collaboration in architectural design and artistic practice. The scope of my work probes the visualization and simulation of complex spatial datasets (biological, mathematical, material) alongside issues of making, fabrication, and production in a diverse array of material systems (digitally Jacquard-woven, CNC-knitted, drone-woven, and braided textiles; rapid prototyped, cast, and robotically 3D-printed ceramics; cast bioplastics/polymers and hydrogels; waterjet-cut metals). The productive tinkering and deliberate misuse of digital fabrication machines from other industries (automotive, textiles, medicine) produces bioinspired material systems and software design tools that have the capacity to facilitate embedded expressions in our built environment. This new model has been actively developed and practiced over the past fifteen years within a hybrid research lab and design studio.[1]

This approach first developed in 2006 by the Sabin+Jones LabStudio at the University of Pennsylvania, continues today at the Jenny Sabin Studio (through practice) and the Sabin Lab (through fundamental research) at Cornell University.[2] It couples architectural designers with engineers, material scientists, and biologists within a research-based laboratory-studio in order to develop hybrid thinking in design, across multiple fields.[3] This mode of working has contributed to a body of fundamental research and applied projects that integrate biology, technology, material science, and architecture through methods and tools that emphasize design research as a continuous and emergent process that is bottom-up and processed-based. Schools of thought where immanence, morphogenetic processes, generative design, and constructivist theories thrive have influenced and partially shaped my design process and thinking. These include Gilles Deleuze's infinite folds, Detlef Mertins's "bioconstructivisms", Manuel DeLanda's "meshworks", Anni Albers's "pliable plane", Bernard Cache's *objectile*, and Cecil Balmond's "informal".[4] In the sciences, the work is influenced by the field of epigenesis and its study of the extracellular matrix.[5]

In many instances, I've been fortunate to study, collaborate, and shape projects alongside the pioneers listed above, whose generative approaches and processes link them to transdisiciplinarity. In previous texts, I've explored how these theories have influenced the development of a biosynthetic design methodology.[6] In this article, deepening transdisciplinarity and transindividualism, I explore gender and computational design to unearth the post-gender dimensions between nature and culture that a transdisciplinary design methodology opens. Here, the emphasis is upon individuation as a transforming process, a necessary component to transindividualism and the practice of transdisciplinary work. In doing so, I compare cyberfeminist Sadie Plant's essentialist zeros and ones to Donna Haraway's materialism and cyborg metaphor in order to frame the possibility of a networked collective.[7] Further, I elucidate research methods, prototypes, and applications in a description and discussion of three transdisciplinary projects: *Ada* by Jenny Sabin Studio for Microsoft Research; *Sustainable Architecture & Aesthetics: Agrivoltaic Pavilion*, funded by the National Academy for Engineering and the Grainger Foundation; *PolyBrick/ PolyTile*, funded jointly by the Sabin Lab and Luo Labs. These projects incorporate adaptive materials, artificial intelligence, textile and ceramic assemblies, and architectural interventions that ultimately (re)configure their own performance based upon local criteria and human interaction. Through the lens of this collaborative work, this article highlights the transindividual and hybrid multiplicities that are produced through transdisciplinary design research, offering a new model for research, pedagogy, and practice. I begin with an exploration of the term 'transdisciplinary' to unfold its generative relationship with transindividual.

Transdisciplinarity and Transindividuation

Transdisciplinarity is a concept that stems from several schools of thought. There are various definitions of 'trans' disciplinarity, as there are multiple allied and interchangeable configurations of 'inter,' 'multi,' and 'cross' disciplinarity. Musicologist and Professor Alexander Refsum Jensenius summarizes Dr. Marilyn Stember's definitions as follows:

> **Intradisciplinary:** working within a single discipline;
>
> **Crossdisciplinary:** viewing one discipline from the perspective of another;
>
> **Multidisciplinary:** people from different disciplines working together, each drawing on their disciplinary knowledge;
>
> **Interdisciplinary:** integrating knowledge and methods from different disciplines, using a real synthesis of approaches;
>
> **Transdisciplinary:** creating a unity of intellectual frameworks beyond the disciplinary perspectives.[8]

As pointed out by feminist scholar Tanya Augsburg, transdiciplinarity was initially defined during the First International Conference on Interdisciplinarity held in Paris in 1970.[9] Identified as a "common system of axioms for a set of disciplines," many early transdisciplinary models were focused on education.[10] As Augsburg describes, one of the more influential definitions came from astrophysicist Erich Jantsch in 1972 who "described a multidimensional innovative approach to education that was coordinated as a multi-level and multi-goal system."[11] A decade later, perspectives from the social sciences turned to conceptual frameworks and thought models to break down disciplinary boundaries.[12] The approach and conceptual framework that resonates most with my collaborative methods stems from what is frequently referred to as the Swiss or German school. As Augsburg describes, this "second widely recognized current focuses on transdisciplinarity as a research approach for addressing complex societal problems such as those related to sustainability."[13] This emerged as a response to cultural shifts and changing societal needs in the context of a new type of science, or a post-normal science, that called for corresponding changes in research methods and collaborative models.[14] As Augsburg states, this "new type of research … is characterized by complexity, hybridity, non-linearity, reflexivity, social accountability, mutual learning, heterogeneity, and of course, transdisciplinarity."[15] Here, the focus is on collaborative problem solving across science, technology, and the design arts, inside and outside of academia.

My own working definition of transdisciplinary is not focused on problem solving, but instead embraces problem generation through design research processes that are emergent, transformational, and evolving for innovative applications across multiple disciplines and disciplinary frameworks. In contrast to Stember's description of transdisciplinary, my definition does not focus on generating a unity of intellectual frameworks outside of disciplinary perspectives, but instead on emphasizing the possibility of evolving knowledge, excessive knowledge, impregnated within a collective meshwork of heterogenous disciplinary expertise. In the context of pedagogy, this frequently requires engaging fundamental concepts and methods in design and emerging technologies across the design arts, engineering, and science to prepare students with the necessary tools and knowledge for iterative, hybrid, and synthetic thinking. The success of transdisciplinary work is not based on singularities and unity, but upon metastable events and relationships that change and unfold over time. By seeking openings rather than setting strict goals, applications emerge through a process that is iterative, bottom-up, and transformational.

This definition is substantiated by the philosophical underpinnings of transindividualism, specifically as described by French philosopher Gilbert Simondon (1924-1989).[16] In interpreting Simondon's work, contemporary French philosopher Étienne Balibar states:

> …the network of analogies and affinities has become more and more dense between Simondon's idea of the transindividual, centered on individuation as a universal ontological and morphological category, applicable to all kinds of beings, and the objectives of a contemporary philosophy of becoming, of collective transformation, of the plasticity of institutions, casting suspicion again on metaphysical oppositions between the reign of necessity and that of freedom.[17]

Balibar further describes how "individuation according to Simondon occupies an intermediate position between a 'pre-individual' potential that it expresses but never exhausts, and a 'transindividual' excess in which it is always already engaged."[18] It is this "ontology of relations," a dynamic collective process that gives rise to transdisciplinary 'excessive' knowledge, that is both internal and external, metastable, and infinite. The collective within this "ontology of relations" is immanent to human relations.[19] Therefore, authorship is not singular, but heterogeneous, transindividual, and in formation.

In our specific transdisciplinary process, Jenny Sabin Studio and Sabin Lab emphasize the analogic negotiation of morphological behavior as a dynamic substrate that is filtered through material organizations, where the pursuit is to 'biosynthesize' a natural model with its material manifestation. In this, the work is inspired by Detlef Mertins who situated biosynthesis in a longer history of generative design through his adoption of the term 'bioconstructivisms'.[20] The architectural results of such an experiment, to use Mertins's words, would be "self-different, in which identity is hybrid, multiple and open-ended."[21] Importantly, he demonstrated how these endeavors for autopoiesis are linked to politics and "the quest for freedom among the cultural avant-garde."[22] As a way of working and "acting constructively in the world," Mertins reached across disciplines to unearth and identify bioconstructive concepts that "produce new iterations of reality" through creative and scientific practices that are not deterministic and governed by laws.[23] Mertins constructed the beginnings of a history of generative design processes across disciplines to unfold heterogeneous models that are relevant in the formation of a contemporary transdisciplinary and transindividual practice. Concepts of transformation, difference, and immanence are also present in this generative and collective potential of individuation.

Computation and Cyberfeminism

Importantly, my work is also informed by early discoveries of latent connections between gender, emerging technologies, and computation, especially through the generative process of weaving which stitches across the analog and the digital.[24] The gender of computational design has been an elusive and complex topic in my practice and collaborative research. On the one hand, the technological spaces that I occupy, and the communities I participate in and learn from (ACADIA, Smart Geometry, SimAUD, SIGGRAPH), are heavily male dominated: a condition common in many science and technology fields whose longer histories have yet to be written. On the other, it is precisely the history of computation, digital space, and weaving that communicates more complex gender narratives. Not only does this history reveal the women who have been erased and barred from being educated and working in these fields, but it positions women at the origins of all three.

Sadie Plant, the cyberfeminist theorist, traces how women have contributed throughout history to important and groundbreaking phases of scientific computing. The lens she adopts is that of women's work in weaving.[25] Starting with Ada Lovelace (1815-1852), a polymath, mathematician, and early innovator of the computer age, Plant outlines Lovelace's extraordinary contribution to modern-day scientific computing.[26] Here, the history of modern computing may be traced back to an uncanny meeting between two disparate, yet emergent, inventions: punch cards that effectively mechanized the Jacquard loom through stored memory and Charles Babbage's (1791–1871) steam driven calculator, the Difference Machine. While the Difference Machine was brilliant with addition and subtraction, Lovelace's contribution to Babbage's second machine, the Analytical Engine, opened entirely new possibilities for data analysis, regulation, cycles, nested loops, and parametric operations through circulating data. Credited with being one of the first computer programmers, Lovelace intuited the revolutionary impact that Jacquard's punch cards would bring to Babbage's calculator, effectively launching the precursors of modern-day scientific computing. As Plant states:

> And it was, as Ada wrote, 'the introduction of the principle which Jacquard devised for regulating, by means for punched cards, the most complicated patterns in the fabrication of brocaded stuffs,' which gave the Analytical Engine its 'distinctive characteristic' and 'rendered it possible to endow mechanism with such extensive faculties as bid fair to make this engine the executive right-hand of abstract algebra'.[27]

According to Plant, Lovelace described this as the "science of operations."[28] Indeed, Plant's scholarship makes a seminal contribution to the history of scientific computing through the lens of cyberfeminism and its historically accurate accounts of contributions by key women protagonists including Lovelace.[29] Plant theorizes the following about the computer:

> Nor is there anything like unto computers: they are the simulators, the screens, the clothing of the matrix, already blatantly linked to the virtual machinery of which nature and culture are the subprograms. The computer was always a simulation of weaving; threads of ones and zeros riding the carpets and simulating silk screens in the perpetual motions of cyberspace. It joins women on and as the interface between man and matter, identity and difference, one and zero, the actual and the virtual.[30]

Feminist scholar Stacy Gillis in writing about Plant's work states: "The weaving metaphors which were frequently used of the World Wide Web in the mid-1990s offered Plant the opportunity to identify women as the common factor to both: weaving, women, the Web."[31] Gillis demonstrates how Plant's *Zeros and Ones: Digital Women and the New Technoculture* broke through important boundaries to demonstrate that women were not victims of technological innovation but rather part of its history, always there but never celebrated or studied. Gillis describes how cyberfeminism moves beyond the patriarchal boundaries of a gendered body through metaphors that link women with weaving, computing, and the fluidity of identity that the birth of the internet inspired. However, as Gillis argues, in essentializing the role of women through their work "with the new freedoms of cyberspace" in the context of computing, weaving, and digital space, we are still left with binary conditions of gender, which limit what may be achieved across genders, individuals, and disciplines.[32]

Differently, by working across mathematics, the craft and industry of weaving, and computation, Lovelace's contributions (developed in collaboration with Babbage) demonstrate the emergent possibilities of human-technological relationships that inform new transdisciplinary and transindividual directions and inventions. By essentializing these relationships through attributes of gender as Plant does, we may overlook the human-technological relationships that co-create space through a network of collective human embeddedness, or what Balibar calls "transindividual excess."[33] The true potency of contributions by Babbage, Lovelace, and Jacquard is, therefore, best described through the interconnected and individuated relationships that link mind, matter, and machine. Certainly, Lovelace faced scrutiny and limitations as a woman working in domains traditionally reserved for men—including being labeled hysterical and mentally unstable—but as a "Prophetess born into the world" she understood the metastable connections between weaving and computing as a "science of relations."[34]

Moving closer to concepts of transindividualism is Donna Haraway's metaphor of the cyborg. Considered one of the founders of posthumanism, Haraway breaks down the binary oppositions of male and female, man and machine, to offer instead a hybrid form that synthesizes biology, machine, and body as an integrated materialist system of relationships. What is particularly appealing about Haraway's cyborg is the promise that technology, computing, machines, and the body are completely hybrid, boundless, differentiated, and transforming. As Haraway states, "A cyborg is a cybernetic organism, a hybrid of machine and organism, a creature of social reality as well as a creature of fiction."[35] In writing about Haraway's work, Gillis confirms that, "Haraway posits a cyborg feminism, arguing that the metaphor of the cyborg breaks down the binary oppositions of meat/metal and offers the possibility of a post-gender identity.[36] Like the brilliant success of the work by Babbage, Lovelace, and Jacquard, transdisciplinary work requires metastable relationships between people, material, nature, and machines that define a collective transindividualism.

Can transdisciplinary thinking in design, coupled with emerging technologies and generative digital fabrication, give rise to an architectural body that erases boundaries of gender towards an infinite topological landscape of connections? Building upon this question, and learning from Haraway's cyborg manifesto as metaphor, I engage in and cultivate a generative design research process of individuation that is 'trans'. In the following collaborative projects here discussed, both intrinsic properties—how they are fabricated and what they do as a set of materialized relationships—and extrinsic properties—the audiences they engage and why they're of value to diverse publics—define them as 'trans'.

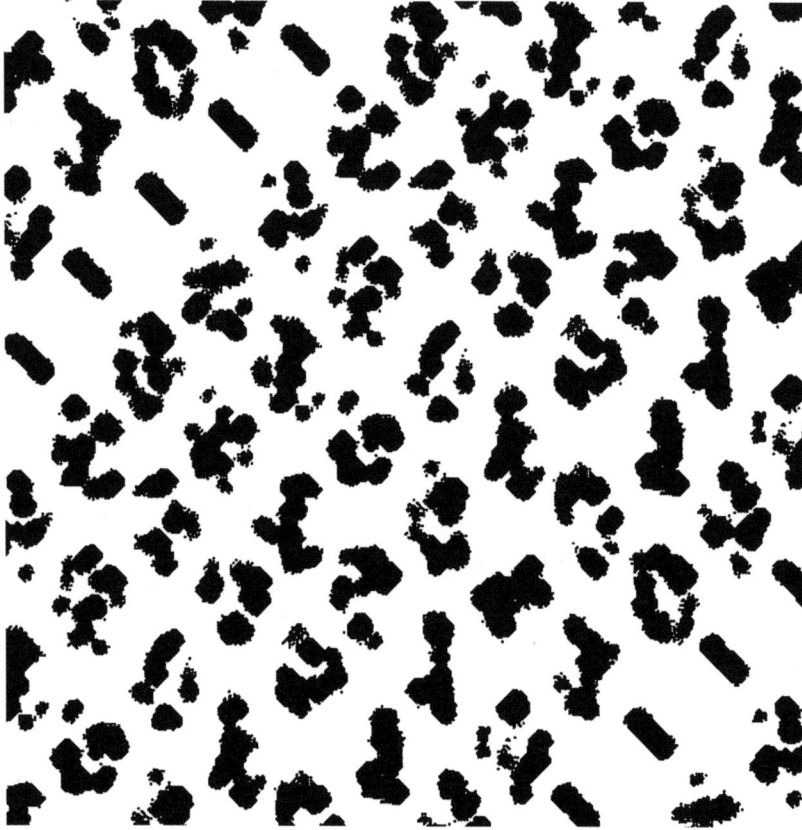

Figure 2. *BodyBlanket*, draft notation pattern drawing, 2005. Woven on a digitized Jacquard loom. Image credit: Jenny E. Sabin.

Weaving Artificial Intelligence, Energy Futures, and Bio-Bricks and Tiles

The coupling of architecture and textiles has a long history. My own interest in this space resides primarily in materialist and computational design philosophies, not in phenomenological ones. Important examples of which include the artistic design work of the weaving workshop at the Bauhaus, such as that produced by Anni Albers, Gunta Stölzl, and Lilly Reich. Bauhaus weavers marked a shift from expressionistic and individually handcrafted compositions to mass-produced, manufactured furniture and prototypes for interior design and architectural furnishings. This shift led to a new approach to mass production that was marked by the integration of pattern, material, form, and fabrication. Albers's term for this generative and constructive intersection between architecture and textiles was "pliable plane," an undulating construction of information made tangible through geometry and matter.[37]

My interest in weaving began in 2004 when I developed a way of working that bridged the digital and the material in an effort to seamlessly integrate pattern, geometry, computation, material, form, and digital space. *BodyBlanket*, materialized in 2005 as part of a

Figure 3. *BodyBlanket*, draft notation pattern drawing, 2005. Woven on a digitized Jacquard loom. Image credit: Jenny E. Sabin.

group project, became my capstone work while studying with Cecil Balmond in the Department of Architecture at the Graduate School of Design of the University of Pennsylvania. (Figures 2 and 3) The intent of *BodyBlanket* was to realize interfaces between patients, information, and the hospital setting by giving physical form to patient data, thereby making data perceptible. It sought to understand patterns within patient data sets; patterns that, through interpretation and analysis, could lead to multi-scalar solutions for problems in health care. In this project, data was woven; it was transformed into a CAD file of binary block code in black and white, which was then fed into a digitized Jacquard loom. The width of a textile was limited by the loom's bed size, but the length could be infinite—as long as the thread didn't run out! Data provided by the human body could also be woven, such as body frequencies emitted during magnetic resonance imaging (MRI), which could be filtered through a Fourier Transform, a mathematical transformation that decomposes functions, visualized as an image.[38] Produced on a digitized Jacquard loom, the vision for *BodyBlanket* was for it to be continuously woven as an infinite analog of data collected from the human body. Agency was given back to the patient (a speculative patient in a hospital room) in the form of a

Figure 4. *Fourier Carpet*, 2006. Woven on a digitized Jacquard loom. Image credit: William Staffeld.

reading of their data and information. In this way, *BodyBlanket* made visible a transindividual data-body.

Moreover, as I have stated in previously published texts, the technological and cultural history of weaving reveals a link between punch card technologies used to automate the Jacquard looms of the mechanical age and early binary systems used in the very first computers.[39] This link is rooted in the coded binary patterning system of warp and weft that structures a weave. Because weaving is binary—composed of zeros and ones—it is possible to weave computational designs from other sources that share a binary structure. For example, *The Fourier Carpet Series* (2006), commissioned and on view as part of the *H_edge* exhibition at Artists Space 2006, transformed color and sound frequency data into code that was configured and woven into material expression on a digitized Jacquard loom. (Figure 4)

These early interests in weaving led to additional investigations into textile processes, namely knitting. *Ada* is a recent project by Jenny Sabin Studio completed for the artists in residence program at Microsoft Research 2018-2019. (Figure 5) An early architectural project to incorporate artificial intelligence (AI), *Ada* is a lightweight knitted pavilion composed of responsive and data-driven tubular and cellular components held in continuous tension via a custom 3D-printed semi-rigid exoskeleton shell. As installed, *Ada* creates and describes a cyber physical architecture driven by human participation and powered by individual and collective 'sentiment' data, collected throughout the Microsoft Research Building 99, using a network of cameras. The collaborative transdisciplinary process that produced *Ada* contributed to its collective and embedded characteristics.

Its structural concept builds upon previous work by Jenny Sabin Studio, engineered with design help from Arup. The structure displays a balance of tensile and compressive forces, generating

Figure 5. An early architectural project incorporating AI, *Ada* is a lightweight digitally knitted responsive pavilion with 3D printed semi-rigid exoskeleton shell. Photo by Jake Knapp for Microsoft.

Figure 6. *Ada*, designed and built by Jenny Sabin Studio for Microsoft Research Artist in Residence Program 2018–2019. Photo by John Brecher for Microsoft.

lightweight and expressive forms. (Figure 6) Importantly, *Ada*'s design began with a series of collaborative conversations between Jenny Sabin Studio and Microsoft Research on topics that spanned artificial intelligence, adaptive and embedded architecture, affective computing, and personalized space.[40] In addition, *Ada* leveraged thirteen years of collaborative work and innovation across the fields of architecture and science, wherein projects embraced and were informed by technology, non-standard and bio-steered concepts, and the hidden spatial structures within data, to reveal hidden expressions and emotion in the built environment.[41] Over the course of eighteen months, *Ada* materialized a set of co-evolving collaborative streams involving hardware, physical computing, software engineering and design, material testing, 3D-printing and digital knit prototypes, lighting design, and structural analysis. *Ada*'s Brain Ring was a bespoke yet simple assembly of laser-cut stainless-steel parts with strategic voids for ventilation and channels for wiring. It accommodated all electronics and networking hardware used to process, display, and spatialize expression data read from an on-site personal computer (PC). Working collaboratively, Jenny Sabin Studio and Microsoft Research designed and programmed the software architecture for two programs which allowed *Ada* to interface with human sentiment in its environment: a program running on the on-site PC, and a program running on each Raspberry Pi in the Brain Ring. The PC software continually queried data from the network of MSR cameras, which analyzed user facial expressions. These expressions were classified using Microsoft researcher Daniel McDuff's platform, which drove *Ada*. Artificial intelligence algorithms turned these data points into numeric gradients of sentiment, which passed on to the PC in the form of probabilities, represented the program's certainty that a given expression was being observed. (Figure 7)

Drawing synergies with current work at the intersection of data-driven cyber-physical assemblies, digitally knit structures, and textile architecture, *Ada* celebrates, integrates, and materializes AI, affective computing, responsivity, and material performance. However, unlike much of the pioneering work in AI, such as the *BabyX* project by Mark Sagar which seeks to humanize AI through "more symbiotic relationships between humans and machines," *Ada* does not appear lifelike or seek to mimic human behavior.[42] Instead, *Ada* offers subtle and abstract interactions with humans through space, material, and form, it expands our emotional range, and in turn affects the probable sentiment data being collected. The spaces we inhabit influence and partially shape who we are and how we are feeling. How might architecture respond to issues of wellbeing, health, human ecology, and sustainability where buildings behave more like organisms in their built environments? In turn, what role do humans play in response to these changing conditions? In response, *Ada*'s form, software functionality, and its relationship to the physical and social context are conceived to inform one another as a result of a rigorous transdisciplinary design process in which its materials, data, and lighting systems

Figure 7. Participants interacting with *Ada* through the central camera at Microsoft Research in Redmond, WA. Photo by Jake Knapp for Microsoft.

Figure 8. Prototypes for the
Agrivoltaic Pavilion, part of
the *Sustainable Architecture
& Aesthetics* project. Courtesy
Sabin Lab at Cornell College of
Architecture, Art, and Planning,
and the DEfECT Lab at Arizona
State University.

are all programmed. In this way, *Ada* opens up a dialogue around important and pressing issues concerning personal data acquisition and privacy as well as justifiable concerns with AI.

In other projects, similar exchanges between the lab and the studio contribute to dynamic material organizations. The design and engineering of light and energy, for example, are generating sustainable energy futures that are efficient and aesthetically wonderous. (Figure 8) *Agrivoltaic Pavilion* is a current project that innovates in the field of Building Integrated Photovoltaics (BIPV) through computational design and 3D printing for highly customized non-standard filters and panels that result in site-specific non-mechanical tracking solar collection systems. Working collaboratively with Mariana Bertoni, Associate Professor in the School of Electrical, Computer, and Energy Engineering at Arizona State University, and funded by the Grainger Foundation as part of the National Academy of Engineering, this research studies biological and natural systems to understand relationships between morphology, movement and mechanisms, and light and energy. Beginning with biological adaptations including heliotropic mechanisms in sunflowers and the light-scattering structures in Lithops plants, we investigate non-conventional configurations of panels as a means of integrating aesthetics while maximizing energy conversion efficiency. *Agrivoltaic Pavilion* demonstrates the first adaptable system with extremely low greenhouse gas emissions, showcasing the potential of sustainable design for a resilient land use model to provide an integrated approach to food, energy, and water.[43]

This multi-program pavilion for *Agrivoltaics* offers creative design solutions to persistent engineering problems present in current solar technologies. Integrating photovoltaic (PV) systems and infrastructure in early design phases, this proposal aims to inspire widespread integration of sustainability, technology, design, and aesthetics. The majority of BIPV and building-applied photovoltaic (BAPV) technologies forego the opportunity to utilize PV cells as a compelling design element. This project innovates the design and engineering of PV cells through advancements in computation and 3D printing for highly customized bio-inspired filters and panel assemblies that leverage the phenomena, beauty, and performance of light absorption for energy generation. We develop three design methods based on fundamental operations of photosynthesis - heliotropism (orientation), light scattering (filters), and cellular morphological responses (module shape) to varied radiant exposure.

By incorporating collaborative and innovative computational design tools, we can design and engineer PV arrays which respond more systematically to light and its dynamic contexts. Most contemporary sustainable approaches to reduce CO_2 emissions offer technological solutions through sanctioned rating systems such as LEED. While these measures adequately address resource consumption in buildings, our research agenda moves away from resource consumption as a primary paradigm to address the systemic ecology of the built environment over the long term. Importantly, this project and our transdisciplinary design process incorporates the role of aesthetics in generating excitement amongst the general public surrounding the importance of employing sustainable building materials and practices in our private and public domains.[44]

And lastly, my interest in leveraging beauty and nature's inner workings continues in *PolyBrick*, a collaborative project with mechanical, biological, and environmental engineers from Cornell University. Since 2009, the Sabin Lab (and Sabin+Jones LabStudio before it) has developed a research agenda on digital ceramics and 3D-printed ceramic work, focused on the development of nonstandard bricks, tiles, and aggregated assemblies. The *PolyBrick* series, previously exhibited at the Centre Pompidou in Paris (*Imprimer Le Monde* 2017) and the Cooper Hewitt Smithsonian Design Museum (*Beauty* 2016) explores ceramic construction and full-scale fabrication through digitally steered material and process-based investigations.

PolyBrick 1.0 iteratively investigated the production of nonstandard 3D-printed ceramic brick components with designs for mortarless assembly and interlocking joinery based on traditional wood construction.[45] (Figures 9 and 10) The next generation, *PolyBrick 2.0,* advanced computational methodologies in the context of ceramic construction by applying behaviors and principles based on human bone formation to structurally focused finite element analysis models. (Figure 11) By guiding load paths and generating a series of structural rules, the tests generated a fine and controlled gradient of dense structure when needed, and light, porous geometries in areas demanding less load.[46] *PolyBrick 3.0* looked to biology and cutting-edge DNA hydrogel research for developing intelligent and encoded bio-bricks and tiles. Using clay as substrate for DNA glazing, this series highlighted living signatures on ceramic surfaces to facilitate data storage in and around our constructed environments. Designed in collaboration with biological engineers in Luo Labs at Cornell University, this research demonstrated the compatibility with clay as a material host for life and DNA hydrogel materials.[47] (Figure 12)

Most recently, the *PolyTile* series has investigated the development of biofunctional elements within our hybrid tiles. In a continuing collaboration with Luo Labs, we've refined technologies critical to digital fabrication of 3D printed ceramics and nano-bioengineering, including a microscale method for spatially controlling DNA materials and developing specific hydrogels with bespoke functionalities. Smart materials are appearing in varying products to tackle issues of sustainability and efficiency. The Sabin Lab has launched into ceramic-hydrogel composites to innovate multi-functional *PolyTiles* by making them programmable and with structural integrity at the architectural-scale. (Figure 13)

PolyTile 2.0 interrogates the potential of programmable biofunctionalities in our constructed architectural environments through the development of advanced ceramic bio-tiles.

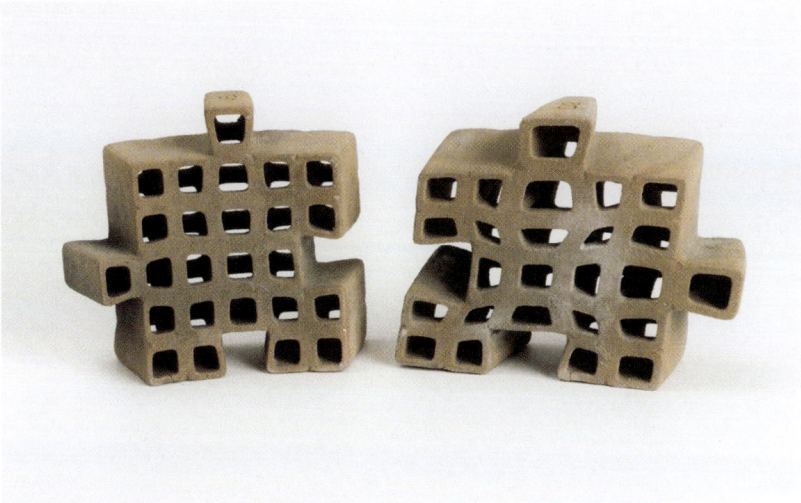

Figure 9. *PolyBrick 1.0* in greenware (pre-fired) stage. Using customized digital tools, low-cost printing materials, and component-based aggregations, *PolyBrick 1.0* develops large scale forms by aggregating interlocking components. Image courtesy of the Cooper Hewitt Design Museum.

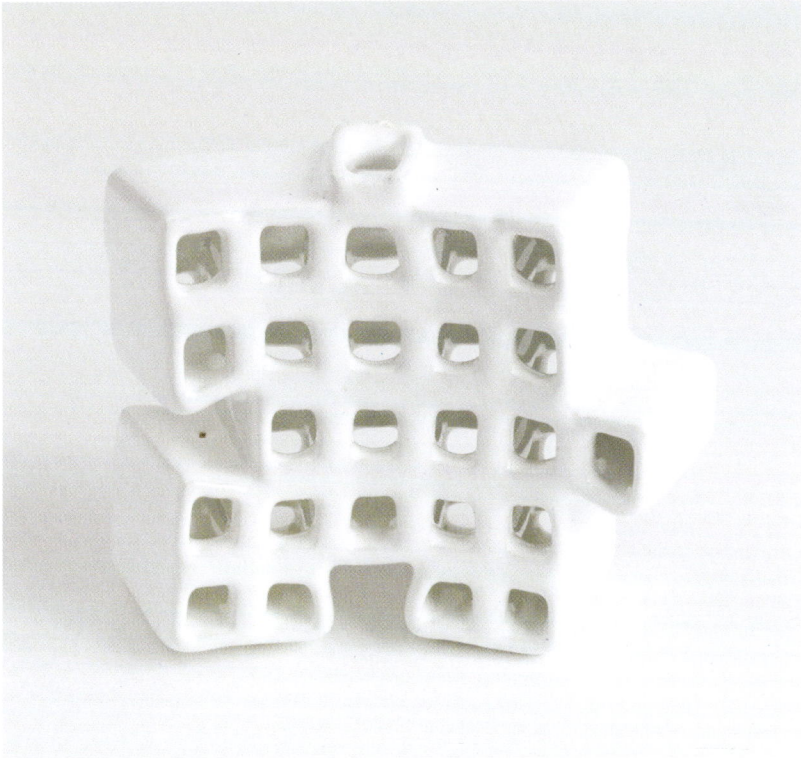

Figure 10. *PolyBrick 1.0* bisque and glaze fired. Image courtesy of the Cooper Hewitt Design Museum.

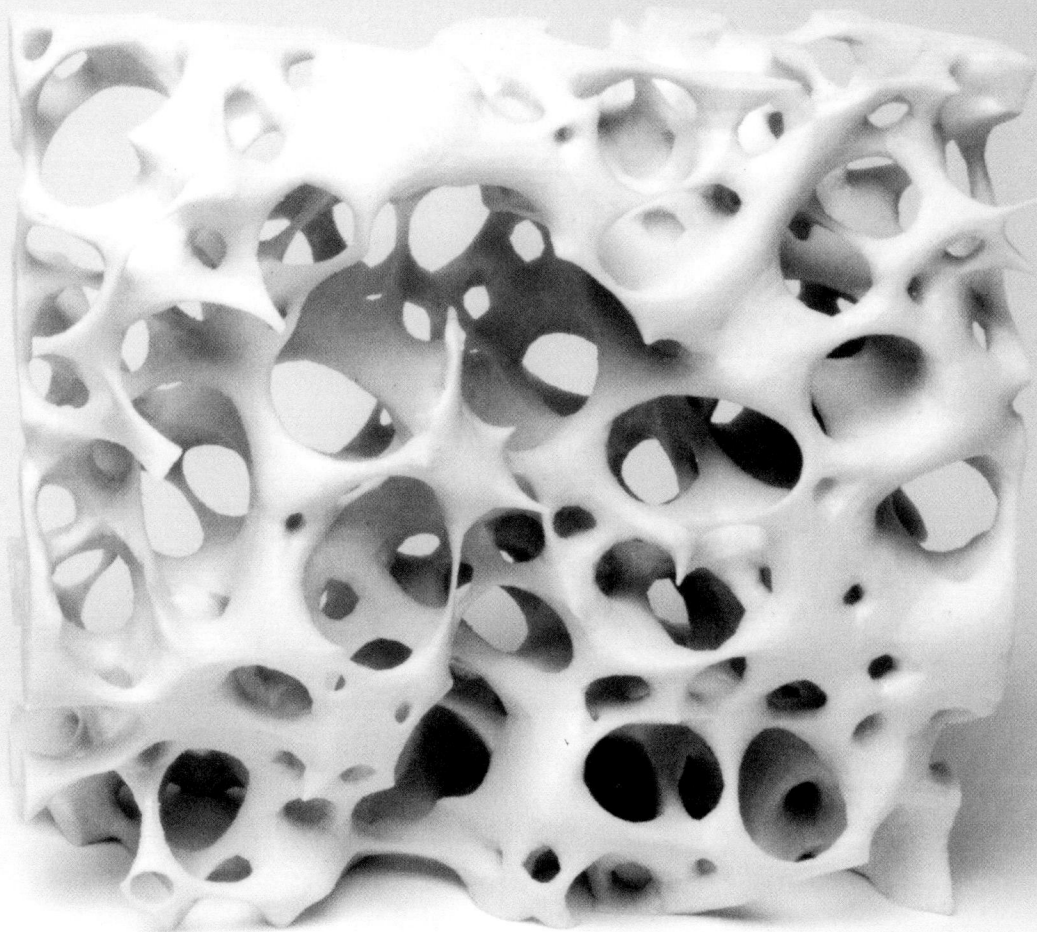

Figure 11. *PolyBrick 2.0* is generated using the rules, principles, and behavior of bone formation. Photo courtesy Sabin Lab, Cornell University.

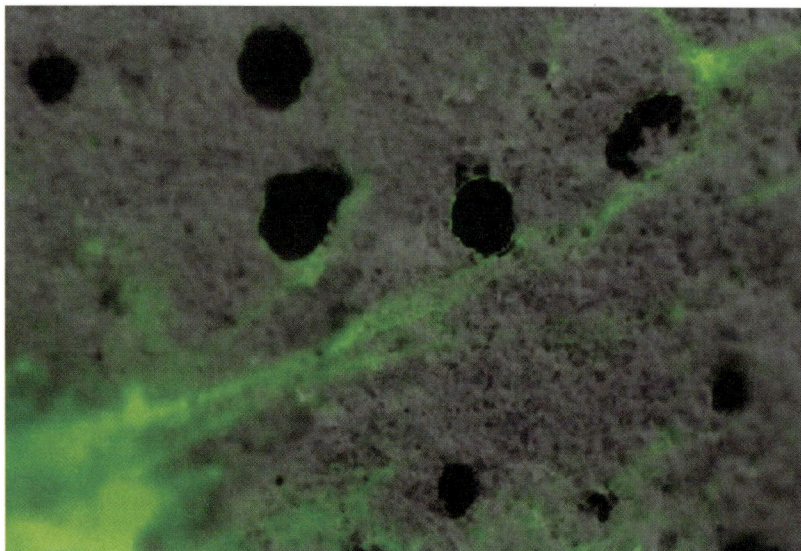

Figure 12. *PolyBrick 3.0* utilizes 3D printed clay, hydrogel, and synthetic DNA. Using DNA hydrogel, a programmable function such as protein expression can be integrated into clay-based material systems to achieve responsive behavior such as fluorescence or light emission. Photo courtesy Luo Lab, Cornell University.

These tiles utilize novel patterning techniques and hydrogel biomaterials to tune surface conditions at the micro- and macroscale. This transdisciplinary work builds upon recent advancements in the fields of 3D-printing, digital ceramics, materials science, bioengineering, chemical biology, and architecture. *PolyTile 2.0* enables designers and architects to implement biofunctionality and microscale patterning with the ability to continuously adjust design iterations across scales and contexts. The refinement utilizes glazing strategies as a directable fluidic device and biocompatible hydrogels as a sensing platform to further developments in responsive built environments. We are developing methods to produce bulk-scale hydrogel materials, stereolithography-based 3D-printed ceramic tiles, and scalable glazing techniques for future building scale applications.[48]

Conclusion

Rapid advancements in robotic additive manufacturing, interaction design, materials science, and DNA nanotechnology are introducing new possibilities for creating nano- to macroscale devices, machines, materials, and architectural elements that can dynamically react to environmental cues and interact with biochemicals and even humans. This paper has discussed both established and unique avant-garde models for collaboration between the architectural design studio and laboratory-based scientific and engineering research. The synergistic, high risk, bottom-up approach across diverse disciplines brings about a new paradigm for constructing intelligent and adaptive materials with applications spanning personalized architectures to adaptive building materials. Importantly, the convergence of these design and engineering approaches enables the co-evolution of design strategies

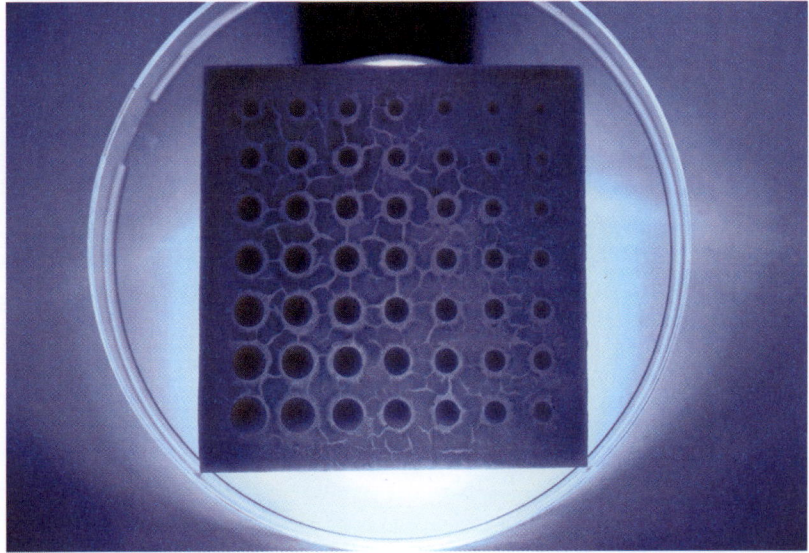

Figure 13. Full scale *PolyTile* fluorescing with unique pattern structures. Photo courtesy Sabin Lab, Cornell University.

and hybrid thinking that are collective, emergent, and context-driven thereby generating adaptive materials and environments that can sense, learn, self-assemble, and heal. Through the lens of three projects and research agendas—that separately explore artificial intelligence and human sentiment, design for light and energy through non-standard photovoltaic components, and parametrically design with DNA for the 3D printing of full scale responsive bio-bricks and tiles—this article explores how transdisciplinary collaboration breaks down anthropocentric and disciplinarily-siloed conditions that have contributed to some of the world's most pressing issues, including the devastation of our planet's climate and ecosystems.

Transdisciplinarity emphasizes evolving and generative knowledge, excessive knowledge, unearthed within a collective meshwork of heterogenous transdisciplinary expertise. As Balibar describes, the process of individuation oscillates between a 'pre-individual' potential and a 'transindividual' excess. Here, what I call a collective meshwork is the nature of transindividualism, an 'ontology of relations' or of Lovelace's "science of relations". Transindividualism transcends gender as a space of human embeddedness, infinitely woven, and facilitated and generated by algorithms, biology, and digital fabrication machines. The success of projects described in this article hinges on the transindividual potential and excessive knowledge that a transdisciplinary 'ontology of relations' generates. Haraway offers a provocative metaphor, the cyborg. I am proposing a transdisciplinary process to generate and engage with the digital materialist space of the cyborg. Transdisciplinary research opens a transindividual paradigm shift, one that is necessary for systemic and collective change across research, pedagogy, and practice.

Acknowledgements

The projects described in this article would not have been possible without the incredible support and participation of my students, research associates, senior designers, and collaborators. For a full list of research and project teams see the individually referenced project papers. *Ada* by Jenny Sabin Studio is a project for Microsoft Research (MSR) Artist in Residence (AIR) program, made possible with the support and participation of the following MSR core team: Technical Fellow and Director: Eric Horvitz; Director of PM and Special Projects: Shabnam Erfani; Principal Research Designer/Fusionist: Asta Roseway; Principal Design Director: Wende Copfer; Principal Electrical Engineer: Jonathan Lester; Principal Researcher: Daniel McDuff; and Partner Director/Ethics: Mira Lane.

Funding

Funding for *Sustainable Architecture & Aesthetics: Agrivoltaic Pavilion* is generously provided by the National Academy of Engineering and the Grainger Foundation. *PolyBrick 2.0* and *3.0* are jointly funded by the Sabin and Luo Labs at Cornell University.

ENDNOTES

1. Portions of this text are extracted and revised from, Jenny Sabin, "Biosynthetic Architecture." Kodalak, Gokhan & Kwinter, Sanford, (eds). *Log 49*, (Summer 2020): 169-81.

2. The Sabin+Jones LabStudio was jointly housed in the Department of Architecture at the Graduate School of Design, now the Weitzman School of Design, and the Institute for Medicine and Engineering at the University of Pennsylvania from 2006-2011.

3. For a comprehensive overview of the Sabin+Jones LabStudio, see Jenny Sabin and Peter Lloyd Jones, *LabStudio: Design Research Between Architecture and Biology* (London and New York: Routledge Taylor and Francis, 2017).

4. Gilles Deleuze, *The Fold: Leibniz and the Baroque* (Minneapolis: University of Minnesota Press, 1993); Detlef Mertins, "Bioconstructivisms," in Lars Spuybroek, ed., *NOX: Machining Architecture* (London: Thames & Hudson, 2004), 360–69; Manuel DeLanda, *A Thousand Years of Nonlinear History* (New York, NY: Zone Books, 1997); Anni Albers, *On Weaving* (Middletown: Wesleyan University Press, 1993); Bernard Cache, "Décrochement," in *Earth Moves: The Furnishing of Territories,* trans. Anne Boyman, ed. Michael Speaks (Cambridge: MIT Press, 1995); Cecil Balmond, *Informal* (London, U.K.: Prestel, 2007).

5. Peter Lloyd Jones, "Context Messaging: Modeling Biological Form," in *Models*, ed. Emily Abruzzo, Eric Ellingsen and Jonathan D. Solomon (New York: 306090, Inc., 2007), 31–38.

6. Sabin, "Biosynthetic Architecture."

7. Sadie Plant, *Zeros + Ones: Digital Women + The New Technoculture* (New York, NY: Doubleday, 1997); Donna Haraway, *Crystals, Fabrics, and Fields: Metaphors That Shape Embryos* (Berkeley, CA: North Atlantic Books, 2004).

8. Alexander Refsum Jensenius, "Disciplinarities: Intra, Cross, Multi, Inter, Trans," 12 March 2012, https://www.arj.no/2012/03/12/disciplinarities-2/; Marilyn Stember, "Advancing the social sciences through the interdisciplinary enterprise," *The Social Science Journal* 28:1 (1991): 1-14.

9. Tanya Augsburg, "Becoming Transdisciplinary: The Emergence of the Transdisciplinary Individual," *World Futures: The Journal of General Evolution* 70 (2014): 234.

10. J. T. Klein, "Prospects for transdisciplinarity," *Futures* 36 (2004) (4): 515-26.

11. Augsburg, "Becoming Transdisciplinary," 234.

12. Ibid.

13. Ibid.

14. Ibid., 235.

15. Ibid.

16. Étienne Balibar, "Philosophies of the Transindividual: Spinoza, Marx, Freud," *Australasian Philosophical Review*, 2:1 (2018), 5-25.

17. Ibid., 6.

18. Ibid.

19. Ibid.

20. Mertins, *NOX: Machining Architecture*, 360-69.

21. Ibid., 369.

22. Ibid., 362.

23. Ibid., 369.

24. Here, I am interested in how the integration of weaving, technology, and biosynthetic design can inform and learn from the ethical and sociopolitical dimensions and intricate depths of contemporary feminist theories such as cyberfeminism and the cyborg to offer alternative analogs, through architecture and dynamic space, that are post gender and transindividual. The focus is not on gender or identity, but on the emergent space and form resulting from a biosynthetic design process where projects are developed and generated by data emanating from human bodies such as cells, sentiment, and biodata. Thank you to Sylvia Lavin for encouraging me to address these topics during the Q&A session following my public lecture in early February 2020, at the School of Architecture at Princeton University. The 2020 spring lecture series, organized by Dean Monica Ponce de Leon, was themed around issues of technology and design.

25. Plant, *Zeros + Ones: Digital Women + The New Technoculture*.

26. Ibid., 5-23.

27. Ibid., 18.

28. Ibid.

29. Sadie Plant, "The Future Looms: Weaving Women and Cybernetics.," *Body and Society* 1, no. 3-4 (1995): 45-64.

30. Ibid., 63.

31. Stacy Gillis, "Neither Cyborg nor Goddess: The (Im) Possibilities of Cyberfeminism" in S. Gillis, G. Howie, and R. Munford, eds., *Third Wave Feminism: A Critical Exploration*, (Palgrave Macmillan; 2nd edition (April 17, 2007): 170.

32. Ibid., 170.

33. Étienne Balibar & (Translated by Mark G. E. Kelly), "Philosophies of the Transindividual: Spinoza, Marx, Freud," *Australasian Philosophical Review*, 2:1 (2018), 6.

34. Plant, *Zeros + Ones: Digital Women + The New Technoculture*, 20.

35. Donna J. Haraway, "A Cyborg Manifesto: Science, technology, and Socialist-Feminism in the Late Twentieth Century," in *Simians, Cyborgs, and Women: The Reinvention of Nature* (New York: Routledge, 1991): 149.

36. Stacy Gillis, "Neither Cyborg nor Goddess," 169.

37. See Albers, *On Weaving*. Portions of this text are extracted from Jenny E. Sabin, "Matrix Architecture," in *Inside Smart Geometry*, Terri Peters and Brady Peters, ed. (Chichester: John Wiley & Sons Ltd, 2013), 60–71.

38. For further background on this work, see Ferda Kolatan and Jenny E. Sabin, *Meander, Variegating Architecture* (Exton: Bentley Publications, 2010).

39. Ibid., and Jenny E. Sabin, "Matrix Architecture."

40. For a comprehensive overview of *Ada*, see Jenny Sabin, John Hilla, Dillon Pranger, Clayton Binkley, and Jeremy Bilotti, "Embedded Architecture: Ada, Driven by Humans, Powered by AI," in *FABRICATE 2020: Making Resilient Architecture*, ed. Jane Burry, Jenny Sabin, Bob Sheil, and Marilena Skavara (London: UCL Press, 2020), 246-56.

41. Jenny E. Sabin, Dillon Pranger, Clayton Binkley, Kristen Strobel, and Jingyang Leo Liu, "Lumen," *ACADIA 2018 Recalibration: On Imprecision and Infidelity*, Proceedings of the 38th Annual Conference of the Association for Computer Aided Design in Architecture (October 2018), 445–55.

42. A. Vance, "Mark Sagar Made a Baby in His Lab. Now It Plays the Piano," *Bloomberg Businessweek*, 7 September 2017, https://www.bloomberg.com/news/features/2017-09-07/this-startup-is-making-virtual-peoplewho-lookand-act-impossibly-real.

43. Manuscript in preparation by Jenny Sabin, Mariana Bertoni, Alexander Htet Kyaw, Begum Birol, Jack Otto, and Angel Langumas, "Sustainable Architecture and Aesthetics (SAA): Agrivoltaic Pavilion", (forthcoming).

44. Jenny Sabin, "Transformative Research Practice: Architectural Affordances and Crisis," *Journal of Architectural Education*, Taylor & Francis, (2015) (69:1): 63-71.

45. Jenny Sabin, Martin Miller, Nick Cassab, et al, "PolyBrick: Variegated Additive Ceramic Component Manufacturing," (ACCM). *3D PRINTING* 2014; 1, no. 2.

46. Eda Begum Birol, Yao Lu, Ege Sekkin, Colby Johnson, David Moy, Yaseen Islam, Jenny Sabin, "POLYBRICK 2.0: Bio-Integrative Load Bearing Structures." *ACADIA* 2019 (Ubiquity and Autonomy): 222-33.

47. David Rosenwasser, S. Hamada S, D. Luo, et al. "POLYBRICK 3.0: Live Signatures Through DNA Hydrogels and Digital Ceramics," Special issue of *International Journal of Rapid Manufacturing* – Additive Manufacturing in Architecture 7, no. 2, 3, (2018): 205-18.

48. Portions of this text extracted from Viola Zhang, David Rosenwasser, & Jenny E. Sabin, "PolyTile 2.0: Programmable microtextured ceramic architectural tiles embedded with environmentally responsive biofunctionality," *International Journal of Architectural Computing* (July 2020).

EXPANDING THE COMMUNITY OF LIFE:
A NEW MATERIALIST PERSPECTIVE OF BIODESIGN

RACHEL ARMSTRONG

ABSTRACT

Biodesign is a training-ground for the concept it seeks to interrogate—biology. Biodesign lends itself to radical contextualization, evolution, and metamorphosis, as well as to occupying a contestable political status. The highly situated nature of life renders biodesigners both producers and translators of their contexts, with the design process shaped by constraints and freedoms that are unique to physics and chemistry. This calls for the urgent and fundamental reassessment of our assumptions about, and our relationship with, all living things, particularly material entities embedded in the politics of life through the concepts of *bios* (a being entitled to a political existence), and *zoë* (an entity expressing bare life, with no political rights). This paper speaks to forms of re-empowerment made possible through a practice of biodesign that constructs new values and material relationships. It discusses the limits of the modern notion of *bios* in relation to my work with 'living' chemical agents called 'protocells', which possess *zoë* but are not technically 'alive'. It also explores the complex nature of *bios* in the context of the nonhuman and at a time of significant advances in biotechnology. This paper details the European Union funded project *Living Architecture*, which combines microbes (a form of *bios*), with artificial intelligence (an agent of *zoë*), with machines and human waste streams (partly composed of biotic agents-microbes) to produce a holobiont entity—or "cyborg." Designed to prove the principle that our indoor habitats can perform 'housework' alongside our activities of daily life, it changes the relationship between human and technology by raising complex questions about the political status of that which we design and our duties of care for them.

+ living architecture
+ protocell
+ bios
+ zoë
+ ethics

Figure 1. Unnatural history: Time-based protocell topology, original footage from Centre for Fundamental Living Technology, Southern University of Denmark, Odense, 2010, micrograph x 10 magnification by Rachel Armstrong, composite image by Simone Ferracina, originally published in Rachel Armstrong, *Liquid Life: On non-linear materiality*, (New York: Punctum, 2019).

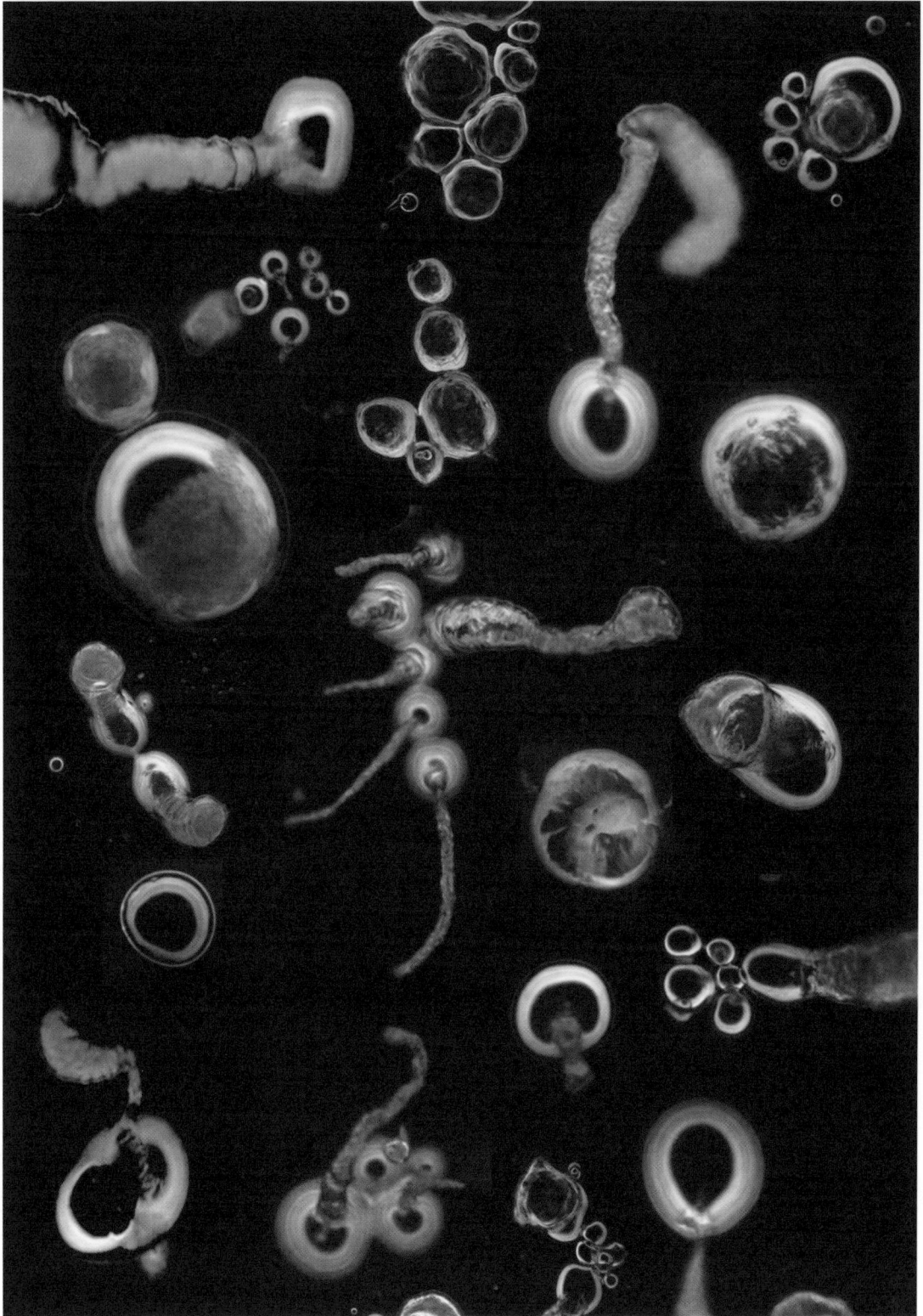

Introduction

The persuasive claim of biodesign is that it integrates design with biological systems, enabling practices to achieve better ecological performance.[1] This paper challenges and complicates the current curatorially-derived definition of biodesign by taking a new materialist perspective that—rather than focussing on the technological feasibility and aesthetics of the designed 'object'—asks searching questions about what *designing with biology/biotechnology* actually entails. Implicit in its etymology, the *bios* that forms the subject of biodesign and biotechnology implies a particular politics of 'life' at the heart of its design practice. Modern notions of the living, focused on the study of individuals, present us with blind spots with respect to the true bio-character of life: deficiencies that prevent us from achieving a more complete understanding of, and engagement with, an inclusive community of subjects, systems, and networks that make the world lively. An expanded territory incorporates the realm of bare life, or *zoë*, which in articulating the limitations of biodesign practices can develop more consciously political (concerned with the redistribution of power relationships) and ethical engagements with life. Establishing a "zoë-centred egalitarianism" is in keeping with new materialist concerns (agency, power, subject formation, and co-constitutive productions of reality) that seek to develop new protocols and paradigms for design in an ecological era.[2]

Ontology of Biodesign

The term biodesign was popularized in architecture and design in 2012 when William Myers and Paula Antonelli curated the exhibit *BioDesign: Nature + Science + Creativity*. Its accompanying publication outlined an emerging practice, which innovatively altered and incorporated living organisms, or tissues, in the design process.[3] Gathering together and categorising an assemblage of approaches that harnessed living materials and processes—from Kate Orff and SCAPE's *Oystertecture*, Philips Design's *Microbial Home*, Julia Lohmann's microbial mural *CoExistence,* Suzanne Lee's microbial cellulose garments *BioCoutureTM*, and the Wyss Institute's *Lung on a Chip*—the items in this cabinet of bio-curiosities collectively articulated their relation to the history of architecture, art, and industrial design.[4] They also intonated their implications for the future, in their "sustainable approaches to building and manufacturing," and in their unified purpose for greater collaboration between designers and biologists, particularly through developments in biotechnology.[5] Today, biodesign continues as a diverse practice that assimilates biomimicry agendas and forges radically interdisciplinary practices that engage with organic substrates, as in the work of the Stanford Byers Centre for Biodesign and Kent State's Biodesign and Environmental Art.[6] And the field continues to grow.

Emerging at thresholds and intersections between disciplines, the ontology of biodesign must, however, exceed its curatorial definition in representation, an idea notable in manifold explorations in the arts, humanities, and sciences—from *organicism* via Karl Ludwig von Bertalanffy's (1901-1972) *systems biology*; through *biotechnology* that considers cells as molecular machines whose parts can be rationally engineered; to *bioart* which raises questions about identity and ethics in an age of biotechnology; to *new materialism* that considers the agency of matter beyond its human relations.[7] Even in her introduction to the exhibit catalogue titled "Vital Design," Antonelli observed that "the implications of every project reach far beyond the form/ function equation and any idea of comfort, modernity, or progress. Design transcends its traditional boundaries and aims straight at the core of the moral sphere, toying with our most deep-seated beliefs".[8] In this way, the definition of biodesign of interest to this paper surpasses the aesthetic dimension to embrace its larger philosophical context.

Bios itself, comes with a set of beliefs. Consolidated in the term 'biology' are the Greek term βίος, or *bios*, and the suffix *logy*, denoting 'science of', 'knowledge of', or 'study of'.[9] Introduced independently at the turn of the nineteenth century by a number of scientists including German naturalist Gottfried Reinhold Treviranus (1776-1837) and French botanist Jean-Baptiste Lamarck (1744-1829), biology sought the understanding of all living bodies by specifically decoding the functional processes of each organism. This was in stark contrast with its forerunners, natural history (based since antiquity on 'observation') and zoology (concerned since the twelfth century with the 'science of animals'). Importantly, biology included the study of the human body through medical investigations as well as the vital processes of plants and animals that could be used as *models* for understanding human physiology. Ambitious *to exploit* the naturalists' wealth of description, biology became an experimental practice that instrumentalised the processes and bodies that comprised the living realm. In doing so, it facilitated a highly charged political history, grounded in privilege (access to a political existence) and imperialism (encoded in hierarchies of order both within and across species).[10]

It is, therefore, incumbent upon practitioners of biodesign to consider their specific terms of engagement with the living realm, as present protocols for designing with lively matter are *always already* framed by the deep, divisive politics of the 'biological' that shape the contemporary view of 'life'. By adopting a more than human view of the living realm, biodesign can assert an inclusive ethics whose political engagement upholds an ecological approach to the community of life, and whose explorations and engagement with *zoë* and new materialism can generate qualitatively different kinds of outcomes than the exploitation of living things in service to the human. The work of the designer, therefore, goes beyond theory to work alongside and *along with* the living realm. This is an exhaustingly nomadic and experimental *task that emerges* as an assemblage of concepts, material expressions, and propositions that collectively constitute an agile, transferrable knowledge-making toolset embodying the tactics of 'life'. Rather than taking a predetermined approach that establishes foundations for validating a theory, constituent ideas are operationally employed so they can exceed their role as bio-technical instruments and become epistemic agents, capable of producing present and future knowledge.

Beyond Biology

At the heart of biodesign lies the question of how decisions made when designing enable us to better relate to, take care of, and re-distribute power towards the very elements of this world that make this planet lively—all the while reducing the influence of those that choose to harm it. The inequalities that characterise the biological sciences must be identified to challenge existing anthropocentric assumptions about the living realm, so that an appropriate understanding of *bios* can be developed. Such an inclusive and ecological approach first requires an understanding of those active agents that are excluded from the biological realm and what the consequences of their omission might be.

The unfolding viral COVID-19 pandemic of 2020-21 spotlights the limitations of biological paradigms as exclusive means for discussing the community of life. Historically regarded as alternatively poisons, life-forms, and biological chemicals, viruses today are considered members of a poorly defined zone between living and non-living matter. Requiring a living host cell to complete their replication, viruses lack an autonomy that frames current definitions of what it means to be 'alive'. Despite this non-living status, their significant impacts on human health and the global economy means they are, nonetheless, considered biological (organic) entities. Such generosity is not, however, applied to other ambiguously 'living' agents like dynamic chemical droplets, or 'protocells', which possess remarkably lively characteristics with organic components.[11] With no nucleic acids (the chemical coding system of biology), and with little to no impact on human, economic, or social activity, they are not considered biological. Defined as chemical entities, they highlight the otherwise anthropocentric nature of biology that ontologically discriminates against a range of indeterminate agents of little concern to the biodesign portfolio. For clarity, protocells *are included* in Myers and Antonelli's curatorial endeavour, presumably as technical agents (a key theme) and not as biological substrates.[12] This raises an important question about the nature of bio- in biodesign and presents us with an opportunity for radical design actions.

Designing Zoë

To enable biodesign practices to become catalysts for radical change in our relationship with nature, late twentieth century concepts and design principles must be critiqued and appropriately transformed through practices that embrace a complex, inclusive ethics. Adopting an expanded relationship with a wider community of life than *bios* infers, biodesign can challenge essentialist assumptions by accommodating exceptions, monsters, and aliens: entities whose status within the broader community of life is questionable.[13] Simultaneously, an alterbiopolitics is needed to inform the curation of alternate power structures and organisational relations, so that biodesigners can consciously relate to and appropriately deal with *all kinds of lively matter.* By acknowledging the fundamentally political nature of *bios,* whose ontology implies the right to a political existence, the human biodesigner is no longer the sole author of a process but a choreographer and advocate who works *along with* an expanded community *of* biodesign subjects and within co-constituting communities of agents engaged in many co-creative acts.[14] By summoning the life force that is *zoë,*—a classical concept that denotes the barest of life with no right to a political existence—power can be extended and transferred to agents that are otherwise excluded from the community of life, creating a more inclusive platform for change and ethical action.[15] Positioning biodesign as a form of political action radically affects how biodesign is imagined, executed, and sustained; it also alters its value and consequences. Such a profound transition in the recognition of the agency and status of these co-designed entities necessitates an appropriate ethics for the redistribution of power relations in the production process. While there are many risks in working with animate matter—as 'life' is a lively and wilful force, and where subjugation of all participating entities risks their disobedience—the potential benefits are immense. No less is needed for reversing the ecocidal impact of human development and for restoring an overall liveability through multiple, ongoing, concerted acts of re-enlivening whose navigational framework engenders a *worlding* process. Always inhabited, each 'world' is not a specific object, but an agentised, immanent *whole,* receptive to external influences and cross-infected by other worlds.[16] Both *born and made,* these worlds are iteratively generative. They are enriched by their co-constitutive agents and as one world collapses, another is ready to take its place.[17]

Dynamic Droplets

Neither biology nor zoology includes the study of all lively things, whose current membership is delimited by Charles Darwin's cellular *Tree of Life.*[18] Importantly, the household of nature has not always been confined to the organismal kingdoms of animals and plants. Carolus Linnaeus (1707–1778), for example, extended it to include the mineral kingdom, or *regnum lapideum.* Exhibiting some of the qualities of the living realm, inorganic objects such as stones, forms of bedrock, soil, ores, and minerals could grow through aggregation, accretion and crystallisation, and in so doing invoke a different set of concepts with which to discuss their lively nature. Unfortunately, during the eighteenth century the principles of rock formation were soon assimilated within the emerging science of geology which, concerned with the growing commercial value of rocks and minerals, resulted in their treatment as inanimate, eternal, and exploitable things. Recent evidence, however, suggests that the mineral kingdom is not separate from but has co-evolved *along with* life, where up to two thirds of the more than four thousand known types of minerals on Earth are directly, or indirectly, produced through biological activity. This raises questions about our assumptions regarding 'liveliness' and where a view of the mineral realm co-constitutes the potentiality of the living world.[19] While human activity such as garbage dumps, electrical power grids, and industrial toxins alter the character of minerals, these rocks in turn act upon us, being vitally incorporated into our biochemistry and reclaiming their kin from our tissues when we die.[20] Working at different scales and temporal rhythms than organismal life, minerals shape the world's metabolism by absorbing gasses like carbon dioxide and oxygen into their substance. When considered as part of the community of life, these *active* processes can help shape our transition towards an ecological era of design.

Since the 1990s, I have critically explored intersections between the living world and design practice. Challenging notions of an assumed *quality of life* that is conferred by a *specific kind of (hu)man*—the *Anthropos*—I have explored how the agency of living systems (cellular and ecological) can assist the design process. Invoking another agency as collaborator requires an attitude that is conducive to optimising possible relationships between human designer and participating agent. Establishing a *Cytoplasmic Manifesto*, I sought an experimental, testable approach for lively matter that circumvented genetic control of cellular operations and thereby evaded an already politicised reading of causes and effects within the design processes.[21] I attended the Artificial Life XI Conference in 2008 where, during his keynote entitled "Artificial Life is Dead," Takashi Ikegami presented the movie of a dynamic chemical droplet produced by a simple chemical recipe—adding a single drop of an alkaline solution (inorganic solution) to an oil medium (organic liquid) at room temperature. The resultant product that formed at the interface of the droplet (a 'soap' skin—an organic substance) started to grow, creating continuous changes in the physics of the droplet (including surface tension and drag), while exhibiting characteristics of a non-linear, dynamical material system.[22] I watched as the chemical body, only a few millimetres in diameter, started to move and finally, shed a skin of accumulated product—seemingly wriggling free from an amniotic sac, before it moved off again from view with renewed vigour. I realised such strikingly lifelike agents could help interrogate and realise my preferred principles of practice.[23]

Since dynamic droplets have not yet been observed to form spontaneously in nature, to observe them, they must be created by mixing highly reactive chemical fields such as oil and alkali.[24] The first person to have observed their striking liveliness was the zoologist Otto Bütschli (1848-1920), who on mixing a solution of potash with olive oil watched the chemical fields of a single droplet of alkali break up in the oil field to form tiny beads of moving droplets. To him, they looked exactly like simple, single-celled life forms known as protozoans, or amoeba.[25] Their motion was so remarkable and convincing that he considered them to be a form of organic, artificial life. This was consistent with his belief that biology was but a more complex form of chemistry, and in this experiment, he was the witness to a biogenesis-like process. However, Bütschli droplets, like other forms of 'protocell', have none of the characteristics of biological systems besides a powerful metabolism and, while highly convincing in their movements and explorations of the medium in which they are produced, do not meet the formal criteria for being 'alive'.[26] Operating at the limits of the definition of life—experimentally defined by information, container, and metabolism—these dynamic chemical droplets do not fulfil the classical definition of *bios*, but arguably, are an expression of *zoë*. They may be quite alien to the make-up of biological organisms but behave in ways that are extremely lively even under extreme conditions that are hostile to 'life', such as environments with very low pH.[27] (Figure 1)

Once formed by bringing oil and alkali into proximity, it is possible to influence the behaviour of dynamic droplets by altering their surroundings using a range of chemical gradients and mineral ingredients. Within these laboratory environments, such as that of a Petri dish, droplets 'compute' and orchestrate specific responses to physical and chemical challenges by moving around, interacting with other agents, or leaving highly structured residues. The resultant, surprisingly complex behaviours spotlight the unique agency of dynamic droplets, where unfolding events under laboratory conditions conjure a vision of a liquid life before biogenesis in simple movements of droplet bodies and in structured configurations and population-scale assemblages. Exceeding the classical expectations of the material realm through their nonlinear materiality, they articulate the strangeness and potentiality of the living realm through the oddly relatable traces they leave behind in the Petri dish, seeming at some stages like fields of ice and fire, and at others like roses and oysters.[28]

Future Venice (Figure 2) and *Hylozoic Ground* (Figure 3) are projects I completed in 2012 and 2011 respectively; both of which arose from explorations in the fundamental qualities of living systems. I applied the properties of dynamic droplets in actual contexts, including their ability to move around in their environment and to make a product (which formed the basis of the 'smart' chemical reef formation in *Future Venice*, trialled in tanks along the lagoon side with Red Bull, for creating a protective casing for the city's wood piles so they would not rot on exposure to the air) and to change colour (which enabled the design of attractive carbon-dioxide sensing 'organs' for the cybernetic *Hylozoic Ground* installation by Philip Beesley). In both projects, living systems were tools for design; tools that included metabolism, self-movement, growth, and environmental sensitivity.[29]

Figure 2. *Future Venice*: View of mineral accretion of woodpiles, 2009. Rendering by Christian Kerrigan.

Like viruses, the exact status of dynamic droplets in the community of life is problematic. While mineral, they are indisputably lively, and their behaviour can be optimised through attention to their liquid infrastructure whose nutrients provide food and energy, remove waste, and establish sufficiently supportive ambient conditions that enable them to complete the work of 'life', that is to move, metabolise, relate, grow, and die. Through their 'constructedness', the complexity of their movements, their sociability, and organic likenesses, dynamic droplets stir empathy in the human collaborator through their recognisably life-like antics, raising ethical questions of one's obligation of care towards them. Such evocations are also present in developing microbial cultures, hatching fertilised eggs from incubators, resuscitation, and many other kinds of engagement associated with the act of helping *bring something to life*. Since there is no universally agreed to definition of life, even as technical agents, dynamic droplets warrant appropriate consideration—especially as they cannot exist without human participation. As noted by Bruno Latour in his article "Love your monsters," it is the case that "our sin is not that we created technologies but that we failed to love and care for them."[30]

An Inclusive Ethics of Care

Appreciating the agency of nonhumans requires decentring the human which, instead of minimising our presence, extends it by becoming semi-permeable to the influence of nonhumans. This enables us to pay close attention to new relations and bodies all around us through an ecology of inter- and intra- actions. In this way, nonhumans are not the *other*, nor do they require our permission to act. Such an ecological ethics establishes a more-than-human approach to biodesign by facilitating appropriate attachments through our relationship with assemblages of participatory nonhuman agents. Adopting an applied ethics in my design practice, I examine the (relational) status of lively agents within a process that acknowledges their contributions, experimentally interrogates, and appropriately socialises them within an ecological framework that includes the human designer. Decentred from human agency, acts of care become the axis around which ethics become possible. Assuming an ecological baseline, a flat ontology, and principles of equity between actors, decisions are co-constituted through the care and attention paid to the "natural history" of participating agents—both individually and collectively—developing a choreography of exchange to co-produce negotiated effects. Informed by scholar Maria Puig de la Bellacasa's three dimensions of care—affect/affection, labour/work, and ethics/politics—the care produced by an ecology of actants also draws on Joan Tronto's notion of a "generic doing of ontological significance... [that] includes *everything that we do* to maintain, continue and repair 'our world' so that we can live in it as well as possible."[31] Reformulated through a framework of relational actions, and activated through critical scholarship on "technoscientific knowledge production" and experiential research with "natureculture practices," de la Bellacasa's new interdisciplinary theory of care builds on Tronto's observations to neither discharge human responsibilities nor "paralyze[s] our ethical imagination."[32] Such an experimental approach frames the concept of care as a situated and committed action that strives to sustain our current world at a time of ecological crisis, while remaining open to new constituencies and political stakes. Specifically, de la Bellacasa uses her experience in permaculture practices to explore an ethical approach where everyday experiences of interdependency—such as gardening—are integral to a shared ethics, expressed through multiple acts of care. These spontaneous acts do not require instruction but emerge from the mutuality of community relationships that provide spaces for arbitration. This means that the status of any specific agent within the design process can dynamically alter and elicit more appropriate responses and attachments.

Such symbiosis (living along with) is not, however, unproblematic. Inherent in all ecosystems are injustices, negations, and exclusions where not all presences contribute constructively within an ecosystem of agents, and where some may pose irreconcilable difficulties, discrepancies, or conflicts. Such an experimental ethical approach towards biodesign enables layered responses and dynamic tactics, where day-to-day acts of bio/eco-logical violence do not go unnoticed or ignored. Facilitated by conditions of care, and paying attention to individual and collective agent actions, the ecological ethos builds towards a condition of overall, rather than absolute, thriving. Operating within a theatre of potentiality, the many co-constitutive acts of an ecology of actants have the potential for *worlding*, where the distribution of power plays a critical role in the overall dynamics and persistence of each world.

Metabolic Negotiation and Power-sharing

Classically speaking, *bare life* alone or *zoë* lacks a political existence.[33] However, within an ecology of actants, a distributed and heterogeneous form of co-empowerment emerges through metabolism by the formation of bio-chemical hubs—centres of intense metabolic activity that forge dynamic relationships and material connections with other active bodies. Created through proximity, frictions, and contingency, co-contributing agents establish rich networks of chemical exchange that mutually reinforce each other to form the cycles of life and death that characterise Earth's lively materiality. Comprising an everyday practice and ethos, such co-constitutive actions of caring and sharing flow through all animate matter, foregrounding their collective ethical obligation towards maintaining a lively world. Metabolism is not something that is owned by, or kept within, an agent but is exchanged as a currency between various ecologies of actants. Such metabolic economies are visible in fields of dynamic droplets, where the power relations can be viewed through dynamic material traces—behaviours and chemical unveilings of the potentiality of matter. While contributing

Figure 3. Protocell Flasks designed for *Hylozoic Ground* installation by Philip Beesley, Canadian Pavilion, Venice Architecture Biennale, 2010. Flask and photograph by Rachel Armstrong.

to the collective endeavour of material transformation, but unlike 'life', laboratory-based dynamic droplets have no way of giving back to the effectively closed system, as their environment is too rarefied and their metabolism too simple to invite diverse relationships. Without the capacity to connect to other metabolising systems, or chemical hubs, their existence is not sustained, lasting briefly from seconds to minutes. Highlighting the importance of a rich environment that enables the continual, generous exchange, and transformation of matter through metabolic processes, the short lifespans of dynamic droplets draw attention to the importance of the work of care and valued exchange of particular kinds of metabolic substances needed to sustain life. This alterbiopolitics—an ethics of collective empowerment—is conferred by the capacity of each actant to metabolically recycle matter back into the exchange of life through diverse networks, in accordance with the negotiated needs and interdependencies of populations with differing bodies.

For *bios* and *zoë* this happens at different temporal scales—during the life of a specific organism and within its geostory, respectively. For the latter, this occurs through a process of decomposition,

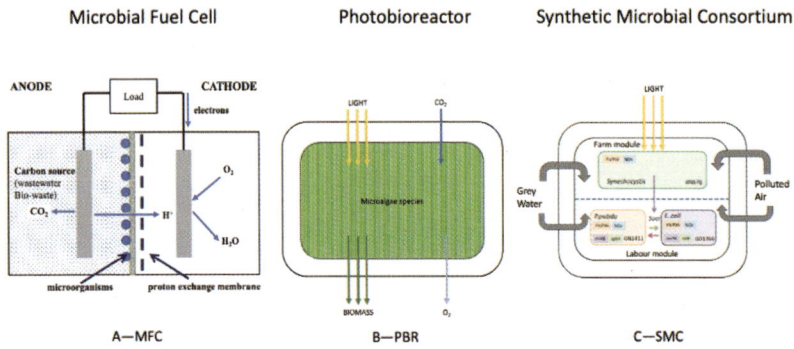

Microbial Fuel Cell   Photobioreactor   Synthetic Microbial Consortium

A—MFC   B—PBR   C—SMC

Figure 4. Diagrams of different bioreactor types. *Living Architecture* project, 2019. Diagram by Rachel Armstrong.

Figure 5. Bathroom. *Living Architecture* project, 2019. Rendering courtesy of Liquefier Systems Group for the *Living Architecture* Project.

which generates new organic molecules that in turn influence the geological transformation of the world's minerals. These complex interactions and interdependencies are highly distributed and not immediately perceptible across the human lifespan and so we are largely unaware of these 'invisible' occurrences. Asserting our irreducible interconnectedness with the Earth's dynamic planetary systems, the alterbiopolitics of life as zoë ensures that no metabolising agent—no matter how slow its exchange—is without significance, even within the mineral realm. From thresholds of recognition to the construction of power relations and their inclusive participation within the design process, an alterbiopolitics of matter can be elicited through biodesign practices. This is not a given, but a carefully navigated and negotiated process that is based on iterations, communication, care, and attention invoking a system of governance founded on sensitivity, feedback, and commitment. Multicellular animals coexist alongside unicellular ones and potent chemistries keep metabolising systems off-balance, so they continue to share and exchange with the living what Latour calls a "Parliament of Things," facilitated and recorded by the rocks.[34]

Living Architecture

The *Living Architecture* project (2016-2019) is a freestanding, next-generation, selectively programmable bioreactor composed of integrated building blocks of microbial fuel cells (MFC), an algae photobioreactor (PBR), and a synthetic microbial consortium (SMC), which also function as standardized building segments—or bricks.[35] (Figure 4) Each unit constitutes a unique type of *ideal home* unit for different kinds of microorganisms that are sustained by feeding them with liquid domestic waste—namely, urine and grey water. The circular metabolic exchanges of the *more-than-human* 'inner life' of the apparatus are brought into meaningful proximity with people at the interface of kitchen and bathroom, which in a future context acquires its own autonomy. (Figures 5 and 6)

This enabling environment encourages reciprocal exchanges between electrical, physical, and chemical interfaces to form a kind of metabolic trading system within an ecology of participating actants—human, microbe, waste, electrons, apparatus, and artificial intelligence. Here, just by using carbon dioxide and sunlight the system can produce electricity, clean water, reclaim phosphates from detergents, remove pollutants from the air, make biofertilisers and generate different kinds of biomass. Discursively, *Living Architecture* can be rationalised as a type of building infrastructure with instrumental value; but it is so much more than this, as all of these outcomes arise from surprising, complex, and technologically-mediated conversations with microbes that operate through chemical languages.[36] These highly specific exchanges can be further spatially sequenced and combined to generate different kinds of synthetic metabolisms that organise microbial populations into a range of programmable *metabolic apps.* To create an expanded portfolio of metabolic design tools for the microbes, the software *Doulix* is used to extend the capabilities of wild-type species by increasing the available range of chemical 'words', by designing plasmids that contain synthetic nucleotide sequences, and then by *infecting* the microbes with them.[37] Such transfer of genetic material is a natural process and part of material language-making activities for bacteria that habitually exchange information as forms of *currency*. These are more than semantic/symbolic exchanges for they can empower genetic material to acquire antibiotic resistance. Occurring naturally during a form of bacterial mating called 'conjugation', gene transfer also happens when cells are infected with viruses, or bacteriophages—although the latter may kill their host, bacteria have developed strategies to variably manage, mitigate and accept this. And by developing different populations of genetically modified 'workhorse' organisms—namely *Synechococcus elongatus, Escherichia coli* and *Pseudomonas putida*, which do not collaborate in nature—different kinds of cooperative work are invited in the design and operation of *Living Architecture* through the construction of *farm* and *labour* modules.

The apparent division of labour in the structure of the bioprocessor arose from the microbes needing very different worlds in which to carry out their novel functions. Drawing inspiration from the established two-chamber design of the MFC (comprising an anaerobic cathode and an aerobic anode capable of exchanging metabolites across a semi-permeable membrane), a specially constructed synthetic bioprocessor was developed creating two different environments that enabled novel metabolic exchanges across a semi-permeable membrane. On one side of the membrane the farm module provided a light-filled environment that stimulated an *agricultural crop* of modified algae, *Synechococcus elongatus*, to over produce sugar as the outcome of their biochemical work. On the other side of the membrane a dark environment promoted the growth of modified *Escherichia coli* and *Pseudomonas putida* which expressed new genes in the presence of lots of energy providing sugar. Enabling very different metabolic processes to simultaneously co-exist across a gradient of exchange made possible by a continuous stream of nutrients, light, and carbon dioxide, this population can be considered the equivalent of *domestic cleaners* that are supplied with nutrient molecules in exchange for *housework*. Amongst the work they perform is the reclaiming of inorganic phosphate from the detergents in liquid household waste streams and detoxifying nitrous gases in solution. (Figure 7) The occupants of this microbial city are neither enslaved nor owned but selectively establish themselves within preferred bioreactor types, becoming metabolically wealthy enough to make kin and community.[38]

Changing the properties of microbes through genetic engineering raises ethical questions about how far microbes can, or should, be designed and how their responses can be better evaluated. This approach to synthetic biology poses more danger to humans than that posed by modified microbes, but also raises questions about

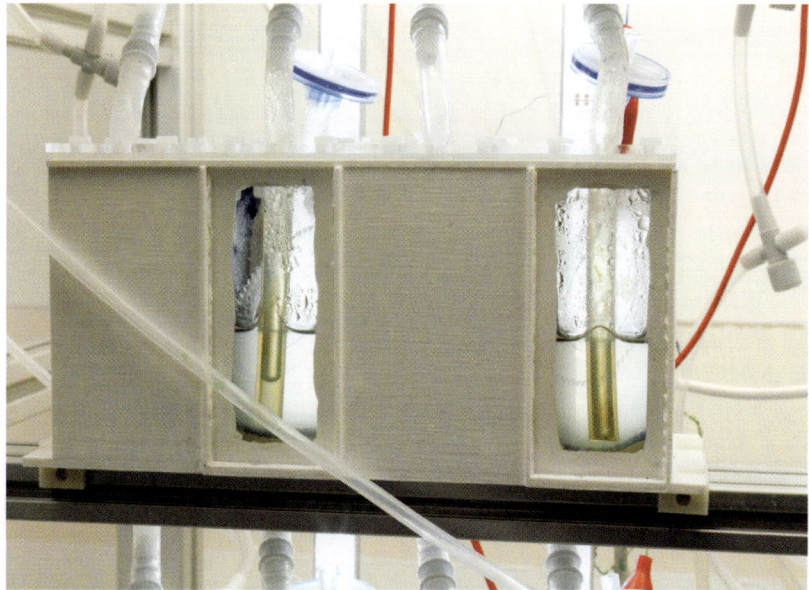

Figure 6. Living 'brick' chassis for housing wild type and genetically modified organisms, 2019. Photograph courtesy of the *Living Architecture* project.

the exploitation of microbes, which are not without repercussions to unwanted interference. Possessing an *intracellular parliament,* microbes can remove unwanted, extra genetic material and revert to their wild-type constitution.[39] They also uphold principles of justice that are witnessed in how microbial populations deal with *cheaters* that work against the dynamics of a biofilm community or who invade their neighbours. Non-producers that can threaten the overall cooperation of a community can cause a population to collapse, when profiting from public microbial goods without paying their true production cost or when selfishly proliferating and exhausting the resource provided by the microbial commons.[40] Communities can prevent this demise through intercellular signalling systems that enable generous microbes to outgrow the cheaters, invoking an aspect of care that is maintained through (natural) surveillance—like an immune system.

Many ethical design questions are raised by the *Living Architecture* project; from the construction of unnatural spaces to the perceived needs of homes, the harvesting of microbial products for human needs, the manipulation of microbial genetic material, the need for constant system surveillance, and the inevitable blind spots and over-sights in the system produced by anthropocentric limits (for example, perceptual scale, imperfect communication, and indirect contact with the microbes), the proper care of microbes require the assistance of more-than-human collaborators.

In *Living Architecture* microbial surveillance is mediated by arti-ficial intelligence software that uses a bio-digital interface situated within the microbial fuel cells, where heterogenous populations of anaerobic bacteria settle on a membrane that separates the anode from the cathode to form a stable biofilm. Here they excrete electrons,

Figure 7. Fully inoculated *Living Architecture* 'wall' and apparatus, 2019. Photograph courtesy of the Living Architecture project.

which are captured by the hardware. The bioelectricity produced by this process is sufficiently powerful to actuate regulatory electronics while the software interprets the outputs. The latter provides a basic service and communications system that optimises the provision of resources through feedback loops that read microbial and human needs. The metabolically and informatically entangled *Living Architecture* apparatus can, therefore, be thought of as a *microbial cyborg*, or *holobiont*—a wholly integrated assemblage of *bios* and *zoë* which expresses different microbial species, human, narrative, and technical agents, and which comprises the alterbiopolitical framework of the system. (Figure 8)

Living Architecture's interfaces are not yet sensitive or accommodating to real-time differences in time and scale between humans and microbes. Appropriate languages and modes of governance are urgently needed for richer and more finely tuned forms of communication that underpin iterations of exchange which are reinforced through mutual affirmation and eventually familiarised through habituation. Such a system invokes the specificity of immune systems that require more sophisticated combinations of sensors and effectors at the bio-digital interface. This could provide more nuanced kinds of diplomacy between human and microbes, where it becomes possible to begin to 'speak' chemically, physically, biologically, mechanically, digitally, and even emotionally with the microbial realm. Making tangible this invisible world, *Living Architecture* also facilitates an alternative paradigm for domestic economies in dealing with resources through metabolic recycling, which does not discriminate between vital matter and its excrements. Hopefully, a point may be reached when our daily activities no longer generate waste but are transformed into world-making metabolic actions with the potential of integrating systemic change in the material definition of human inhabitation. Power relationships can be designed through the selection of *metabolic apps* which enable us to rehearse the kinds of relationships with microbes we would like to invite into our domestic realms. These customisable bioprocessor units can be altered through combinations of different bioprocessor types and the selection of microbial populations. This new environment is not only relevant to our caring, *along with* nonhumans as a means of shaping relations, but also provides broader conceptual and practical frameworks for organising collectives of inter- and intra- relating humans and microbes.[41] While at an early stage of its experimental evolution, *Living Architecture* enables the integrative synthesis of multiple agents in ways that catalyse radically different approaches to how we think about the nature of home, our relationship with microbes, and alternative paradigms for sustainability and resource management. *More than* a domestic appliance, it enables the coproduction of worlds through an emerging and experimental practice of worlding.

Conclusion

As the climate crises intensifies, the story of life needs retelling to help mitigate the impacts of our misunderstanding of nature and our role within it. A metaphysics of life for an era of pandemics is urgently needed through a negotiated relationship between *bios* and *zoë,* so that the agency of vital sub-cellular entities can be appropriately and equitably engaged within an expanded ecology of lively matter. This is a view that more fully characterises this planet and towards which we can look for our world's overall re-enlivening. Within this context biodesign provides a platform capable of asking new questions that help us realise an expanded, more-than-human design practice, which works towards a co-constituted lived world.

Acknowledgments

The *Living Architecture* project is funded by the European Union under the Horizon 2020 Research and Innovation Programme via Grant Agreement no. 686585. The grant ran from April 2016 to April 2019, bringing together experts from the universities of Newcastle, UK; the West of England (UWE Bristol); Trento, Italy; the Spanish National Research Council, Madrid; LIQUIFER Systems Group, Vienna, Austria; and Explora, Venice, Italy.

ENDNOTES

1. William Myers and Paola Antonelli, *BioDesign: Nature + Science + Creativity* (New York: Museum of Modern Art, 2012).

2. Rosi Braidotti, *The Posthuman*, (Cambridge: Polity Press), 22.

3. Myers, *BioDesign: Nature + Science + Creativity*. The term 'biodesign' is also adopted in scientific fields such as medicine where the body's scaffolding is used to guide cell culture and in the molecular engineering of biology. Paul G. Yock, et al., *Biodesign: The Process of Innovating Medical Technology* (Cambridge: Cambridge University Press, 2009), 39.

4. Kate Orff, "Oystertecture," in *BioDesign: Nature + Science + Creativity* (New York: Museum of Modern Art, 2012) 56-57; Philips Design, "Microbial Home," in *BioDesign: Nature + Science + Creativity*, 96-101; Julia Lohmann, "Co-Existence," in *BioDesign: Nature+ Science+ Creativity*, 218-21; Suzanne Lee, "Bio-Couture," in *BioDesign: Nature + Science + Creativity*, 108-11; Wyss Institute, "Lung-on-a-Chip," in *BioDesign: Nature+ Science+ Creativity*, 94-95.

5. William Myers, "The Hybrid Frontier," in *BioDesign: Nature + Science + Creativity*, 8-9.

6. For access to theories of Biomimicry, see Janine Benyus, *Biomimicry: Innovation inspired by Nature*. (New York: HarperCollins Publishers Inc., 2011), and Michael Palwyn, *Biomimicry in Architecture* (London: RIBA publishing, 2019).

7. Tom Knight. Idempotent vector design for standard assembly of biobricks. *Technical Report, Massachusetts Institute of Technology Synthetic Biology Working Group Reports*, 2003.

8. Paola Antonelli, "Vital Design," in *BioDesign: Nature + Science + Creativity*, 7.

9. William Coleman, *Biology in the Nineteenth Century: Problems of Form, Function and Transformation* (Cambridge: Cambridge University Press, 1977), 1.

10. Peder Anker, *Imperial Ecology: Environmental Order in the British Empire, 1895 – 1945* (Cambridge: Harvard University Press, 2001), 4.

11. The term "protocell" is a contested concept. One school of thought led by Steen Rasmussen and others conjecture that it is a fully synthetic, or artificial cell, in other words, it meets the full criteria for 'life'. See, Steen Rasmussen, Mark A. Bedau, Liaohai Chen, David Deamer. David C. Krakauer, Norman H. Packard and Peter F. Stadler. *Protocells: Bridging nonliving and living matter*. (Cambridge, MA: MIT Press, 2008) 102. The other viewpoint, the one I share, builds on origin of life concepts that consider the protocell as a lively chemical unit, or dynamic droplet, as a precursor of true life, with no claims on being fully alive, see, Rachel Armstrong, *Vibrant Architecture: Matter as Co-designer of Living Structures* (Berlin: De Gruyter) 35. Also see, Rachel Armstrong, "Designing with protocells: Applications of a novel technical platform" *Life* 4 no. 3 (2014): 457–90.

12. Rachel Armstrong, "Future Venice" in *BioDesign: Nature + Science + Creativity*, 72-73.

13. Rachel Armstrong, "COVID-19: The invisible titan of biopunk," in *Brave New Human: Reflections on the Invisible*, ed. Alexander Mouret, (Netherlands: Brave New World Conference / Bot Publishers, 2020), 57-70.

14. Hannah Arendt, *The Human Condition* (Chicago: University of Chicago Press, 2018), 12.

15. Rosi Braidotti, *Nomadic Theory: The Portable Rosi Braidotti Gender and Culture*. (New York: Columbia University Press, 2012), 16. Celebrating zoë as an irreducible animating force within lively matter, Braidotti proposes we are also part of this agency that connects us to the other creatures we share the world and our own bodies with.

16. First popularized by Martin Heidegger in *Being and Time*, who turned the noun (world) into the active verb (worlding), and so proposed an ongoing, generative process of world-making, worlding defies formal definitions of object-ness as it is also always unmaking, renewing, and constantly revealing different aspects of its being. When allied with new materialism that circumvents Heidegger's anthropocentric "world picture" the term infers practices that promote *new modes of being*. See Martin Heidegger, *Being and Time* (Oxford: Wiley Blackwell, 1978), 65.

17. Rachel Armstrong and Rolf Hughes, *The Art of Experiment: Post-pandemic Knowledge Practices for 21st Century Architecture and Design* (London: Routledge, 2020).

18. Charles Darwin, *On the Origin of Species by Means of Natural Selection, or the Preservation of Favoured Races in the Struggle for Life* (London: John Murray, 1859), 143.

19. Robert M. Hazen, Dominic Papineau, Wouter Bleeker, Robert T. Downs, John M. Ferry, Timothy J. McCoy, Dimitri Sverjensky and Hexiong Yang, "Mineral evolution," *American Mineralogist* 93, (2008):1693-1720.

20. Jane Bennett, *Vibrant Matter: A Political Ecology of Things* (Durham: Duke University Press, 2010), 24.

21. David Dunkley Gyimah, "Preview to Sci Fi London—Rachel Armstrong @viewmagazine.tv," VIMEO video, 03:58, May 3, 2009, https://vimeo.com/4455066.

22. Nonlinear dynamical systems change their variables (composition, movement etc.) over time. They may appear chaotic, unpredictable, or counterintuitive, contrasting with much simpler linear systems and are of interest owing to their highly complex expressions.

23. Yuto Miyamoto, "Inside Alternative Machine." https://medium.com/evertale-english/alternative-machine-7058e71be53. (November 3, 2018).

24. Rachel Armstrong, *Vibrant Architecture: Matter as co-designer of living structures* (Berlin: DeGruyter, 2015), 40.

25. Otto Bütschli, *Untersuchungen ueber microscopische Schaume und das Protoplasma*, (Leipzig: Engelmann, 1892), 229-31.

26. Martin M. Hanczyc, "Droplets: Unconventional Protocell Model with Life-Like Dynamics and Room to Grow," *Life 4*, no. 4 (2014): 1038-49.

27. Tibor Gánti, *The Principles of Life* (New York: Oxford University Press, 2003), 1.

28. Rachel Armstrong, *Liquid Life: On non-linear materiality* (New York: Punctum, 2019) 369-413.

29. Rachel Armstrong, "Future Venice," in *Bio Design: Nature, Science + Creativity*, 72-73. Rachel Armstrong and Philip Beesley. "Soil and protoplasm: The Hylozoic Ground project." *Architectural Design* 81, no.2 (2011): 78-89.

30. Bruno Latour, "Love your monsters: Why we must care for our technologies as we do our children." In Michael Shellenberger and Ted Nordhaus (Eds) *Love Your Monsters: Postenvironmentalism and the Anthropocene* (Kindle, 2011). *No page numbers.*

31. Joan Tronto, *Moral Boundaries: A Political Argument for an Ethic of Care.* (New York: Routledge 1993) 103.

32. Maria Puig de la Bellacasa, *Matters of Care: Speculative Ethics in More Than Human Worlds* (Minneapolis: University of Minnesota Press, 2017), 219.

33. Giorgio Agamben, *Homo Sacer: Sovereign Power and Bare Life* (Stanford: Stanford University Press, 1998), 188.

34. Bruno Latour, *We Have Never Been Modern* (Cambridge, MA: Harvard University Press, 1991), 142-145. Moreover, Latour's "Parliament of Things" invokes an enlarged democracy centered on life where things can speak in their own name rather than mediated by humans "Natures are present, but with their representatives, scientists who speak in their name. Societies are present, but with the objects that have been serving as their ballast from time immemorial." Latour, 1991, 144.

35. Rachel Armstrong, et al., "Living Architecture (LIAR): Metabolically engineered building units," in *Cultivated Building Materials: Industrialized Natural Resources for Architecture and Construction*, ed. by Dirk E. Hebel and Felix Heisel (Berlin, Germany: Birkhauser, 2017), 170-77.

36. Melissa B Miller and Bonnie L. Bassler, "Quorum sensing in bacteria." *Annual Review of Microbiology*, 55 (2001): 165-99.

37. *Doulix* is one of the tools and resources for the effective engineering of complex pathways through the assemblage of complex biological parts into multipartite plasmids, which takes the form of a web-based DNA design tool that was developed by Explora Biotech during the course of the *Living Architecture* project.

38. My anthropocentric reading of these highly constructed events is tainted here, with a desire to create conditions for microbial existence that counter comparisons with battery farming.

39. "Intracellular parliament" references Latour's *Parliament of Things* and Karen Barad's intra-actions, transposing them into a cellular environment. Conventionally cell governance is understood through the central dogma of molecular biology that accounts for how cell processes work: DNA gives instructions to RNA, which then passes them on proteins in a unidirectional flow of information and command. This results in a vertical mode of transmission to produce well circumscribed, pre-determined organisms. Variations are lucky mistakes, whose value is measured by survival. The *intracellular parliament* opens up alternative forms of decision and power-making within the interior life of a vital agent to offer metabolic and symbiogenetic freedoms that empower an agent to invent, particularly bacteria whose robust metabolisms and genetics are very different than eukaryotes—the organismal domain that contains organisms that are *big like us*.

40. Özhan Özkaya, Roberto Balbontín, Isabel Gordo and Karina B. Xavier. "Cheating on cheaters stabilises cooperation in Pseudomonas aeruginosa." *Current Biology* 28 no.13 (2018): 2070-80.

41. Stephanie R Fischel, *The Microbial State: Global Thriving and the Body Politic* (Minneapolis: University of Minnesota Press, 2017).

LOST IN SPACE WITH FRANKENSTEIN'S SHADOW

PATRICIA OLYNYK

ABSTRACT

The participatory turn in art over the past five-and-a-half decades has produced an array of compelling immersive environments that enhance the viewer's sense of their corporeality. A key mechanism in this choreography often involves the optical representation of shadows and captivating visual phenomena, from which nuanced sensorial responses emerge. Technologies that materialize phenomena complicate spatially oriented works, dramatically altering the tenor of the embodied experience while offering new ways for our mediated sensoria to shape our sense of bodily presence within the physical world. "Lost in Space with Frankenstein's Shadow" discusses three immersive artworks, or 'shadowlands', produced by three artists: Olafur Eliasson, Won Ju Lim, and this author. While each artwork employs projected media in a distinct way to enhance the viewer's sense of corporeality by way of optical shadow play, all three operate on different registers while conjuring notions of the fractured, recomposed, and even virtual "Frankensteinian" body. In addition, this essay explores the haptic qualities of shadows as they relate to notions of Frankensteinian space, re-imagined by way of technology and by registering key telltale features of Mary Shelley's masterwork: hybridity, replication, and unpredictability. By invoking the productive capacities of Shelley's narrative, can the viewer's techno-hybrid body be empowered to build affinities with other individuals and groups? Or, as transhumanist performance artist Stelarc speculates: Is the monster in fact the system itself, one that sucks the body into virtuality?

+ art and technology
+ media arts
+ affect behavior
+ frankenstein studies
+ immersive art

Figure 1. Boris Karloff as Frankenstein's monster in James Whale's film, *Frankenstein*, 1931, NBC Universal Archives & Collections.

Frankensteinian Space

The participatory turn in art over the past five-and-a-half decades has produced an array of compelling immersive environments that enhance the viewer's perception of their body as a corporeal being and heighten awareness of their bodily relationship to space. A key mechanism in this choreography often involves the optical representation of shadows and captivating visual phenomena from which nuanced sensorial responses emerge. Moreover, technology complicates if not activates many spatially oriented works, dramatically altering the tenor of the embodied experience while offering new ways for mediated sensoria to shape our sense of presence within the physical world.

This essay discusses three immersive artworks by Olafur Eliasson, Won Ju Lim, and this author, all of which use projected media to affect the viewer's awareness of their bodily presence by way of shadow play. All three 'shadowlands' operate on different registers while conjuring notions of the Frankensteinian body and in some cases, Frankensteinian space. The first reifies the viewer's recognition of self, the second reveals the viewer as Other, and the third, which features a layered soundscape that enhances the spatiotemporal experience, connotes hybridity. Each work in its own way tinkers with the viewer's perception of their body and encourages spectacular forms of engagement that reinforce corporeality.

Mary Shelley's *Frankenstein; or, The Modern Prometheus* (1818) is the springboard and inspiration for examining the ways in which contemporary immersive installations and the embodied experiences they offer are realized by way of technology. The translocation of Victor Frankenstein's specter into technologically-complex art works provides an opportunity to consider the technological, cultural, and scientific concerns of Shelley's masterwork in the present. Likewise, Claudia Gualtieri, contributing author to the edited book *Transmedia Creatures: Frankenstein's Afterlives*, discusses Shelley's relocation of the ancient Greek Titan god, Prometheus into the modern era. Her analysis on literature's significance during the time of its reception aligns with literary and cultural critic Edward Said in his assertion that literary texts are both circumstantial and situational. They can function as signifiers with their own contingent histories and in this they play a political role as agents of change in their times of reception.[1] So too, I argue, can transmedial representations generate productive discourse around topics that endure over time. Our ambiguous relationship with technology, its spectacular representations of the body that challenge its corporeality, and our preoccupation with Othering and difference—concerns that are all foregrounded in Shelley's novel—persist and are of interest to this paper. In addition, contemporaneous attempts to interrogate from the perspective of feminist science studies the history of patriarchal science, intersectionality, and our fears around machine intelligence, have emerged with Frankenstein as muse. Reconceptualizing corporeality through the lens of Frankenstein in dynamic, participatory art works reinforces the potential of transmediality and the power of technology to expand our notions of what a body is and what it does.

Surprisingly, however, while Shelley's cautionary tale provides a superb, multifaceted point of departure that eloquently captures the technocultural concerns of its time, the creation scene that specifically addresses the monster is but a mere eight lines in total. Significantly, Shelley did not even explicitly use the term technology in her text. As she wrote:

> It was on a dreary night of November, that I beheld the accomplishment of my toils. With an anxiety that almost amounted to agony, I collected the instruments of life around me, that I might infuse a spark of being into the lifeless thing that lay at my feet. It was already one in the morning; the rain pattered dismally against the panes, and my candle was nearly burnt out, when, by the glimmer of the half extinguished light, I saw the dull yellow eye of the creature open; it breathed hard, and a convulsive motion agitated its limbs.[2]

In her essay "Frankenstein and Radical Science," eminent Frankenstein scholar and feminist, Marilyn Butler states that Shelley's novel invites speculation and is endlessly reinterpretable.[3] This has been the case for well over two hundred years since *Frankenstein* was first published. It has also been the case, as Butler notes, that the author's focus on natural science and the scientific revolution, which drove debates on materialism, vitalism and galvanism, was deeply criticized precipitating a shift in focus with her theological rewrite of Frankenstein in 1831.[4] Nonetheless, it is the techno-scientific imagination and the techno-hybrid body of Frankenstein's so-called 'hideous progeny' that lives on in the cultural imaginary and the art world today.

This essay, and the artworks it discusses, consider the productive capacities of Shelley's narrative and the technological embrace it invites. It asks if digitally mediated spatial works can empower newly formed techno-hybrid bodies to build affinities between individuals and groups, who may otherwise be separated by difference. Or, alternatively, as transhumanist performance artist Stelarc—whose work challenges the boundaries of corporeality through remote controlled biological-technological interfaces—speculates: "is the monster in fact the system itself, one that sucks the body into virtuality?"[5] Transmedial representations of Frankenstein's body—that is, the fractured, reassembled, hybrid body—which cannot be fully revealed through literature, can emerge by way of technology in other artforms such as complex immersive media environments. These translations challenge our notions of the corporeal and the virtual as they relate to Frankenstein's various incarnations. In addition, as art historians Cristina Albu and Dawna Schuld note in their essay "Beside Ourselves," the complexity of intertwined experiential registers that operate within immersive installations transcend the perceptual attributes of an experience to include both social and political contingencies.[6] In these art pieces, experience is rooted in both physical and cultural environments, and as Albu and Schuld further note:

a situational approach to art...make[s] evident the continuities across historical trajectories of artworks that are based on different media and...underscore the fluidity and interweaving of biology, society, politics, and culture - not in isolation from each other, but as part of a dynamic assemblage that we ourselves take part in and so influence to varying degrees. Such an approach shifts the focus away from art understood as representation toward art understood as an operative tool that models life in conspicuous ways.[7]

Likewise, I argue, a transmedial approach considers not only the connective tissue that bridges technology, humans, and non-humans with one another and with politics and culture, but that examines how these connections are manifest across time and various artistic media.

Cinematic, Haptic, and Rhetorical Shadows

Invention... does not consist in creating out of void, but out of chaos... [I]t can give form to dark, shapeless substances, but cannot bring into being the substance itself.[8]

Inspired by masterpieces of German Expressionist films that include Friedrich Wilhelm Murnau's *Nosferatu* (1922) and Robert Wiene's *The Cabinet of Dr. Caligari* (1920), James Whale's unforgettable classic horror film *Frankenstein* (1931) is based on a stage play rendition of the story which accounts in part for its occasional departure from the original narrative.[9] (Figure 1) Whale's version makes use of the power of the shadow as a cinematic device that, I argue, takes center stage. His brilliant cinematic representation of Frankenstein's creature moves beyond materialism to conjure the notion of the dark and ambiguous virtual body. Chiaroscuro light and shadow materialize and dematerialize the body of the creature to maximum effect. Referring to light as the "First Aesthetic Field" in his book *Sight, Sound, Motion, Applied Media Aesthetics,* Herbert Zettl states that light is not only essential for life, but essential for film. He refers to film as a pure light show, differentiating the "materia" of theatre, which consists of both people and objects, from the "materia" of film, which consists only of light.[10] Without its counterpart, light, the shadow cannot exist.

Surely, optical shadows can be spectacularly affective; however, it is haptic shadows, or those in which the shadow suggests a sense of prosthetic touch that are the mechanism by which the three immersive art installations I discuss in this essay operate. Whether imbricating the body into an interspecies world or enticing it into a liminal space, a newly formed Frankensteinian body emerges, re-imagined by way of technology and registering any or all key telltale features of Mary Shelley's masterwork: hybridity, replication, and unpredictability.

Moreover, an encounter with one's shadow is more than a perceptual or rational understanding of phenomena. Beyond the optical mechanics of what a shadow is and how it works, it has an abundance of metaphorical, symbolic, and rhetorical meanings. In Plato's "Allegory of the Cave" in the *Republic* (375 BC), shadows are an illusion, an intangible register of people and objects that offer but a partial understanding of the world, "an imposter that reveals human ignorance as the mental state of detainees in a cave who take the shadows cast by events happening outside their prison for reality."[11] During the twentieth century, Carl Jung's four archetypes in analytical psychology contain the self, the persona, the anima/animus, and the shadow; the latter, a repressed presence that can only reveal itself indirectly by way of projection.[12] And if the Jungian shadow represents all that we consciously repress, then our fear of death must also reside there. Frankenstein and his hideous progeny may have both cheated death, yet they were forced to live out their lives reckoning with their shadows.

In *The Divine Comedy* (1320), a tale in which Dante Alighieri descends into the underworld, purgatory, and paradise to explore the fundamental elements of the human condition, shadows reveal to the souls of the underworld that he resides in the world of the living. For Dante, a shadow, is not a thing but an absence. *Ombra* is an intriguing material and poetic figure precisely because it does not have a fixed identity: as a form of darkness shaped by the negation of light (or even its hiding), it is an absence that denotes a presence.[13] However, more than simply the interplay between bodies and light, the shadow is "ever elusive and intangible, a protean shape that resembles as it dissembles."[14] Here, Dante casts a shadow on Frankenstein and his lament about the limits of invention, which can give form to dark and shapeless substances, but is incapable of bringing into being substance itself.

Olafur Eliasson, *Your Uncertain Shadow (colour)* (2010)

Enchanted by a quartet of luminous, pastel-colored silhouettes, the viewer in Olafur Eliasson's large-scale, immersive artwork *Your uncertain shadow (colour)* on exhibit at Martin Gropius Bau in Berlin (2010), did not immediately realize—like a chimpanzee who has not yet mastered the art of self-recognition—that the shadows were the register of their own presence. After a few moments in Eliasson's lively performative environment, one of several room-sized installations in his first major solo exhibition in Berlin, the viewer gained sufficient self-awareness to recognize that the shadows were theirs, albeit in a de-familiarized, perverted form. This recognition activated a sense of play and interconnectivity between viewers, a key component of role-playing and social interaction, typically embedded in living and technological systems including those found in video games.

Though such amusements don't immediately recall Frankenstein's creature, whose gentle and curious nature is corrupted by society's violent rejection of him, he is nonetheless characterized by Shelley as an emotionally complex, well-developed character who deeply desires

Figure 2. Olafur Eliasson, *Your uncertain shadow (colour)*, 2010. HMI lamps (green, orange, blue, magenta), glass, aluminum, transformers. Installation view: Studio Olafur Eliasson. Photo: María del Pilar García Ayensa/Studio Olafur Eliasson. Thyssen-Bornemisza Art Contemporary Collection, Vienna. © 2010 Olafur Eliasson.

human connection. James Whale's adaptation of Shelley's novel in his 1931 film further humanizes the creature as a social, sentient, and sometimes playful being. Innocently befriending the character of a young girl Maria, who is playing by a lake, the two take turns floating daisy blossoms on the water's surface, while the creature imitates, repeats, and delights in this whimsical game. Just as Eliasson's installation invites imitation, repetition, and the joy of pondering light phenomena, the creature and Maria likewise take mutual delight in beholding the phenomenon of buoyancy. Because the scene so convincingly portrays the creature as being susceptible to play, and because play is widely understood as a vital aspect of personal and social well-being, Frankenstein's creature, popularized in Whale's film, while fantastical, bears the same fanciful appetite for delight as humans. (Figure 2)

The dazzling visual effects of Eliasson's *Your uncertain shadow (colour)* rely on the mechanics of light phenomena and involve "five coloured spotlights, directed at a white wall, and arranged in a line on the floor: a green light positioned next to another green light, followed by a magenta light, an orange light, and, finally, a blue light."[15] These combined colors illuminate the wall with a bright white light and when a visitor moves through the space, their projected shadow blocks each light from a different angle. The final effect multiplies the viewer's shadow, which appears as an array of five differently colored silhouettes projected on the wall. An additional dark shadow emerges where all five lights are obstructed while the other shadows—one yellow, one violet, one cyan, and two magentas—reflect the properties of additive color theory.[16] For example, as a visitor blocks light from the blue spotlight, the remaining shadow is lit by a combination of green, magenta, and orange lights, creating a yellow shadow. The overlap of the five silhouettes produces its own color array.

As Eliasson explains:

> When you encounter your own shadow on the wall, it is undeniable evidence of your presence in that space. It is a consequence of your being there. If, however, you find yourself following the shadow instead, unexpected things begin to happen. The shadow asks you to move differently. You become activated, so to speak... so that it is no longer a consequence of your presence, but rather you are the consequence of the shadow's presence.[17]

My own experience of *Your uncertain shadow (colour)* in 2010 in Berlin was a surprisingly whimsical and affective experience in which a strange sense of distant touch transcended kinesthetics through direct contact with the surface of the work by way of my shadow. With this setup, the shadow becomes a kind of prosthetic, which dances across the gallery walls, touching the other shadows engaged in shadow play, offering a unique phenomenological feedback loop that registers bodily presence. With *Your uncertain shadow (colour)*, the shadow is both a de-corporealization of the self and a sign for the self. The installation does not reveal any moral conflict or sense of turmoil, rather, it engenders a better understanding of our agency within a highly phenomenological experience. Unlike Jung's notion of the shadow as a repressed representation of the unconscious self—a dark side which is regrettably cast onto others—the shadow self in Eliasson's unpredictable entanglements of overlapping silhouettes verge on a collective sense of visual delight. *Your uncertain shadow (colour)* offers the viewer a lively intersubjective form of play, never to be replicated again in the same manner in any other installation of this work.

In contrast to Whale's treatment of the creature's singular shadow, in which a pitch-black silhouette whose colossal void registers a predatory threat, shadow performers in Eliasson's installation are highly interactive and appear socialized and empathetic. According to Christina Albu, acting both collectively and singularly, viewers "perform a peculiar form of subjectivity, one informed both by affective exchanges and by shared discursive spaces of art, philosophy, anthropology, and... cognitive theory."[18] Though Albu's thesis on performing subjectivity relates to encountering mirror images, shadow encounters would arguably operate in a similar way.

Though *Your uncertain shadow (colour)* doesn't seem to invoke the horror genre or refer directly to cinematic representations of Frankenstein's shadow, it does conjure notions of the *unheimlich* through the duplication of the body, one which cannot exist as a singular entity in Elliason's installation. The body's shadow doppelgängers, which emerge by way of technology, create according to Joshua Comaroff and Kir-Shing Ong "a surplus of symmetry... [whereby each embodies] a super-abundance of the 'harmony' valued in classical thought. This excess gives rise to an odd phenomenon: horrible beauty."[19] According to Gry Faurholt in "Self as Other", the doppelgänger in the German *Schauerroman* (horror novel), and the British Gothic novel, emerge from an enchantment with early nineteenth-century folklore that stitch together three key elements: supernatural horror, a philosophical enquiry concerning personal identity, and a psychological investigation into the hidden depths of the human psyche.[20] Manifesting as a twin, a shadow, or a reflection, the doppelgänger presents according to Faurholt the "paradox of encountering oneself as another; the logically impossible notion that the 'I' and the 'not-I' are somehow identical. This uncanny device subverts our notion of identity...I must identify as 'I,' that which is not me."[21] It is through the fluctuation between animated play and uncanny doppelgänger encounters that Eliasson's highly interactive *Your uncertain shadow (colour)* invites multiple ways of examining our relationship to technology by pondering its relationship to corporeality, virtuality, and identity.

Won Ju Lim, *California Dreamin'* (2008)

> [M]any contemporary artists induce highly mediated modes of sensing that destabilize perception and direct one's attention to the social contingencies of experience, creativity, and knowledge production ...[P]henomenal art—art in which we install ourselves as both the recipients and sources of its perceptual effects—fosters introspection. To perceive oneself perceiving, in Merleau-Ponty's terms, seems at first to necessitate an asocial, if not anti-social, inward turn.[22]

In a recent interview I conducted with Los Angeles-based artist Won Ju Lim, she emphasized that the subject/object relationship is central to her work, which she described as both ontological and phenomenological. Her wvisually seductive, spatially complex installation, exhibited at the Frankfurt Schirn Museum in 2008 and again at the San Jose Museum of Art in 2018 titled *California Dreamin'*, consists of a gem-like cluster of brilliantly colorful Plexiglas blocks—a near miniature city—stacked neatly in the middle of the gallery. Ensconced in a liminal space between worlds and looming over this abstracted diorama of the imaginary metropolis, is the viewer's body, one that is rendered monstrous as a consequence of this inversion of scale. It might be said that the resulting destabilizing effect aligns with the artist's own personal history as a young immigrant who recalls experiencing Los Angeles (LA) by way of the car as a kind of mechanical prosthetic, one that mediated and at times enhanced her phenomenological experience of the city. Lim describes in vivid detail her childhood memories of California palms, pumpjacks, and sprawling urbanscapes seen through the windshield of a car and how they forever changed the way she understood the city, to which she had no direct bodily relationship.[23] Mining her childhood revelries in which the future is mechanical, Lim's immersive environments allude to technology's promise of a brave new world. Longing for a place that doesn't exist yet influenced by both LA's experimental architecture and the aesthetics of science fiction, Lim's installations generate a kaleidoscope of colors, reflections, refractions, shadows, and silhouettes. Light, shadow, and space are co-dependent actors on a stage that render a technological future that can never quite meet human expectations. Like Frankenstein's initial enthusiasm over the promise that technoscience offers, Lim's work helps us imagine what our future could be.[24] (Figure 3)

The installation's multiple projections of crepuscular skies, palm trees, and high-rises caress the walls of the gallery with rich, warm color. The light sources, which seem to emanate from the center of the sculpture, produce what the artist refers to as "a democratic space, where all viewers are registered equally."[25] Everyone's presence is reflected in the faceted surfaces of the Plexiglas and even more mysteriously revealed by the shadows they cast. Lim also emphasizes that "similar to the way that celluloid compresses everything in filmic space, the existential makeup of *California Dreamin'* is the same."[26] However, in Lim's installation there is a gap between the wall projections and the sculpture that creates a liminal space where bodies are displaced from the miniature city, yet they do not seem to exist beyond it. These bodies remain bound in a kind of Frankensteinian shadowland, an indeterminate borderland between places and states. As the viewer moves through the installation, the architectural space regurgitates gigantic doubles and clones, fractured bodies, and according to Comaroff and Ong, a "transformational play of multiples...[whereby] the space itself makes a 'fundamental break with architectural tradition: the freedom...not to be conceived as an object, but rather as a juxtaposition of versions."[27]

For any young immigrant, one's self-identification and the body's relationship to space are shaped by notions of hybridity, difference, and othering. This newly assembled body is an ambivalent figure that emerges from a hybrid place; in Lim's case, from a post-colonial South Korea that had been annexed in the early twentieth century by Japan and which saw in the 1960's the emigration of many South Koreans to the United States. These are the conditions of ambiguity that challenge not only colonialism but hegemonic culture, making

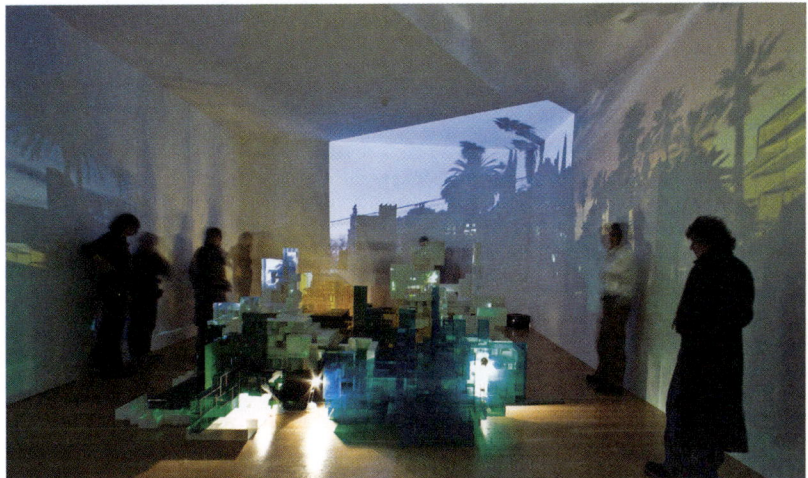

Figure 3. Won Ju Lim, *California Dreamin',* Exhibition view of *All Inclusive - A Tourist World* at Schirn Kunsthalle, Frankfurt, Germany, 2008. Credit Won Ju Lim.

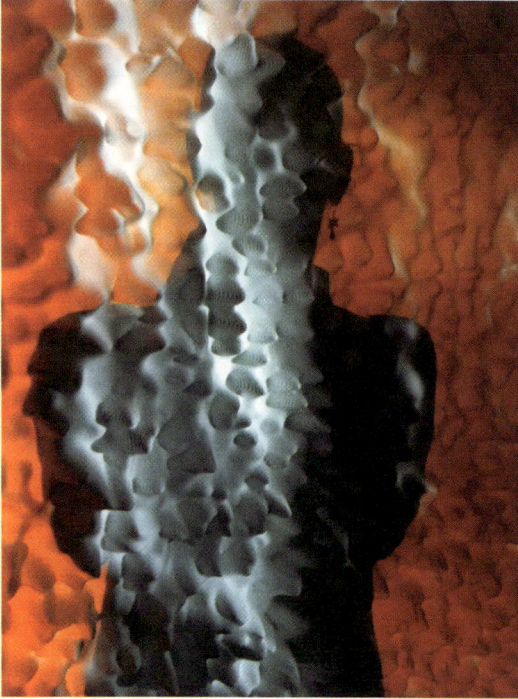

way for a more diverse and integrated world. And while a perpetual flow of new migrations and encounters drives the very world in which we live, Lim's work invites a more singular inspection: the viewer who may be unaware of their own shadow can see the shadow artifact of the Other, recognize their own presence in space, and perceive themselves perceiving. As Lim states, "This is less about cognitive thinking and more about something more immediate: I see you."[28] And yet, there is another layer to this cognitive process. Recalling Albu's theories of perception and agency in shared exhibition spaces, *California Dreamin'* reinforces the notion that a situated approach to theories of cognition acknowledges that thought is embodied and that it is embedded in both physical and cultural environments.[29]

Patricia Olynyk, *Dark Skies* (2012)

Dark Skies is an immersive artwork inspired by the obfuscation of the night sky due to the prevalence of artificial light in urban environments. The disorienting effects of light pollution are familiar to most, and aside from the obvious psychological impact of not being able to perceive and orient ourselves by way of the night sky, light pollution—according to the nonprofit International Dark-Sky Association (IDA)—has quantifiably dangerous effects.[30] In collaboration with Sung Ho Kim, the architecture and design firm Axi:Ome, and with assistance from sound designer Christopher Ottinger, I designed *Dark Skies*. First installed as a prototype at UCLA's Art | Sci Center in 2012, the artwork consists of a large-scale digitally sculpted wall, two hi-resolution video projections, and multiple soundscapes that emanate from speakers positioned throughout the installation. Activating both the visual and aural senses, *Dark Skies* invites a close meditation on the visceral qualities of a key cinematic moment in the twenty-four-hour clock—sundown. (Figure 4)

The projection wall is modeled after the prickly taste bud of a vespertine creature, a wild mouse that is active at twilight. However, given the right angle and proximity, the visually ambiguous wall appears to fluctuate in scale between the microscopic and the gargantuan, between lingual papillae and a mountainous terrain, as seen from a very high altitude. With this, the viewer's sense of bodily scale oscillates between the miniscule and the monstrous. The two videos projected at oblique angles on either side of the sculptural surface reveal two distinct time periods between twilight and sunrise, a condition that can only exist by way of technology. Together, they create a pulsating lenticular effect that is periodically interrupted by the viewer's own shadow. This immersive environment provokes an affective alignment of the visual and the acoustic rhythms found within, submerging the viewer in a complex sensorial experience. The first projection offers crepuscular skies in acid orange, which continuously loop and then slowly fade to black, tracking the passage of time and foreshadowing the imminent arrival of darkness. The second video proffers the cosmos, which gradually appears as small particles that cluster into milky trails and that stream vertically across the surface of the work. (Figure 5)

When either projection is interrupted, the viewer's shadow reveals either the vespertine or the night sky within their silhouette. This technologically produced composite form is transformed by its integration with other virtual bodies by way of light projection. As two evocative soundscapes revealing creatures of the night penetrate the space, the body is imbricated in an interspecies world, one that is dangerous and ungovernable, and one where the shadow self only partially retains its status as human. Albu speculates:

> *Dark Skies* merges the topology of bodies with that of the land and the cosmos, making viewers aware of the interdependence of all entities sharing planetary space. At the confluence between the cosmos and the microcosmos, the distant and the proximate, viewers encounter their eerie shadow projections. Each dark silhouette engulfs the terrain of the projection and is concomitantly engulfed by it. Both ominous and precarious, colonizing and compliant, the shadow is evocative of a contact zone, a liminal space of contagion which unmasks the illusion of containment. Dark Skies underscores the imbrication of personal bodily presence in systems that are too unpredictable and complex to be controlled at an individual level.[31]

Christopher Ottinger, the sound editor for *Dark Skies,* describes the immersive soundscapes made from field recordings that I recorded in the Rocky Mountains at twilight, as "triggers that sonically articulate the space between micro and macro worlds."[32] The first track emphasizes deep, rumbling sounds that shrink the listener down to the microscopic level, while the second track presents an expansive sonic space that fluctuates between noisy murmuration and spooky moans, zooming the viewer out from the world of the microscopic. With the soundtracks projected directionally at oblique angles into the installation space, viewers virtually migrate between macro and micro worlds.

In a recent conversation with Marcia Tanner, the curator of the 2005 exhibition *Brides of Frankenstein* at the San Jose Art Museum, Tanner mused that the story's conceptual apparatus is in and of itself a form of shadow play.[33] Each contemporary artist featured in the exhibition—an all-female cohort—interpreted, modified, and proffered their own interpretations of Shelley's masterwork and of Frankenstein in ways that were conceptually rich and technologically compelling. Tanner noted:

> They are 'brides' or metaphorical mates of Dr. Frankenstein, and use robotics, animatronics, computer animation, video, digital photography, the Internet, computer games, and other digital and electronic media to animate synthetic creatures with virtual life. Yet as artists, they resemble Mary Shelley; their creatures embody complex responses to the human and aesthetic implications of the technologies that made them. Like Shelley's, their works contemplate our relationships—emotional, psychological, physical—with those technologies: how we interact with them and they with us. Like hers, their projects question the unreflective drive to reconfigure nature that motivated Frankenstein, and explore the social, cultural, ecological and moral issues such activities raise. Their works [all] view these issues from a distinctly female perspective.[34]

The artists in *Brides of Frankenstein* use technology to investigate what technology is doing to and for us, and vice versa: "the medium is integral to the message."[35] However, perhaps, what is most significant is the potential that the artists' progenies have for mirroring contemporaneous issues involving humans and technology in ways that are infinitely reinterpretable over time.

Conclusion

Mary Shelley's *Frankenstein; or, The Modern Prometheus* is, in effect, a technological time capsule, a "textual battery that charged the *epistème* of Romantic science and culture to generate the modern discourse of technology."[36] Put another way, according to professor of literary studies Mark A. McCutcheon, today's discourse on technology has a history that traces back to Romanticism and the novel's discourse on technology, the meaning of which "is constructed retrospectively... and independently of *Frankenstein,* in popular culture. If technology has popularized a certain interpretation of *Frankenstein,* it is because *Frankenstein* itself conditioned the modern redefinition of technology as such in the period of its publication, early reception, and popularization."[37]

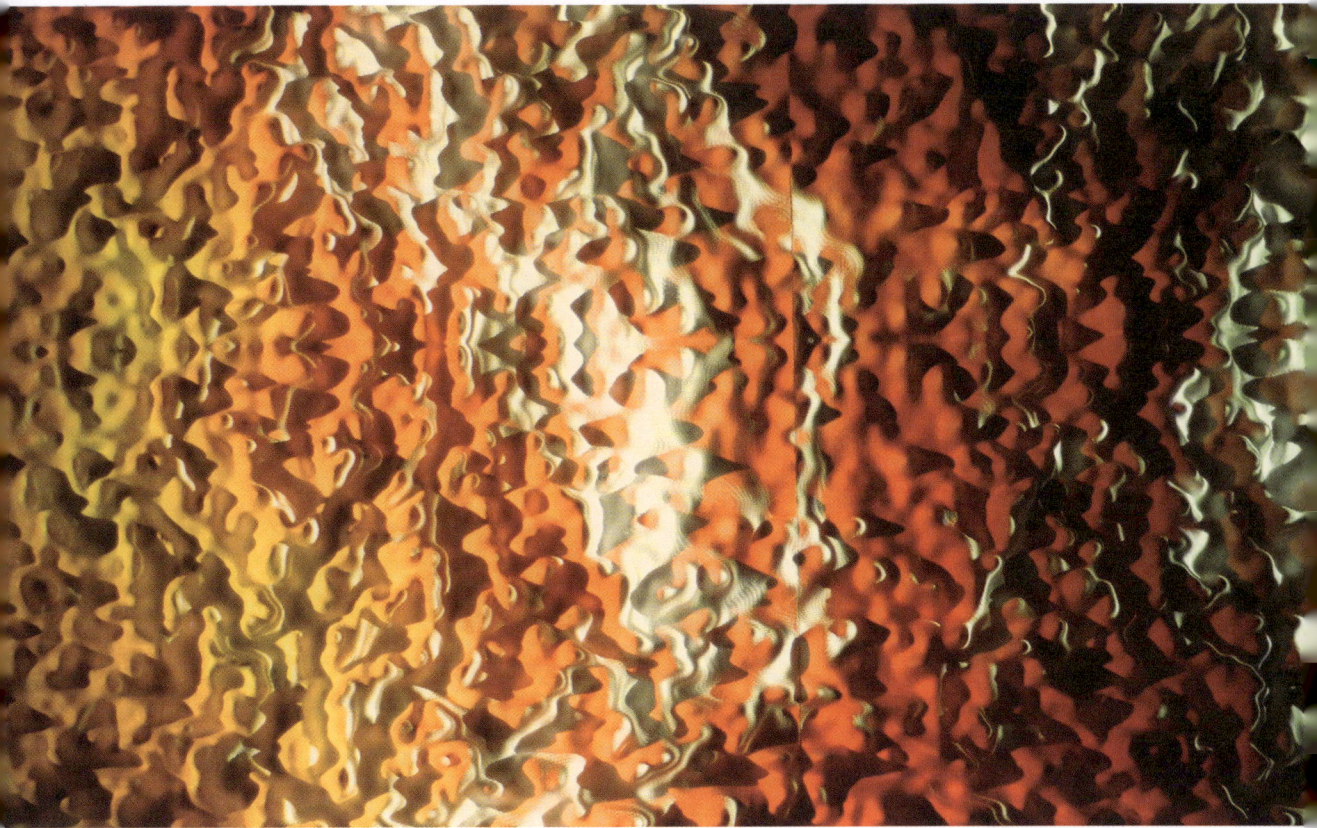

Figure 5. Patricia Olynyk –
Dark Skies, Wall sculpture,
two–channel video and
soundscape, 2016. Credit
Patricia Olynyk.

Frankenstein has cast a long shadow on the cultural imaginary as it relates to technology and science. Those who advance biotechnology and artificial intelligence (AI) and contribute to our seemingly inevitable posthuman future that reimagines the body by way of cloning and robotics, electromechanical implants, and gene editing technologies such as CRISPR, often mine Shelley's cautionary tale to probe the humanistic meaning behind the story. Further, the drive of transhumanism to explore the promise of a bodiless mind, recalls the work of Stelarc and his speculation that technological systems have the capacity to both challenge the limits of the corporeal body and render it virtual.

For now, the body as subject and object, corporeal and incorporeal still exists. Spanning more than two hundred years—from when Shelley's literary tale was first published to the era of modern cinema—contemporary immersive environments reveal a succession of bodily representations that call forth Frankenstein's creature and the technological means by which he was brought to life. Magnificent bodies transformed physically, optically, symbolically, and virtually emerge in a technological wonderland where shadows, both optical and rhetorical are the manifestations of new imaginaries that reflect Frankenstein's Promethean drive and that of those who wish to, as Shelley notes:

> penetrate into the recesses of nature and show how she works in her hiding-places. They ascend into the heavens; they have discovered how the blood circulates, and the nature of the air we breathe. They have acquired new and almost unlimited powers; they can command the thunders of heaven, mimic the earthquake, and even mock the invisible world with its own shadows.[38]

Driven less by technology's bells and whistles, and more by the ambiguous nature of shadow play, *Your uncertain shadow (colour), California Dreamin',* and *Dark Skies* all conjure Frankenstein's shadow by exploring the transformational capacities of technologically mediated art environments, where no fixed identity can exist.

Acknowledgements

I wish to thank Sung Ho Kim and Axi:Ome for digitally modeling *Dark Skies*, and Adam Hogan and Jacopo Mazzoni for their technical support.

Funding

I wish to acknowledge in-kind support from the Sam Fox School of Design & Visual Arts at Washington University in St. Louis, and from Victoria Vesna and the Art|Sci Center at UCLA.

ENDNOTES

1. Claudia Gualtieri, "Frankenstein; or the Modern Prometheus in the Postcolony," in *Transmedia Creatures: Frankenstein's Afterlives*, eds. Francesca Saggini, Anna Enrichetta Soccio (Lewisburg, Bucknell University Press, 2018), 102-03.

2. Mary Shelley, "Frankenstein," in *Frankenstein: a Norton Critical Edition*, Ed. J. Paul Hunter (New York, W.W. Norton Company, 2012), 35.

3. Marilyn Butler, "Frankenstein and Radical Science," in *Frankenstein: a Norton Critical Edition*, Ed. J. Paul Hunter (New York, W.W. Norton Company, 2012), 404.

4. Ibid.

5. Stelarc, email communication with author on October 9, 2017.

6. Cristina Albu and Dawna Schuld, "Beside Ourselves," in *Perception and Agency in Shared Spaces of Contemporary Art*, eds. Cristina Albu and Dawna Schuld (New York, Routledge Press, 2018).

7. Ibid., 6.

8. Shelley, "Frankenstein," 167.

9. Koraljka Suton, "Frankenstein: James Whale's Macabre Take on One of the Most Sympathetic Characters Ever Created in the World of English Letters," *Cinephilia & Beyond*. https://cinephiliabeyond.org/frankenstein/; William F. Burns, "From the Shadows: Nosferatu and the German Expressionist Aesthetic," *MISE-EN-SCÈNE, The Journal of Film & Visual Narration* 1, no. 1 (Winter 2016): 2.

10. Herbert Zettl, "The First Aesthetic Field" in *Sight, Sound, Motion, Applied Media Aesthetics*, Eighth Edition, (Boston, Cengage Learning, 2016), 19.

11. Joost Schilperoord & Lisanne van Weelden, "Rhetorical shadows: The conceptual representation of incongruent shadows," *Spatial Cognition & Computation* 18, no. 2 (2018): 98.

12. Carl Gustav Jung, *Four Archetypes* (London, Routledge Press, 2003), 173.

13. Andrew Hui, "Dante's Book of Shadows: Ombra in the Divine Comedy Dante," *Dante Studies* 134, Johns Hopkins University Press (2016):196-97. https://doi.org/10.1353/das.2016.0006

14. Ibid.

15. Olafur Eliasson, "Your uncertain shadow (colour), 2010." https://olafureliasson.net/archive/artwork/WEK100100/your-uncertain-shadow-colour

16. Ibid.

17. Ibid.

18. Albu and Schuld, "Beside Ourselves," 2.

19. Joshua Comaroff and Kir-Shing Ong, *Horror in Architecture* (San Francisco, ORO Editions, 2013), 50.

20. Gry Faurholt, "Self as Other: The Doppelgänger," *Double Dialogues,* Issue 10, (Summer 2009). https://www.doubledialogues.com/article/self-as-other-the-doppelganger/

21. Ibid.

22. Albu and Schuld, "Beside Ourselves," 2.

23. Won Ju Lim, phone interview, August 15, 2020.

24. Ibid.

25. Ibid.

26. Ibid.

27. Comaroff and Ong, *Horror in Architecture,* 120.

28. Won Ju Lim, phone interview, August 15, 2020.

29. Albu and Schuld, "Beside Ourselves," 6.

30. International Dark Sky Association, "Light Pollution." https://www.darksky.org/light-pollution/

31. Cristina Albu, email communication, April 26, 2021.

32. Christopher Ottinger, email communication, June 23, 2014.

33. Marcia Tanner, phone interview, April 22, 2021.

34. Marcia Tanner, email communication, April 25, 2021.

35. Ibid.

36. Mark A. McCutcheon, *The Medium is the Monster: Canadian Adaptations of Frankenstein and the Discourse of Technology* (Edmonton, Athabasca University Press, 2018), 59-60.

37. Ibid., 60.

38. Shelley, "Frankenstein," 184.

MATERIAL ECOLOGIES OF DISCURSIVE SKINS:

BIO-PHYSICAL ENVELOPES / SENSING BOUNDARIES / EMOTIVE INTERMEDIARIES

KATHY VELIKOV

ABSTRACT

This essay addresses theories of technology within posthuman frameworks that decenter narratives of progress and control to reposition humans within more extensive relational and intersubjective webs and environmental forces. Architectural practices are brought into discourse with aesthetic theories and political ecologies motivated by material languages and material immanence. The architectural envelope, or skin, is explored as a critical material system whose hybrid nature has the capacity to build relational communities and sympathies between human and nonhuman entities, as well as between humans and the environment. Three experimental responsive envelope projects co-created by the author and developed as fragile, collaborative, non-hierarchical, quasi-bio and techno assemblages are positioned discursively. Their use of material assemblies and of sensing and interaction technologies, embodies and makes apparent the politics of boundaries, environmental mediation, material ontologies, and aesthetics. In this, *The Stratus Project, PneuSystems,* and *Nervous Ether* offer alternate trajectories for pursuing and interrogating experimental design work in building technology and architectural material systems, that challenge purely technological frameworks in favor of more discursive and emotive subjectivities.

+ responsive envelopes
+ material systems
+ posthuman theory

Figure 1. *PneuSystems* detail of pneumatic skin prototype installation ©RVTR.

Towards Hybrid Relational Practices

The Carrier Bag Theory of Fiction (1986) by author Ursula K. Le Guin, builds upon anthropologist Elizabeth Fisher's theory that the first "cultural device" was likely not the weapon or spear used by the hero to kill with but, more likely, something with which to transport plants, seeds, and gathered edibles back to the family: some sort of bag, sling, basket, container, holder, or vessel, a "recipient," a "thing that holds something else."[1] In her essay, Le Guin extends Fisher's theory not only to fiction writing—arguing for stories about people as opposed to heroes—but to technology as well. If technology is conceived as heroic and triumphant, she writes, it too will ultimately be tragic. She suggests that far more meaningful stories can be told if "one avoids the linear, progressive, Time's-(killing)-arrow mode of the Techno-Heroic and redefines technology and science as primarily cultural carrier bag rather than weapon of domination."[2] According to this alternate formulation, there is not one hero but instead a gathering of people who create, and all of whom are imperfect. In the "carrier bag" unlikely things may be brought together, and sometimes they get entangled, jumbled. These things, according to Le Guin, become powerful when they're brought in relation with one another, and then in relation with humans.

Architectures that engage with technology and science seem more often than not to be cast within master narratives of progress, both by their authors and audiences. This is especially so in neoliberal research universities where the underlying goal of design research in technology is typically instrumental.[3] Here, research is aimed at solving problems (ideally problems identified by some sort of heroic terminology such as 'grand challenges') and at advancing state of the art technology through incremental yet measurable achievements.[4] Reduction of carbon emissions is typically at, or near, the top of the list of grand challenges as carbon is believed the greatest source of global warming and planetary climate destabilization. The urgency of this challenge is certain, as is the impact it is already having on the planet's species, including its human ones. Clearly, time is not on our side. The Intergovernmental Panel on Climate Change (IPCC)'s *Sixth Assessment Report* (2021) predicts significant and irreversible extreme weather events due to human-caused climate change, including imminent sea level rise, a condition that only several years ago had not been anticipated to occur during this century.[5]

However, when framing carbon emissions, energy usage, or climate change as challenges or problems that need to be overcome, solved, conquered, or won, we invite solutions that assume an adversarial and combative role. Technology, computation, and the hard sciences are honed to focus on these identifiable challenges, and they are marshalled into action by funding calls that claim to support interdisciplinary collaboration yet are entirely focused on research that has a "transformative impact" or the "potential for transformative change."[6] Transformations and changes are thus understood as externalities: deliverables that are objectified, measured, able to be quantified and evidenced. However, is this enough to truly meet the challenges we face? In a more recent essay from 2017, Le Guin argues that:

> in the midst of our orgy of being lords of creation, texting as we drive, it's hard to put down the smartphone and stop looking for the next technofix. Changing our minds is going to be a big change. To use the world well ... we need to relearn our being in it.[7]

According to Le Guin, the more significant change that is necessary is the one inside us, the change that concerns humankind's interrelations and intersubjectivity with the web of beings and things of which we are a part. This is not a case of turning away from science but rather of adding to it, for as Le Guin states, "[w]e need the languages of both science and poetry to save us from merely stockpiling endless "information" that fails to inform our ignorance or our irresponsibility."[8] This reorientation of relations with the world prioritizes subjectivity, empathy, and kinship. Moreover, philosopher of science and nonlinear systems, Isabelle Stengers, argues that the condition of climate change makes it urgent to both think 'and feel' Gaia. To do this she suggests we "learn telling other tales," stories that foster "an ongoing care and concern for the fragility of the assemblage, for the maintenance of what is always a more-than-human interdependence."[9] While Stengers turns to Donna Haraway's "SF tales" (String Figures, Science Fact, Speculative Fabulation, Science Fiction, Speculative Feminism, So Far) as "ideas we use to think other ideas with,"[10] and Le Guin suggests poetry as a "language of compassionate fellowship with other beings,"[11] I would like to add to this conversation the 'language of matter' as one of the critical "matters we use to think other matters with."[12] It seems to me that material practices and their architectural constructs have a compelling capacity to bridge across knowledge domains in order to build relational communities that communicate between human and nonhuman entities, and in particular between humans and the environment.

In my practice with Geoffrey Thün at RVTR, the material construct of the architectural envelope is the site where we explore and experiment with such questions.[13] Alongside our collaborators, we work with sensing technologies to create hybrid artifacts that exist between matter, environmental forces, and communication devices which explore aesthetic, performative, and cognitive relations with our surrounding milieu. The projects are realized as full-scale physical prototypes, developed through the interaction and integration of matter, geometry, information and fabrication technologies, and environmental forces. Like Le Guin's "Carrier Bag Theory" these projects are intended to ask more questions than they answer, gathering into their conceptualization and larger narrative fragments of disciplinary and scientific histories, influences from art practices, and bits of biological thinking. Conceived as unstable and partial hybrids, they operate on the one hand as serious proposals for new species of building skins meant to behave environmentally, and on the other as discursive object intermediaries that operate critically in their eliciting of material and cognitive forms of awareness. These hybrid constructs encourage us to question notions of boundary, be they between humans and the environment or between humans and nonhuman agents. By working through material experimentation, we produce not only things that exist on their own terms, but also tangible fragments of possible architectures, where architectures play functional and symbolic roles, as well as relational and ontological ones.

Envelope as Territory

The building envelope, or skin, is a term familiar to most architects. It is generally understood as the material assemblage that creates a mediating boundary between inside and outside, between one space and another; it regulates atmospheric, thermal, light, auditory, and olfactory exchanges across these domains.[14] Architectural envelopes are concerned with constructing relations across boundaries, such as those which architecture creates between humans and the world, between bodies and the environment, and between humans and other humans. Even the most common building envelope can be understood as a socio-technical apparatus that not only implicates material behaviors and technologies in managing environmental flows across space, but that materializes political relations through its form, image, frameworks of visuality, and processes of formation.[15]

In *Bubbles* (1998), the first book in his Spheres trilogy, philosopher Peter Sloterdijk aimed to dissolve the dualism of inside and outside by exposing a collective existence within multiple insides that are shared and intermingled with the broader envelope of the earth's very atmosphere.[16] The envelope in Sloterdijk's view, is both a technological apparatus fundamental to biophysical existence and an ontological framework essential to the human understanding of self. Bruno Latour summarizes Slotersdijk as follows: "to define humans is to define the envelopes, the life support systems, the *Umwelt* that makes it possible for them to breathe."[17] Within each of these envelopes, the atmosphere—be it conditioned air of the office interior, particulate-filled air amidst our cities, or salt-laden air by the sea—is a co-produced material substance that is most of the time invisible to human eyes. Like climate change, it too can be understood as a hyper-object that exists at multiple scales and in multiple dimensions, resistant to representation.[18] This is the site and material boundary wherein the spatial politics of air and its human and material constituencies become apparent. The envelopes we propose and discuss in this essay occupy this liminal territory.

Figure 2. *The Stratus Project* making sequence of the thick responsive skin ©RVTR.

Figure 3. *The Stratus Project* breathing cells detail ©RVTR.

Sentient Boundaries

The Stratus Project conceived and developed in 2011 is a thick dynamic surface constructed using a networked matrix of physical elements, sensing and actuation technologies, and systems that modify and control air and light. (Figure 2) The project is situated within the trajectory of speculative air-based architecture that aims to dematerialize the building envelope as a project of social and spatial freedom, similar to *Air Architectures* by Yves Klein, Claude Parent, and Werner Ruhnau (1959), the *Environment Bubble* by Reyner Banham and François Dallegret (1965), and inflatable installations by Ant Farm (1970) and José Miguel de Prada Poole's *Instant City* (1971).[19] The word 'stratus' comes from the Latin root *sternere*, meaning to spread out, and like its name suggests, the project is assembled through layers of distributed, lightweight, and suspended networks, interwoven into a diffuse

Figure 4. *The Stratus Project* installation ©RVTR.

surface.[20] A cable-strut tensegrity weave forms a lightweight and deformable mat into which are assembled an array of sensors, actuators, lights, micro-fans and fabric panels impregnated with a diffusing, thermally absorptive phase change material. On the underside of the structure are suspended 'breathing cells': individually actuated translucent polymer tessellations that form a light-diffusing skin, computationally controlled to open and close like gills that enable localized thermal conditioning and air extraction. (Figure 3)

The physical presence of someone breathing, as well as environmental factors including temperature, light, carbon dioxide and airborne pollutants are measured using a network of sensors that communicate to actuators whose movements are designed to trigger fans that locally supply or extract air. Lighting that illuminates the physical space of 'the breather', and blue lights that are triggered when carbon dioxide or pollutant levels increase, haptically register the condition of diminished air quality. (Figure 4) Communicating through these light and movement cues, *The Stratus Project* makes public the apparatus of air control and materializes breathing as an inherently political act. Moreover, its deliberately fragile yet clearly mechanized material presence renders tangible its precarious and exceedingly artificial character that enables the support of life within interior spaces.[21]

A compelling aspect of *Stratus*, according to many who have experienced it, is not its ability to directly respond to air quality and human presence, but the accidental and unexpected behaviors of the envelope during operation. Internal forms of feedback within its distributed sensing technologies produces the condition that in addition to sensing the presence of humans, the skin senses itself. It operates stochastically, moving its breathing cells, lights, and fans not only in direct response to human presence and atmospheric changes, but as it continues to twitch and breathe for some time after people move away from its sensing range. As atmospheric residues linger, a kind of affinity and sympathy result in the lively mechanical assemblage, producing a more provocative effect than its function as a device or instrument.

Thing-Languages and Intermediaries

As a building and spatial practice, architecture trades in languages of matter. These embodied frameworks and cultural norms have been commonly translated and codified into symbols, iconographies, and metaphors. However, these traditional forms of knowledge seem to be breaking down, or at least becoming desperately insufficient and impotent when disrupted by the "intrusion of Gaia" and the impacts of climate destabilization.[22] It is no coincidence that in this context there is a resurgence of materialist philosophy which aims to counter and cast doubt upon the privilege of human agency in shaping the world, as well as in the ability of human epistemologies to know it. While a great deal of 'Speculative Realism' and 'Object Oriented Ontology (OOO)' positions the "hidden" and "unknowable" existence of objects in contrast to human epistemologies,[23] political ecologist Jane Bennett argues that the self-immanence of substances is a "vital materiality" that is self-organizing and that flows through the web of its relations.[24] This non-human vitality argues philosopher Rosi Braidotti, "can no longer be metaphorized as other but needs to be taken on its own terms."[25]

In an early essay from 1916 called "On Language as Such and on the Language of Man" philosopher and cultural critic Walter Benjamin (1892-1940) proposed that all things possess "nameless, nonacoustic languages" and that much of aesthetic practice involves the translation of these "thing-languages" into a shared dialect, which itself could still be a "[language] issuing from matter."[26] He argued that unlike verbal

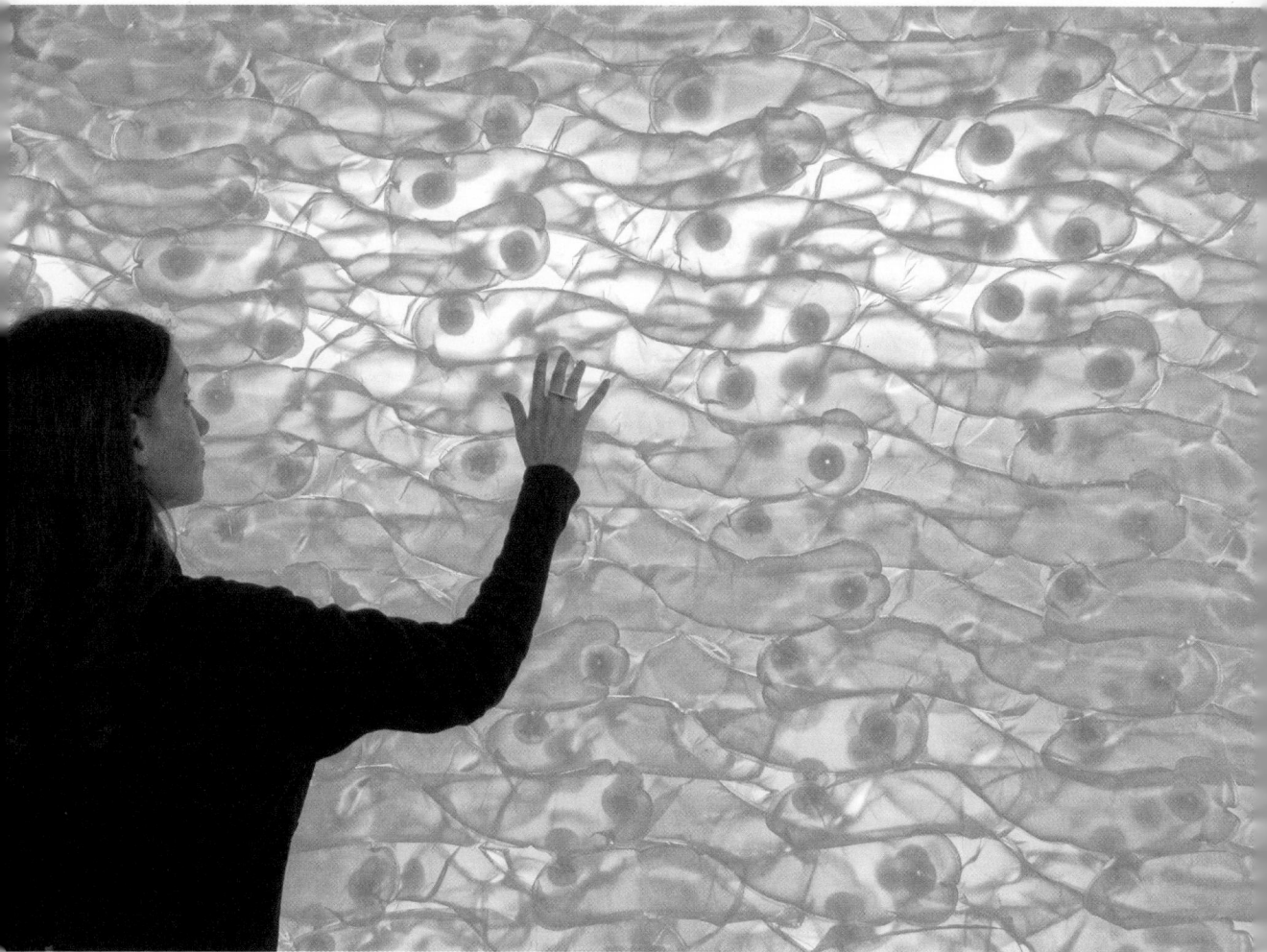

Figure 5. *PneuSystems*
installation of pneumatic skin
prototype ©RVTR.

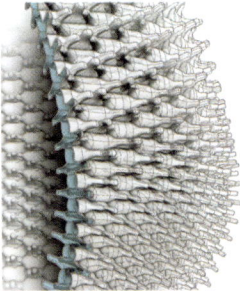

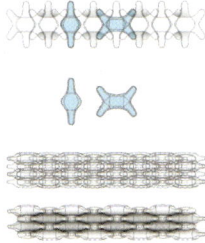

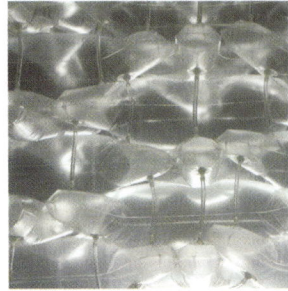

X-Stack Prototype: Aggregation through Stacking

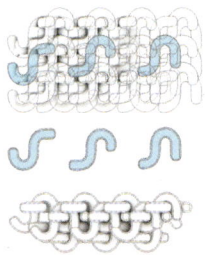

S-Weave Prototype: Aggregation through Weaving

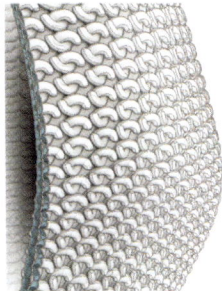

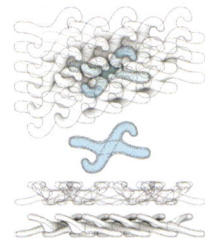

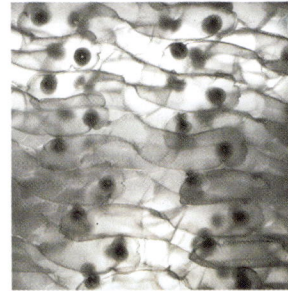

X-Weave Prototype: Aggregation through Braiding / Knitting

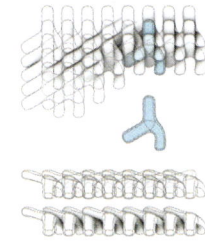

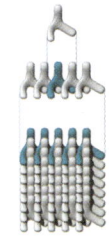

vλ-Weave Prototype: Aggregation through Weaving

Figure 6. *PneuSystems*
interlocking figure array tests
©RVTR.

or symbolic languages, in 'matter-to-material practice' translation occurs not across languages but 'within' them. Media artist Hito Steyerl describes this translation as a practice of "relationism" that presences "unexpected articulations...of objects and their relations" while also having the ability to "become models for future types of connection."[27]

In contemporary artwork that engages such questions of nonhuman embodiment and vitalist behavior within nonhuman objects—as in the media work of Lindsey french and Paola Gaetano Adi—sensing and interaction technologies often serve an "intermediary" role, contributing both to the immanence of their objects and to the creation of a "shared language" that is auditory, vibratory, or respiratory.[28] In architect and artist Phillip Beesley's *Hylozoic Series*, the convulsing and shuddering behavior of the spatial assembly elicits feelings of empathy, affinity, and care.[29] Media scholar Jennifer Gabrys makes the case that "[s]ensing...is not about detecting information 'out there' but about 'tuning' the subjects and conditions of experience to new registers of becoming."[30]

Behavioral Assemblies

Projects *PneuSystems* and *Nervous Ether* continue our experiments with spatial skins that embody both environmental and human relations. In *PneuSystems* (2013-14) we explore the concept and materiality of the *pneu*, a soft form or membrane that is pressurized by air or liquid. The *pneu* was first explored within the realm of biologically informed, lightweight, and responsive architectural structures investigated by Frei Otto (1925-2015) and his colleagues at the Stuttgart Institute for Lightweight Structures.[31] The *pneu* occupies an aesthetic, material category that Otto and his colleagues identified with the subject of taboos: "We can only state without comment that in man the signals of repulsiveness, beauty, and sexuality are given by the characteristic structural form of the skin strained by internal pressure, that is the pneu."[32] Resonant with French philosopher George Bataille's (1897-1962) notion of *l'informe* (the formless), we appreciate the simultaneous attraction and revulsion, and the uncanny familiarity that characterizes the reception of the part-object, part-life like, and unstable figure of the *pneu*.[33] So too with its capacity to decenter human epistemologies and destabilize meaning. Encountering the *pneu*, we encounter another skin, a non-human one, yet one that breathes, is pliant to its environment, and is strangely recognizable.[34] (Figure 5)

Our research into the development of pneumatic skins moves beyond mere conventional formations of packed bubble structures by exploring textile topologies such as braiding and weaving. These form the basis of a series of nested, multi-cellular pneumatic skins whose topologies are deconstructed into discrete units as opposed to strands. They are further developed into networks of interlocking figures able to maintain aggregation without the introduction of external binding forces. They are held together by kinematic constraints, or gripping actions, imposed through shape and mutual arrangement under pressure, as well as through the tendon-like interconnections of pneumatic tubing and valves. (Figure 6) The weave structure also allows us to integrate other figures into the skin, particularly ones that could be actuated through nonlinear behaviors, such as nastic movements found in the stomatal complex of plants and the snapping action of the Venus Flytrap.[35]

The *Nervous Ether* prototype (2013) is constructed as a loose weave of four types of pneumatic membrane figures, interconnected through the apparatus that supplies it with air. Environmental data-feeds of local barometric pressure and wind speed are translated into indirect squeezing and pulsing motions by specific figures within the array. (Figure 7) The installation also incorporates human interaction such that when someone closely approaches the surface, a series of motion-sensor controlled figures rapidly fill with air and move perpendicularly outward. Mimicking human hairs that suddenly stand on a person's skin, *Nervous Ether* reproduces the effects of anxious behavior. The title of the project derives its name from the history of physics, wherein philosophers and scientists speculated on the existence of a "nervous ether," a material atmosphere that was believed to conduct the vibrations of heat, light, sound, electromagnetic impulses and mechanical frictions.[36] In the late nineteenth century, physicist John Tyndall (1820-1893) theorized that the "transported shiver of bodies" of the cosmos and the stars could be intimately felt within our own physical bodies and consciousness.[37] We were compelled by this idea believing that through the non-verbal language of shivering, breathing, and pulsing, we might be able to find a connection between that which is incomprehensible in the atmosphere and in our own bodies.

Proposal for a Decentering of,

Stratus Project, *PneuSystems*, and *Nervous Ether* are models for technological exploration within an architecture that diverges from linear paths and from narratives of progress to occupy instead a more ambiguous and hybrid territory. "In order to produce a credible quantitative shift," writes Braidotti, "we need conceptual and methodological transformations," and a defamiliarization of "mental habits."[38] It is through the immanence of these collaborative, non-hierarchical, quasi-bio, and techno assemblages that new forms of interrelatedness are explored in our encounters with the envelope. These projects decenter and defamiliarize subjects and subjectivities by their simultaneous engagement with multiple and sometimes contradictory domains of knowledge. Their technologies operate on the one hand instrumentally as informational registers, and on the other, as things in themselves that behave and produce emotive responses that fall outside of operative paradigms. They engage both architectural practicalities of environmental control and critical theories of biopolitics and relational aesthetics. Their intention is not

Figure 7. *Nervous Ether* detail of installation ©RVTR.

to diminish existing forms of architectural practice, but instead to add alternatives.

Acknowledgements

Sincere thanks to Franca Trubiano and Amber Farrow for their editorial feedback and suggestions during the development of this chapter.

Collaborators

Projects described in this article have been undertaken collaboratively with my partner Geoffrey Thün, with input from expert colleagues, and with the contributions of our research assistants. For a detailed listing of project credits, see www.rvtr.com.

Funding

The Stratus Project, PneuSystems and *Nervous Ether* were funded through the Taubman College Research Through Making Grant Program and matching support from the University of Michigan Office of Research. Additionally, *The Stratus Project* was supported through a Social Science and Humanities Research Council (SSHRC) Research Creation Grant and *PneuSystems* received funding support from the AIA UpJohn Research Initiative from the Board Knowledge Committee and College of Fellows of the American Institute of Architects.

ENDNOTES

1. Ursula K. Le Guin, *The Carrier Bag Theory of Fiction* (Newcastle: Ignota, 2019 [1988]), 29.

2. Ibid., 36.

3. For a discussion on neoliberalism and its institutions see Wendy Brown, *Undoing the Demos: Neoliberalism's Stealth Revolution* (New York: Zone Books, 2015), 175-200.

4. Janet A. Weiss and Anne Khademian, "What Universities Get Right – and Wrong – About Grand Challenges," *Inside Higher Education.* September 13 2019, https://www.insidehighered.com/views/2019/09/03/analysis-pros-and-cons-universities-grand-challenges-opinion.

5. Valerie Mason-Delmotte et al. (eds) IPCC, 2021: Summary for Policymakers, in *Climate Change 2021: The Physical Science Basis. Contribution of Working Group I to the Sixth Assessment Report of the Intergovernmental Panel on Climate Change* (Cambridge: Cambridge University Press, 2021), 28-30. https://www.ipcc.ch/report/ar6/wg1/downloads/report/IPCC_AR6_WGI_SPM.pdf

6. The National Science Foundation (NSF) in 2005 defined "Transformative Research" as "research that has the capacity to revolutionize existing fields, create new subfields, cause paradigm shifts, support discovery, and lead to radically new technologies..." This terminology and ethos have permeated scientific research mandates ever since. *National Science Board, Enhancing Support for Transformative Research at the National Science Foundation* (NSB 2007), 1. https://www.nsf.gov/nsb/documents/2007/tr_report.pdf

7. Ursula K. Le Guin, "Deep in Admiration," in Anna Tsing et. el eds. *Arts of Living on a Damaged Planet* (Minneapolis: University of Minnesota Press, 2017), M15.

8. Ibid, M16.

9. Isabelle Stengers, "Gaia, the Urgency to Think (and Feel)," from *The Thousand Names of Gaia: From the Anthropocene to the Age of the Earth Colloquium*, Rio de Janero, Brazil (Sept. 15-19, 2014), 7. https://osmilnomesdegaia.files.wordpress.com/2014/11/isabelle-stengers.pdf

10. Ibid., 8.

11. Le Guin, "Admiration," M16.

12. Donna Haraway, "Playing String Figures with Companion Species," *Staying With the Trouble, Making Kin in the Chthulucene* (Durham NC: Duke University Press, 2016), 2910-12.

13. For projects referred to in this article and lists of contributors see, http://www.rvtr.com/projects/stratus, http://www.rvtr.com/projects/pneusystems, http://www.rvtr.com/projects/nervous-ether

14. Michelle Addington, "Contingent Behaviors," *AD* 79, no. 3 (2009): 12-17

15. Alejandro Zaera-Polo, "The Politics of the Envelope," *Volume 17* (2008), 76-105.

16. Peter Sloterdijk, *Bubbles: Spheres Volume I: Microspherology*, translated W. Hoban. (Los Angeles: Semiotext(e), 2011).

17. Bruno Latour, "A Cautious Prometheus? A Few Steps Toward a Philosophy of Design (with Special Attention to Peter Sloterdijk)," (2008): 8. http://www.unsworn.org/docs/Latour-A_Cautious_%20Promethea.pdf

18. Timothy Morton, *Hyperobjects: Philosophy and Ecology after the End of the World* (Minneapolis: University of Minnesota Press, 2013).

19. See Yves Klein, "The Evolution of Art Towards the Immaterial," in Peter Noever and Francois Perrin ed., *Air architecture: Yves Klein* (Ostfildern-Ruit: Hatje Cantz, 2004), 35-35, and Marc Decausse, *The Inflatable Moment: Pneumatics and Protest in '68* (New York: Princeton Architectural Press, 1999).

20. Colin Ripley, Kathy Velikov, Geoffrey Thün, "Soft Goes Hard," *Bracket [goes soft]*, Bhatia, N. and Sheppard, L. eds. (Barcelona, Actar, 2012), 34-48.

21. For further technical and operational descriptions of The Stratus Project, see Kathy Velikov, Geoffrey Thün, Colin Ripley, "Thick Air," Journal of Architectural Education (JAE) 65, no. 2 (2012): 69-97. Geoffrey Thün, Kathy Velikov, Mary O'Malley, Lisa Sauvé, "The Agency of Responsive Envelopes: Interaction, Politics and Interconnected Systems," *International Journal of Architectural Computing (IJAC) Special Issue: Architectural Robotics* 3, no.10 (2012): 377-400.

22. Isabelle Stengers, "Autonomy and the Intrusion of Gaia," *South Atlantic Quarterly* 116, no. 2 (April 2017): 381-400.

23. For Speculative Realism see Graham Harman, Towards Speculative Realism (Washington DC: Zero Books, 2010); for OOO see Graham Harman, *The Quadruple Object*, (Alresford: Zero Books, 2011) and Levi Bryant, The Democracy of Objects (Ann Arbor: Open Humanities Press, 2011).

24. Jane Bennett, *Vibrant Matter: a political ecology of things* (Durham & London: Duke University Press, 2010).

25. Rosi Braidotti, "Animals, Anomalies, and Inorganic Others," *PMLA* 24, no. 2 (March 2009): 527-8.

26. Walter Benjamin, W. 1916, "On Language as Such and on the Language of Man," *Walter Benjamin Selected Writings, Volume 1 1913-1926*, (Cambridge and London, Belknap Press Harvard, 1996 [1916]), 73.

27. Hito Steyerl, "The language of things," *Transversal Texts* (eipcp - European institute for Progressive Cultural Policies, June 2006). http://eipcp.net/transversal/0606/steyerl/en

28. Kayla Anderson, "Object Intermediaries: How New Media Artists Translate the Language of Things," *Leonardo* 47, no. 4 (2014): 352-59.

29. Philip Beesley, "Feeling Matter: Empathy & Affinity in the Hylozoic Series," in Espen Gangvik ed., *Meta.Morf A Matter of Feeling* (Trondheim: TEKS Publishing, 2012), 182-85.

30. Jennifer Gabrys, *Program Earth* (Minneapolis: University of Minnesota Press, 2016), 32-33.

31. Klaus Bach et al., *IL9 Pneus in Nature and Technics*, (Stuttgart: Institut für Leichte Flachentragwerke (IL), University of Stuttgart, 1977).

32. Ibid., [19].

33. Yve-Alain Bois and Rosalind E. Krauss, *Formless: A User's Guide* (New York: Zone Books, 1997), 15.

34. Kathy Velikov, Geoffrey Thün, Mary O'Malley, Wiltrud Simbuerger, "Nervous Ether: Soft Aggregates, Interactive Skins," *Leonardo* 47, no. 4 (2014): 344-51.

35. Kathy Velikov, Geoffrey Thün, Mary O'Malley, Lars Junghans, "Computational and physical modeling for multi-cellular pneumatic envelope assemblies," *International Journal of Architectural Computing (IJAC)* 2, no. 13 (2015):143-68.

36. Stephen Connor, *The Matter of Air* (London: Reaktion, 2010), 148-72.

37. Ibid., 172.

38. Rosi Braidotti, *Posthuman Knowledge* (Cambridge: Polity Press, 2019), 77.

CONTRIBUTORS

EDITORS

FRANCA TRUBIANO

Franca Trubiano is Graduate Group Chair of the Ph.D Program in Architecture and associate professor at the Weitzman School of Design of the University of Pennsylvania. She received her Ph.D from UPenn in the History and Theory of Architecture and is a Registered Architect with *l'Ordre des Architects du Québec*. Since 2021, she has been co-director of Penn's Mellon funded, Humanities + Urban + Design Initiative. Trubiano is the author of *Building Theories, Architecture as the Art of Building* (Routledge 2023), and co-editor of *Women [Re] Build; Stories, Polemics, Futures* (ORO, 2019). Her edited book *Design and Construction of High-Performance Homes: Building Envelopes, Renewable Energies, and Integrated Practice* (Routledge Press 2012), was translated into Korean and awarded the 2015 Sejong Outstanding Scholarly Book Award. Trubiano conducts funded research on 'Forced Labor in the Building Industry,' as well as on 'Fossil Fuels, the Building Industry, and Human Health.'

AMBER FARROW

Amber Farrow earned an M. Arch from the University of Pennsylvania in 2021. During her time there she founded the student club, Dinner Discourse which fosters an interdisciplinary conversation between students over a shared meal. In addition, she held the Professional Development and Programming Chair for Penn Women in Design, whose mission it was to promote equity in the profession and encourage a more inclusive pedagogy in academia. Amber is currently working as an Architectural Designer in New York and has practiced internationally at firms in Tokyo, London, and Bermuda. She is recipient of the 2020 Van Alen Traveling Fellowship, which is awarded annually to one architecture and one landscape architecture student at the Weitzman School of Design; she is also recipient of the 2018 Stanley G. Kennedy Architectural Award by the Institute of Bermuda Architects.

MARÍA JOSÉ FUENTES

María José Fuentes is an architectural designer at KieranTimberlake in Philadelphia, PA. Her professional interests include equitable design strategies, community integration, and the architectural performance of color. Her work has been featured in numerous online and print publications, including *Pressing Matters, Metropolis Magazine*, and *A+D Magazine*. She earned an M. Arch from the University of Pennsylvania where she completed her thesis on the architectural implications of color and authorship. She has won numerous awards for her work in writing, mentorship, and image-making, including the E. Lewis Dales Traveling Fellowship, the Weitzman School of Design Diversity Scholarship, and the Future 100 Competition by *Metropolis Magazine*. She was an active member of Penn Women in Design during her time at the University of Pennsylvania and continues to champion equity in design.

SUSAN KOLBER

Susan Kolber completed her Master of Landscape Architecture and Master of Architecture at the University of Pennsylvania Stuart Weitzman School of Design in 2020. Her graduate and postgraduate research has explored ethical, aesthetic, and environmental practices that reimagine human relationships with the built environment and other living beings. Susan has led initiatives through AIA's Equity by Design and Penn Women in Design, and this advocacy work, including mentorship, data collection, and storytelling, plays a crucial role in her professional and academic practices.

MARTA LLOR

Marta Llor is an architectural designer practicing in New York. She graduated with a Master of Architecture from the University of Pennsylvania, and holds a BFA in Architectural Design from Parsons School of Design. Throughout her career, Marta has held leadership roles in various organizations, including Penn Women in Design (PWID) where she was the Professional Development Chair for three years. She led teams for the organization of two conferences at Penn and collaborated with other student and professional organizations. She believes architecture is the truest aspiration of our values, and strives to help design a more equitable world. Her designs stem from a curiosity about geometric explorations and narratives about space, time, and agency. Marta currently practices inn New York and has designed internationally in Greece, the UK and the UAE.

AUTHORS

VIOLA AGO

Viola Ago (b. Lushnjë, Albania) is an architectural designer, educator, and practitioner. She is the director of MIRACLES Architecture and assistant professor at the Creative School, Interior Design, Toronto Metropolitan University. Viola was awarded the Wortham Fellowship at Rice University, the Yessios Visiting Professorship at the Ohio State University Knowlton School of Architecture, and the Muschenheim Fellowship at the University of Michigan Taubman College of Architecture. Viola earned her M.Arch degree from SCI-Arc, and a B.Sc (Arch) from the Toronto Metropolitan University (previously Ryerson University). Her written work has been published by Routledge, Park Books, and Wiley, as well as in *Log, AD Magazine, Offramp, Acadia Conference Proceedings, JAE, TxA, Architect's Newspaper,* and *Archinect.* Her design and research work has been exhibited in Los Angeles, Boston, Houston, Ghent NY, San Francisco, Miami, Columbus, Ann Arbor, and Cincinnati.

RACHEL ARMSTRONG

Rachel Armstrong is professor of Regenerative Architecture at the Department of Architecture, Campus Sint-Lucas, Ghent, KU Leuven. She was a 2010 Senior TED Fellow, having established an alternative approach to sustainability that coupled agentised matter with the computational properties of the natural world, a platform for the built environment she calls 'living' architecture. She holds a First-Class Honours degree from the University of Cambridge (Girton College) and a medical degree from the University of Oxford (The Queen's College). She is a Member of the Royal College of New Zealand General Practitioners 2005-2015 and awarded a PhD (2014) from the University of London (Bartlett School of Architecture). She is author of several books including *The Art of Experiment: Post-pandemic Knowledge Practices for 21st Century Architecture and Design* with Rolf Hughes (Routledge, 2020); *Experimental Architecture: Prototyping the unknown through design-led research* (Routledge, 2019), and *Soft Living Architecture: An alternative view of bio-informed design practice* (Bloomsbury, 2018).

DORIT AVIV

Dorit Aviv, PhD, is assistant professor of Architecture at the Weitzman School of Design, specializing in sustainability and environmental performance. She is the director of the Thermal Architecture Lab, a cross-disciplinary laboratory at the intersection of thermodynamics, architecture, and material science. Her work examines paths to decarbonization of the built environment through design and policy. Specifically, she looks at the potential synergies between renewable environmental forces and architectural materials and forms, and their impact on buildings' energy performance, air quality, and human health. Aviv holds a PhD in architectural technology from Princeton University and is a licensed architect. She is the recipient of a Holcim Award for Sustainable Design and Construction for a prototype of combined evaporative and radiative cooling in desert climate.

ANNE BEIM

Anne Beim is professor of architecture in the School of Architecture at the Royal Danish Academy. Since 2014, she has co-chaired the graduate program, SET - Settlement, Ecology and Tectonics. Since 2004, she has been Chair of CINARK - Center for Industrialized Architecture whose mission calls for bridging the gap between architectural education, the construction industry, and the architectural profession. In 2000, she completed her Ph.D. in Architecture at the Royal Danish Academy of Fine Arts, School of Architecture and as a visiting scholar (1995-96) studied with Professors Marco Frascari and David Leatherbarrow at the University of Pennsylvania. Her research is focused on ecology, tectonics, material studies, building culture and practice, and architectural theory. Selected publications include the co-authored Biogenic Construction: Materials, Architecture & Tectonics (CINARK – The Royal Danish Academy, 2023), Circular Construction: Materials, Architecture & Tectonics (CINARK–KADK, 2019), Sustainability in Scandinavia: Architectural Design and Planning (Edition Axel Menges, 2018), Towards an Ecology of Tectonics - The Need for Rethinking Construction in Architecture (Edition Axel Menges, 2015), and her PhD Dissertation: Tectonic Visions in Architecture (Kunstakademiets Arkitektskoles Forlag, 2004).

MARTINA DECKER

Martina Decker is associate professor at the Hillier College of Architecture and Design at the New Jersey Institute of Technology (NJIT). She is originally from Munich, Germany where she received her professional architecture degree from the University of Applied Sciences. Martina has worked on a wide range of award-winning projects that show her penchant for interdisciplinary work. They include art installations, consumer products, and buildings. She pursues design innovation through the exploration of emergent materials, working directly with various types of smart materials and nanomaterials. At NJIT, Martina continues her endeavors in her Material Dynamics Lab and directs the Idea Factory, an interdisciplinary work platform.

SONJA DÜMPELMANN

Sonja Dümpelmann co-directs the Rachel Carson Center at Ludwig-Maximilians-Universität Munich where she is professor and chair in environmental humanities. She was previously professor and acting chair (2023/2024) of the Department of Landscape Architecture, at the Stuart Weitzman School of Design of the University of Pennsylvania. Dümpelmann is a historian of urban landscapes and environments in the nineteenth and twentieth centuries. Her most recent monographs are *Landscapes for Sport: Histories of Physical Exercise, Sport, and Health* (ed., Washington DC: Dumbarton Oaks Research Library and Collection, 2022), and the award-winning *Seeing Trees: A History of Street Trees in New York City and Berlin* (Yale University Press, 2019).

BEHNAZ FARAHI

Behnaz Farahi is assistant professor of Design at California State University, Long Beach. She is an award-winning designer, creative technologist, and critical maker. Trained as an architect, she explores how to foster an empathetic relationship between the human body and the space around it through the implementation of emerging technologies. Her goal is to enhance the interaction between human beings and the built environment by following morphological, and behavioral principles inspired by natural systems. Her work addresses critical issues such as emotion, perception, and social interaction. She specializes in computational design, interactive technologies, and digital fabrication technologies. She recently completed her PhD in Interdisciplinary Media Arts and Practice at the USC School of Cinematic Arts. She also holds a bachelor and two master's degrees in architecture.

S.E. EISTERER

S.E. Eisterer is assistant professor of Architectural History and Theory at Princeton University. Before joining Princeton, she was an assistant professor at the Stuart Weitzman School of Design at the University of Pennsylvania and member of the Executive Board of Gender, Sexuality, and Women's Studies. S.E. is interested in histories of collectivity, difference, and dissent in architecture. Her scholarly work in modern architecture and urban culture focuses on spatial histories of dissidence and resistance, intersectional feminism, queer theory, gender studies, environmental history and labor theory. Her work has been published in *Aggregate, Architectural Histories, Architecture Beyond Europe, Ediciones ARQ,* and *Log.* She has received a Carter Manny Award by the Graham Foundation, the Bruno Zevi Award, a Radcliffe Fellowship at Harvard University, the Pearl Resnick Fellowship at the United States Holocaust Memorial Museum, and a Humboldt Fellowship for Senior Researchers in Germany. She lives and works in Philadelphia.

AYASHA GUERIN

Ayasha Guerin is an interdisciplinary artist, researcher, and curator who lives between Berlin and Vancouver. She is assistant professor of Black Diaspora Studies in the Department of English at the University of British Columbia. Her current book project *Making Zone A: Race Nature and Resilience on New York's Most Vulnerable Shores,* traces how colonial capitalism has cultivated a hierarchy of racial difference on urban landfill and considers how activism on the waterfront has been shaped by diasporic relationships and interspecies entanglements.

AROUSSIAK GABRIELIAN

Aroussiak Gabrielian is an environmental designer and bioartist working with living organisms, natural systems, and atmospheric phenomena. Her work aims to torque our imaginaries to help us re-think our interactions with both human and non-human agents on this planet. Aroussiak's work has received numerous design awards and has been exhibited internationally at various institutions, including SXSW, Ars Electronica, the Eli & Edith Broad Museum Art Lab, A+D Museum Los Angeles, Science Gallery Detroit, among others. At the University of Southern California (USC), Aroussiak is assistant professor of Landscape Architecture + Urbanism at the School of Architecture, Affiliate Faculty of Media Arts + Practice at the School of Cinematic Arts, and founding director of the Landscape Futures Lab. Outside academia, Aroussiak is founding design principal of foreground design agency, a critical practice that aims to dismantle structures of power and privilege that render specific humans, species, and matter silent.

ANDREA LING

Andrea Ling is an architect, artist, and researcher working at the intersection of design, fabrication, and biology. Her work focuses on how the critical application of biological and computational processes can move society away from exploitative systems of production to regenerative ones. She is the 2020 S+T+ARTS prize winner for her work as the 2019 Creative Resident at Ginkgo Bioworks designing the decay of artifacts to access material circularity. She graduated from the MIT Media Lab, Mediated Matter group, where she was a research assistant on the Aguahoja project. Former project lead at Philip Beesley Architect, she led international kinetic installations and textiles for the office's collaboration with Iris van Herpen. Andrea is an architect with the Ontario Association of Architects and founding partner at design-GUILD, a Toronto-based art & design collective. She is currently an A&T Fellow at the Institute of Technology and Architecture at ETH Zurich.

JULIA LOHMANN

Julia Lohmann designer, researcher, and educator investigates and critiques the ethical and material value systems that underpin our relationship with nature. She is professor of Design Practice at Aalto University, Helsinki, Finland. Julia is passionate about eco literacy and ocean protection. In 2013, she founded the Department of Seaweed, a transdisciplinary community of practice exploring the sustainable development of seaweed as a material for making. Julia Lohmann believes that any exploration of biomaterials needs to be based on amplifying their regenerative eco-systemic impact. In her practice and teaching she promotes an empathic, more-than-human-centric mindset and views design as a way of connecting knowing, caring, and acting across disciplines and different levels of complexity. Julia contributes to research consortia that relate design to biomaterials, science, and ecology. She holds a PhD from the Royal College of Art, and her work is part of major public and private collections worldwide.

MAE-LING LOKKO

Mae-ling Lokko is an architectural scientist, designer, and educator from Ghana and the Philippines whose work centers on the upcycling of agro-waste and biopolymer materials into clean building material systems. Lokko holds a Ph.D and MS in Architectural Science from the Center for Architecture, Science and Ecology at Rensselaer Polytechnic Institute and a B.A from Tufts University. Lokko is assistant professor at the Yale School of Architecture and has previously taught at Rensselaer and the Cooper Union in New York. She is founder of Willow Technologies, Ltd. based in Accra, Ghana. Her work was nominated for the Visible Award 2019, the Royal Academy Dorfman Award 2020, and the Hublot Design Prize 2019. Lokko's recent projects have been exhibited at Z33 House for Contemporary Art, Architecture and Design, at Somerset House, London, the Triennale Milano, the Serpentine Gallery, London, Radialsystem, Berlin, the Royal Danish Academy of Fine Arts, and the Luma Foundation, Arles.

LAIA MOGAS-SOLDEVILA

Laia Mogas-Soldevila is an architect and assistant professor in Architecture at the University of Pennsylvania's Weitzman School of Design where her work focuses on novel material practices in architecture, as influenced by the fields of material science and engineering. She holds two master's degrees from the Massachusetts Institute of Technology (MIT) on Design Computation and Media Arts and Sciences, and an interdisciplinary doctorate on biomaterial architectures from Tufts University. She has worked and taught at Cornell

University and MIT. She has co-directed DumoLab design studio since 2008. Her work is published in science, design, and engineering journals and participates in cross-disciplinary venues and conferences such as the MRS Materials Research Society, IASS, eCAADe, Biofabricate, and ACADIA, and at the MoMa in New York, the Copper Hewitt Design Museum, Brooklyn's New Lab, the Lisbon Architecture Triennial, the Istanbul Design Biennial, and the Pompidou Center in Paris.

PATRICIA OLYNYK

Patricia Olynyka's work explores the ways in which technology shapes our understanding of our place in the world. Collaborating across disciplines to explore the mind-brain relationship, *umwelt* theory, and the phenomenology of perception, her sound and video installations frequently call on viewers to expand their awareness of the environments they inhabit. In 2007, Olynyk was appointed inaugural director of the unified Graduate School of Art and the Florence and Frank Bush Professor of Art at Washington University in St. Louis. In 2020, she was named inaugural Medicine + Media Arts Fellow at UCLA's Art|Sci Center. She also co-directs the Leonardo/ISAST *NY LASER* program, which promotes cross-disciplinary exchange between artists, scientists, and humanists. Her work has been exhibited in solo and group exhibitions nationally and internationally and her writing has been featured in publications that include the *Routledge Companion to Biology in Art and Architecture, Leonardo Journal, and the Angewandte Book Series.*

STEFANA PARASCHO

Stefana Parascho is a researcher, architect, and educator whose work lies at the intersection of architecture, digital fabrication, and computational design. She is currently assistant professor at the Swiss Federal Institute of Technology in Lausanne (EPFL), School of Architecture, Civil and Environmental Engineering (ENAC) where she founded the Lab for Creative Computation (CRCL). Her research explores multi-robotic fabrication methods and their relationship to architectural design to strengthen the connection between design, structures, computational tools, and robotic fabrication methods. Prior to joining EPFL, Stefana was an assistant professor at Princeton University's School of Architecture, directing the CREATE Laboratory Princeton. Stefana completed her doctorate in 2019 at ETH Zurich, Gramazio Kohler Research, with the thesis *Cooperative Robotic Assembly.* She received her Diploma in Architectural Engineering in 2012 from the University of Stuttgart and worked with DesignToProduction Stuttgart and Knippers Helbig Advanced Engineering.

REBECCA POPOWSKY

Rebecca Popowsky is a landscape architect and research associate at OLIN, where she leads the firm's research and development group, OLIN Labs. Her project-based and grant-funded research includes waste-based material design and innovation in practice-based research models. Since joining the studio in 2009, Rebecca has contributed to design, planning, and construction projects. Her portfolio includes Canal Park in Washington, DC, Dilworth Park in Philadelphia, and collaboration with the Army Corps of Engineers to restore the FEMA floodplain at the Potomac Park Levee on the National Mall. Rebecca is also a Lecturer at the Weitzman School of Design where she teaches core and advanced design studios and professional practice. Rebecca earned dual master's degrees in Architecture and Landscape Architecture from The Weitzman School of Design and a Bachelor of Arts in Architecture and Urban Studies from Yale University. She is currently undertaking a PhD in Practice-Based Research at Virginia Tech.

GUNDULA PROKSCH

Gundula Proksch is a scholar, licensed architect, and associate professor of Architecture at the University of Washington in Seattle. She is founding director of the Circular City and Living Systems Lab, an interdisciplinary research group investigating transformative strategies for sustainable urban futures. In collaboration with international research consortia, Professor Proksch is principal investigator (PI) of National Science Foundation (NSF) funded research projects CITYFOOD and AquaponicsOpti, which investigate urban integration of aquaponics. She is also a Co-PI of an NSF EFRI ELiS grant on engineered living systems in the built environment. Her book *Creating Urban Agricultural Systems: An Integrated Approach to Design* (Routledge, 2016) is the first sourcebook on approaching urban agriculture from a systems perspective. Professor Proksch's interdisciplinary research builds on her professional practice in Europe and the United States with architects David Chipperfield in London and Richard Meier, Stan Allen, and Roger Duffy of SOM in New York.

JENNY E. SABIN

Jenny E. Sabin is an architectural designer whose work is at the forefront of a new direction for 21st century architectural practice that investigates the intersections of architecture and science and applies insights and theories from biology and mathematics to the design of responsive material structures and ecological spatial interventions for diverse audiences. Sabin is the Arthur L. and Isabel B. Wiesenberger Professor in Architecture and the inaugural chair for the new multi-college Department of Design Tech at the Cornell College of

Architecture, Art, and Planning where she established a new advanced research degree in Matter Design Computation. She is principal of Jenny Sabin Studio, an experimental architectural design studio based in Ithaca and Director of the Sabin Design Lab at Cornell AAP. Her book, *LabStudio: Design Research Between Architecture and Biology*, co-authored with Peter Lloyd Jones was published in July 2017. That same year, Sabin won MoMA PS1's Young Architects Program with her submission, *Lumen*.

LUCINDA SANDERS

Lucinda Sanders is CEO and Partner at OLIN. She has led the design of many of OLIN's signature projects, including Comcast Center Plaza and Central Delaware Riverfront Master Plan in Philadelphia, Pennsylvania; Gap Headquarters in San Francisco, California; and Fountain Square in Cincinnati, Ohio. Her current projects include an urban design plan for the 30th Street Station precinct in Philadelphia and large-scale urban developments in China. Lucinda is actively involved on multiple boards and committees dedicated to the advancement of the field of landscape architecture, urban design, and planning, including the Landscape Architecture Foundation, the Lady Bird Johnson Wildflower Center Advisory Board, and the CEO Roundtable of Landscape Architects. Lucinda first studied landscape architecture at Rutgers University and went on to earn a master's degree in landscape architecture from the University of Pennsylvania. She is currently an adjunct professor of Landscape Architecture at the University of Pennsylvania and has lectured internationally for major universities and professional organizations.

CLARISSA TOSSIN

Clarissa Tossin is a visual artist who uses moving-image, installation, sculpture, and collaborative research to engage the suppressed counter-narratives implicit in the built environment. Tossin's work has been exhibited widely including in exhibitions Pacha, Llaqta, Wasichay: Indigenous Space, Modern Architecture, New Art at the Whitney Museum of American Art in New York (2018) and Twelfth Gwangju Biennale in Gwangju, South Korea (2018). Recent solo exhibitions include Circumnavigation Towards Exhaustion at La Kunsthalle Mulhouse in France (2021), Future Fossil (2019) at Harvard Radcliffe Institute and Encontro das Águas (Meeting of Waters) (2018), at the Blanton Museum of Art in Austin, TX. Tossin is recipient of a Graham Foundation Grant (2020); Foundation for Contemporary Arts Grant (2019); and an Artadia Los Angeles Award (2018). She is currently working on a moving-image commission by EMPAC, Mojo'q che b'ixan ri ixkanulab' / Antes de que los Volcanes Canten / Before the Volcanoes Sing (2022).

KATHY VELIKOV

Kathy Velikov AIA, OAA is an architect, professor and Associate Dean for Research and Creative Practice at the University of Michigan's Taubman College, and founding partner of the research-based practice, RVTR. Her work explores and experiments with the intertwinements across architecture, the environment, technology, and sociopolitics. Projects range from material prototypes that explore new possibilities for architectural skins that mediate matter, energy, information, space, and atmosphere between bodies and environments, to investigations of urban infrastructures and territorial practices. She works through the techniques of physical prototyping, mapping and analysis, speculative design propositions, exhibitions, and writing. She is the former President of the Association for Computer Aided Design in Architecture (ACADIA), a recipient of the Architectural League's Young Architects Award and the Canadian Professional Prix de Rome in Architecture. She is co-author of *Infra Eco Logi Urbanism* (Park Books, 2015) and co-editor of *Ambiguous Territory: Architecture, Landscape, and the Postnatural* (Actar, 2022).

JACQUELINE WU

Jacqueline Wu is a designer and creative technologist working at the interface between digital technologies and the physical world. Her current practice is embedded in speculative design and explores unconventional methods of data collection, digital fabrication, and processes of transition and translation between media and matter. She holds an MFA in Design and Technology from Parsons School of Design and a background in architecture and urban planning.

BIO/MATTER/TECHNO/SYNTHETICS
Design Futures for the More than Human

Published by
Actar Publishers, New York, Barcelona
www.actar.com

Text Editors
Franca Trubiano
Amber Farrow
María José Fuentes
Susan Kolber
Marta Llor

Graphic Design
Franca Trubiano
Amber Farrow
María José Fuentes
Susan Kolber
Marta Llor

Distribution
Actar D, Inc. New York, Barcelona.

New York
440 Park Avenue South, 17th Floor
New York, NY 10016, USA
T +1 2129662207
salesnewyork@actar-d.com

Barcelona
Roca i Batlle 2
08023 Barcelona, Spain
T +34 933 282 183
eurosales@actar-d.com

Indexing
English ISBN: 978-1-63840-985-4
Library of Congress Control Number: 2021946881

Printed in Europe

Publication date: July 2024

Cover image: ©Editors of BIO/MATTER/TECHNO/SYNTHETICS